The Comprehensive Textbook of Clinical Biomechanics

For Elsevier
Commissioning Editor: Rite Demetriou-Swanwick/Poppy Garraway
Development Editor: Joanna Collett/Veronika Watkins
Project Manager: Andrew Riley
Designer: Paula Catalano
Illustrator Manager: Amy Faith Heyden

The Comprehensive Textbook of Clinical Biomechanics

Second Edition

JIM RICHARDS BEng, MSc, PhD

Professor of Biomechanics
Lead for the Allied Health Research Unit
University of Central Lancashire
Preston, UK

ELSEVIER

ELSEVIER

First edition 2008
Second edition 2018

Print with eLearning Course ISBN: 978-0-7020-5489-1
eBook with eLearning Course ISBN: 978-0-7020-6495-1
Text-only ISBN: 978-0-7020-5490-7
Text-only eBook ISBN: 978-0-7020-7468-4
Previous edition ISBN: 978-0-443-10170-0

Notices

Knowledge and best practice in this field are constantly changing. As new research and experience broaden our understanding, changes in research methods, professional practices, or medical treatment may become necessary.

Practitioners and researchers must always rely on their own experience and knowledge in evaluating and using any information, methods, compounds, or experiments described herein. In using such information or methods they should be mindful of their own safety and the safety of others, including parties for whom they have a professional responsibility.

With respect to any drug or pharmaceutical products identified, readers are advised to check the most current information provided (i) on procedures featured or (ii) by the manufacturer of each product to be administered, to verify the recommended dose or formula, the method and duration of administration, and contraindications. It is the responsibility of practitioners, relying on their own experience and knowledge of their patients, to make diagnoses, to determine dosages and the best treatment for each individual patient, and to take all appropriate safety precautions.

To the fullest extent of the law, neither the Publisher nor the authors, contributors, or editors, assume any liability for any injury and/or damage to persons or property as a matter of products liability, negligence or otherwise, or from any use or operation of any methods, products, instructions, or ideas contained in the material herein.

CONTENTS

EDITOR

JIM RICHARDS BEng, MSc, PhD

Jim Richards was appointed Professor in Biomechanics and research lead for Allied Health Professions at University of Central Lancashire in 2004. Professor Richards' work includes the clinical application of biomechanics, the development of new assessment tools for chronic disease, conservative and surgical management of orthopaedic and neurological conditions, and development of evidence based approaches for improving clinical management and rehabilitation. The focus of Professor Richards work is to encourage inter-professional research and to develop direct parallels with research to the 'real world' of allied health work. Professor Richards holds a number of International Professorial Chair positions and is an associate editor for "The Knee" and on the editorial board for "Gait and Posture". Professor Richards has authored over 130 peer reviewed journal papers, over 170 conference papers and written and edited a number of textbooks including *Biomechanics in Clinic and Research* (2008) and the 5th edition of *Whittle's Gait Analysis* (2012). He has also contributed to *Tidy's Physiotherapy* (2003, 2008, 2012), the 10th edition of *Mercer's Textbook of Orthopaedics and Trauma* (2012), *Experimental Research Methods: A Guidebook for Studies in Trauma Care* (2015) and *Patellofemoral Pain: A Clinical Guide* (2017).

CONTRIBUTORS

NACHIAPPAN CHOCKALINGAM BENG, MSC, PHD, CENG, CSCI, PFHEA
Professor of Clinical Biomechanics in the School of Life Sciences and Education at Staffordshire University, UK

PAOLA CONTESSA BSC(ENG), MSC(ENG), PHD
Research Scientist at Boston University, Research Scientist at Delsys Inc., USA

CARLO J DE LUCA, PHD
was Professor of Biomedical Engineering, Founder and Director of the NeuroMuscular Research Center, Research Professor of Neurology, Professor of Electrical and Computer Engineering, Professor of Physical Therapy and Founding President of Delsys Inc. In 2015 he was appointed Professor Emeritus of Boston University College of Engineering, USA

AOIFE HEALY BSC, MSC, PHD
Senior Research Officer in the School of Life Sciences and Education at Staffordshire University, UK

SARAH JANE HOBBS BENG(HONS), PHD
Reader in Equine and Human Biomechanics and research lead for Centre for Applied Sport and Exercise Sciences at University of Central Lancashire, UK

DAVID LEVINE, PT, PHD, DPT, OCS
Professor and Walter M. Cline Chair of Excellence in Physical Therapy at The University of Tennessee at Chattanooga, USA

RICARDO MATIAS PT, PHD
Researcher at the Neurobiology of Action Group, Champalimaud Foundation and Neuromechanics of Human Movement Research Group, University of Lisbon, Portugal

ROBERT NEEDHAM BSC, MSC, FHEA
Lecturer in Biomechanics in the School of Life Sciences and Education at Staffordshire University, UK

SERGE ROY SCD, PT
Director of Research, Delsys Inc, Natick, MA; and Adjunct Research Professor at Sargent College of Health and Rehabilitation Sciences at Boston University, USA

JIM RICHARDS BENG, MSC, PHD
Professor in Biomechanics and research lead for Allied Health Professions at University of Central Lancashire, UK

JAMES SELFE DSC, PHD, MA, GDPHYS, FCSP
Professor of Physiotherapy in the Department of Health Professions at Manchester Metropolitan University, UK

JONATHAN SINCLAIR BSC (HONS), PHD
Senior Lecturer and Course Leader for MSc Sport & Exercise Sciences, Sport, Exercise & Nutritional Sciences at University of Central Lancashire, UK

DOMINIC THEWLIS BSC (HONS), PHD
Associate Professor of Biomechanics and NHMRC
    R.D. Wright Career Development Fellow at Centre
    for Orthopaedic and Trauma Research, University
    of Adelaide, Australia

NATALIE VANICEK BSC (HONS), MSC,
PGCHE, PHD
Reader in Biomechanics focusing on preventing falls
    and improving musculoskeletal function at the
    University of Hull, UK

For Jackie, Imogen and Joe

ACKNOWLEDGEMENTS

I owe an enormous debt of gratitude to all my colleagues and students past and present. Particularly I wish to thank my contributors for their tireless work and contributions.

I would also like to take this opportunity to say farewell to my friend and mentor Professor Carlo John De Luca who passed away on July 20, 2016 at the age of 72. Carlo has been an inspiration to me and to countless others. Throughout his life he challenged the status quo which led to significant breakthroughs on the frontiers of neuromuscular control, signal processing and EMG sensor technology. This is an example to us all. Carlo you have left an unrivalled legacy in your field, you will be missed.

INTRODUCTION

Over the years many clinicians have commented about the increase in the need for what is described as Evidence Based Practice or Evidenced Based Medicine. The users' guides to evidence-based medicine (Journal of the American Medical Association, 1992) states that;

- The understanding of basic mechanisms of disease are **not** sufficient guides for clinical practice alone.
- **Systematically recorded observations** and **reproducible measurements** are needed to study the effectiveness of clinical practice.

Two of the challenges in clinical practice are, the reproducibility of measurements of effectiveness and the clinical relevance of the measures made. Two questions that have become commonplace in clinical research in the last few years are:

- What are Minimal Clinical Important Differences (MCIDs)? These can be defined as smallest measureable differences between the patient and a defined "normal".
- What are Minimal Clinically Important Changes (MCICs)? These can be defined as the smallest change in score in the domain of interest due to a treatment which patients perceive as beneficial.

So the questions that need to be asked of biomechanics are:

- Can biomechanics offer new and sensitive measures of assessment?
- Can biomechanics assess the effectiveness of different treatments?
- Can biomechanics offer immediate, informed and direct feedback to clinical practice?

This book covers the concepts and theory necessary to understand the nature of biomechanical measurements, and the methods available to collect, analyse, and interpret biomechanical data in a clinically meaningful way. This includes: the mathematical and mechanical concepts necessary for the understanding of the musculoskeletal system and the interpretation of biomechanical measurements, the variety of methods available for biomechanical measurement, and the biomechanics of conservative management of musculoskeletal and neurological pathologies. This book also covers the biomechanics of prosthetics and orthoses and the biomechanics of common movement tasks used in clinical assessment. This therefore should allow undergraduate and postgraduate allied health professionals to advance their biomechanical knowledge and understanding in a way relevant to both training and clinical practice.

A substantial interactive virtual learning environment and teaching resource runs parallel with this book. The virtual learning environment contains lessons relating to the material covered in the paper text book. This also includes many interactive questions to help the learner determine the level of their understanding as they proceed. This virtual learning environment "course" is highly illustrated and contains animations which describe the mathematical and mechanical concepts needed to understand biomechanics. These animations demonstrate the theory covered, and allow the user to control animations of the various clinical case studies included in the text. This aims to provide a stream of online information on biomechanics in a modular format for teaching and learning, and builds understanding and application of biomechanics at a steady pace. This structured approach is designed to act as a companion to undergraduate and postgraduate courses featuring clinical biomechanics.

BOOK STRUCTURE

This book is divided into three sections: Section 1 Mechanics and Biomechanics Theory, Section 2 Methods of Measurement and modelling and Section 3 Clinical Assessment. This structure also allows the lecturer to plan their teaching in relation to specific learning outcomes, and aims to help both lesson delivery and the development of structured courses.

Section 1 Mechanics and Biomechanics Theory

Chapter 1: Maths and Mechanics

Chapter 1 covers the basic mathematics and mechanics needed to understand the much more complicated problem of the mechanics of the human body. This chapter shows how problems may be broken down into separate parts. The techniques covered in the chapter aim to make the more advanced biomechanical problems covered later much easier to solve.

Chapter 2: Forces, Moments and Muscles

Chapter 2 considers the use of mathematics and mechanics techniques in relation to the musculoskeletal system in more detail. Using these techniques and the study of the properties of the body segments, the joint moments, muscle forces, and joint reaction forces in upper limb and lower limb are also considered.

Chapter 3: Ground Reaction Forces and Plantar Pressure

Chapter 3 considers the use of ground reaction forces and the various measures that may be drawn from them. This covers ground reaction forces during postural sway, walking and different running styles and the methods of measuring Impacts, Impulse and Momentum. In addition, this chapter considers foot pressure measurements.

Chapter 4: Motion and Joint Motion

Chapter 4 covers the basic methods of gait assessment through to the description and discussion of the involvement of the three-dimensional movement of the foot, ankle, knee, hip and pelvis and some more advanced methods of analysing the function of walking.

Chapter 5: Work and Power During Human Movement

Chapter 5 covers the concepts of linear and angular work energy and power and how these can be determined from force data, and demonstrates the concept of how angular work and power can be used to analyse the action of joints and muscles during walking and running. In addition the method of calculating the energies involved in the movement of body segments is also considered.

Chapter 6: Inverse Dynamics Theory

Chapter 6 covers the concept of inverse dynamics. This is an important link between the more basic biomechanical models considered so far and includes examples of how we can consider and calculate dynamic joint moments and forces and considers the consequences of not considering dynamic forces.

Section 2 Methods of Measurement and Modelling

Chapter 7: Measurement of Force and Pressure

Chapter 7 covers the measurement of Force and Pressure. This includes the different methods of assessing force and pressure and the identification of a variety of measurements that are commonly used in research and clinical assessment.

Chapter 8: Methods of Analysis of Movement

Chapter 8 covers strengths and weaknesses of different methods of movement analysis from the use of camera technology through to inertial measurement units. This includes the processes required to collect and analyse movement data and the consideration of possible errors.

Chapter 9: Anatomical Models and Marker Sets

Chapter 9 covers different marker sets that can be used in movement analysis. This includes both modelling of the foot, lower limb, spine and shoulder joint. The nature of six degrees of freedom measurement is considered and the associated errors encountered when considering different models and coordinate systems.

Chapter 10: Electromyography

Chapter 10 covers the nature of an electromyographic (EMG) signal and the different methods of measuring muscle activity using EMG. This includes the setup and use of EMG and considers standard data processing techniques used in EMG and which factors affect the quality of the EMG signal. In addition, this chapter also covers some more recent advancements in EMG data collection and processing that allows individual motor units to be measured.

Section 3 Clinical Assessment

Chapter 11: The Biomechanics of Clinical Assessment

Chapter 11 covers the biomechanics of common movement tasks used in clinical assessment of the lower limb. This includes step and stair ascent and descent, sit to stand, timed up and go, gait initiation, and squats and dips. In addition the use of isokinetic and isometric testing are also covered and how these can relate to different aspects of muscle function and physiological cost.

Chapter 12: Biomechanics of Orthotic Management

Chapter 12 covers the biomechanics of orthotic management of the lower limb. This includes the theoretical mechanics of indirect and direct orthotic management and clinical case study data of the use of the devices covered.

Chapter 13: Biomechanics of the Management of Lower Limb Amputees

Chapter 13 covers the compensatory kinematic and kinetic strategies used by lower limb amputees during level walking and stair climbing in relation to their level of amputation and the effect of prosthetic componentry.

GLOSSARY OF TERMS

AAR: Active Angle Reproduction

accelerometer: electro-mechanical devices which measure acceleration

AFO: Ankle Foot Orthosis

ambulation: Walking

angle of gait: The angle of foot orientation away from the line of progression

angular displacement: The rotational component of a body's motion

angular velocity: The rate of change of angular displacement

anterior: The front of the body or a part facing toward the front

anthropometry: The study of proportions and properties of body segments

biomechanics: The study of mechanical laws and their application to living organisms, especially the human body and its movement

CAST: Calibrated Anatomical System Technique

cadence: The number of steps taken over a period of time, usually steps per minute

cardan sequence: Ways of defining one local (LCS) or segment coordinate system (SCS) relative to another

centre of mass: The midpoint or centre of the mass of a body or object

centroid: The two-dimensional coordinates of the centre of an area

clusters: Rigid plate with four or more reflective markers

CKC: Closed Kinetic Chain

concentric: Where the muscle shortens as it contracts under load

coplanar: Lying or acting in the same plane

coronal plane: Frame of reference for the body – viewed from the front (see frontal plane)

direct linear transformation (dlt): The common mathematical approach to constructing the three-dimensional location of an object from multiple two-dimensional images

dorsiflexion: To flex backward, as in the upward bending of the fingers, wrist, foot, or toes

double support: The stance phase of one limb overlaps the stance phase of the contralateral limb creating a period during which both feet are in contact with the ground

eccentric: Where the muscle lengthens under load

electrogoniometer: A device for measuring changes in joint angle over time using either a potentiometer or strain gauge wire

E_m: Energy expenditure per metre (J/kg/m)

EMG or electromyography: The study of the electrical activity of muscles and muscle groups

E_s: Instantaneous energy of any body segment

E_w: Energy expenditure per minute (J/kg/min)

extension: A movement which increases the angle between two connecting bones

filtering: The process of manipulating the frequencies of a signal through analogue or digital processing

flexion: A movement which decreases the angle between two connecting bones

Foot angle: The angle of foot orientation away from the line of progression

force platform: A device for measuring the forces acting beneath the feet during walking

force twitch: The force generated by the contraction of a single motor unit

frontal plane: Frame of reference for the body – viewed from the front

g: Acceleration due to gravity
gait: The manner of walking
gait analysis: The study of locomotion of humans and animals
gait initiation: To start walking
GCS: Global coordinate system
global frame of reference: A set of orthogonal axes, one of which is parallel with the field of gravity
goniometer: a simple hand-held device for measuring joint angles
GRF or ground reaction force: The reaction force as a result of the body hitting or resting on the ground
gyroscopes: electro-mechanical devices which measure angular velocity

habituation: Becoming accustomed
hemiplegia: Paralysis of one side of the body
h_s: Height of the centre of mass above the datum

IMU: inertial measurement unit
impulse force: Area under a force-time curve
instantaneous power: Power at a particular moment in time
I_s: Moment of inertia about the proximal joint
Isometric: Where the joint angle and muscle length do not change during contraction
Isotonic contraction: Where the tension in the muscle remains constant despite a change in muscle length

JCS: Joint coordinate system

KAFO: Knee Ankle Foot Orthosis
KAM: Knee Adduction Moment
kinematics: The study of the motion of the body without regard to the forces acting to produce the motion
kinetic energy: The energy associated with motion, both angular and linear
kinetics: The study of the forces that produce, stop, or modify motions of the body
k_s: Radius of gyration of body segment

LCS: local coordinate system
linear displacement: Distance moved in a particular direction
linear velocity: Speed at which an object is moving in a particular direction
loading response: Period immediately following the initial contact of the foot

markers: small reflective balls that are used to track movement
mid stance: The period from the lift of the contralateral foot from the ground to a position in which the body is directly over the stance foot
mid swing: This is the period of swing phase immediately following maximum knee flexion to the time when the tibia is in a vertical position
Monopodal: Standing on one leg
moment of inertia: The rotational inertial properties of an object
motion or movement analysis: A technique of recording and studying movement patterns of animals and objects
motoneuron: neurons that originate in the spinal cord
motor unit (MU): functional unit of muscles comprising of a single motoneuron and all the fibers innervated by the motoneuron.
motor unit action potential (MUAP): response of all single muscle fibre action potentials belonging to one motor unit
m_s: Segment mass

non-collinear: Points that do not lie in a straight line

obliquity: Pelvic movement when viewed in the coronal plane
OKC: Open Kinetic Chain

PAR: Passive Angle Reproduction
pascals: Units of pressure
pedotti diagram: Ground Reaction Force vector diagram
plantarflexion: The downward bending of the foot or toes
posterior: The back of the body or a part placed in the back of the body

potential energy: The energy associated with the vertical position of the centre of mass of an object

power: The rate of performing work

pressure: force divided by area

pressure time integral: area under the pressure-time graph

preswing: The period immediately before the lifting off of the stance foot

pronation: To rotate the foot by abduction and eversion so that the inner edge of the sole bears the body's weight

radius of gyration: This is a fictitious distribution of the mass around the centre of mass

range of motion: The angular excursion through which a limb moves

rehabilitation: Restoring a patient or a body part to normal or near normal after a disease or injury

relative velocity: A measure of velocity in terms of the height of the individual. The units reported are statures/s

RMS: Root Mean Squared

r_s: Position of the centre of mass from the proximal joint

SACH: A type of prosthetic foot with a solid ankle and a cushioned heel

sagittal plane: Frame of reference for the body – viewed from the side

SCS: Segment coordinate system

single support: The period during the gait cycle when one foot is in contact with the ground

SNR: Signal-to-noise ratio

spatial: Distance

stance phase: The period when a foot is contact with the ground

step length: Distance between two consecutive heel strikes

step time: Time between two consecutive heel strikes

stride length: Distance between two consecutive heel strikes by the same foot

stride time: Time between two consecutive heel strikes by the same foot

supination: To rotate the foot by adduction and inversion so that the outer edge of the sole bears the body's weight

superposition: when one wave is superimposed 'sat on top of' another

swing phase: Period when a foot is not in contact with the ground

temporal: Timing

total support: The total time the body is supported by one leg during one complete gait cycle

translation: Movement in a particular direction

transverse plane: Frame of reference for the body – viewed from above

TTDPM: Threshold to detect passive motion

v_s: absolute velocity of the centre of mass

varus/valgus: Angle of the ankle or knee joint viewed in the coronal plane

work done: product of a force and displacement

walk mat and walkway: A device to measure the temporal and spatial parameters of gait

$\dot{x}$, $\dot{y}$ and $\dot{z}$: Linear velocities in the x, y, and z directions

ω_s: Absolute angular velocity of segment

Section 1

MECHANICS AND BIOMECHANICS THEORY

1

MATHS AND MECHANICS

JIM RICHARDS

This chapter covers the key terminology, basic mathematics and mechanics needed to understand the much more complicated problem of the mechanics of the human body. It illustrates how problems may be broken down into separate parts and shows the techniques used for the more advanced biomechanical problems covered later.

AIM

To consider and describe the maths and mechanics necessary to build and understand more complex biomechanical concepts.

OBJECTIVES

- To describe key terms used when describing the body
- To describe how vectors can be resolved
- To explain how Newton's laws relate to the human body and the difference between mass and weight
- To explain how the action of force vectors on joints can be explored
- To explain frictional forces under the foot
- To explain what is meant by a turning moment.

1.1 KEY TERMINOLOGY

1.1.1 Units – System International

The system of units we use for measurement is the Système International (SI), which was devised in 1960. It defined a system of units to be used universally. The system of measurement units was based on the MKS (metre, kilogram, second) system. These and only these units should be used when working out problems, do not use pounds and feet! If you do not use SI

units in your calculations then problems become a lot harder to solve, and you don't want that, do you?

Some of the common SI units that are used in biomechanics are given later. Many of these units have a close relationship with one another, which aims to make problems easier to solve (Table 1.1).

1.1.2 Indices

Indices are a way of expressing very large or very small numbers without including lots of zeros (Table 1.2). For example, 100 000 m may be written as 100 km, and a pressure of 10 000 000 pascals may be written as 10 MPa. This can be very useful in biomechanics, in particular for pressure measurement when the values can be very large.

1.1.3 Introduction to Anatomical Terms

The motion of the limbs is described using 3 perpendicular planes: sagittal, coronal and transverse. The sagittal plane can be described as a view from the side, the coronal (sometimes called the frontal) plane is a view either from the front or the back and the transverse plane is a view from above or along the long axis of a body segment. Therefore, flexion and extension is described as movement in the sagittal plane (or dorsiflexion and plantarflexion when considering the ankle), abduction and adduction is described as

TABLE 1.1
Units

Quantity	Name of Base	SI Unit Symbol
Length	Metre	m
Mass	Kilogram	kg
Time	Second	s
Area	Square metre	m^2
Volume	Cubic metre	m^3
Velocity	Metre per second	m/s
Acceleration	Metre per second squared	m/s^2
Force	Newton	N
Pressure	Pascal	N/m^2
Energy	Joule	J
Power	Watt	W

TABLE 1.2
Indices

Multiplication Factor	Prefix Symbol
1 000 000 000	10^9 giga G
1 000 000	10^6 mega M
1000	10^3 kilo k
100	10^2 hecto h
10	10^1 deka da
0.1	10^{-1} deci d
0.01	10^{-2} centi c
0.001	10^{-3} milli m
0.000 001	10^{-6} micro μ
0.000 000 001	10^{-9} pico p

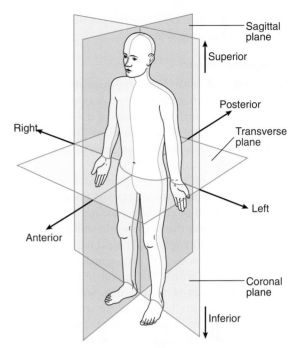

FIGURE 1.1 ■ Anatomical planes *(From Levine, Whittle's Gait Analysis, Churchill Livingstone, 2012.)*

movement in the coronal plane, and internal and external rotation in the transverse plane. Anatomical terms are also used to describe relationships between different segments relative to the centre of the body. These include anterior (front), posterior (back), superior (above), inferior (below), medial (towards the midline of the body), lateral (away from the midline of the body), proximal (towards the rest of the body) and distal (away from the rest of the body) (Fig. 1.1).

1.2 MATHS

The nature of biomechanics means that an understanding of basic mathematics is absolutely essential.

Although the problems themselves may get a lot harder as we progress through the chapters, the basic principles remain consistent throughout. Many clinicians have difficulties with the maths element of biomechanics due to its abstract nature, so I have tried to present this material with reference to anatomy and clinical assessment to show its relevance.

1.2.1 Trigonometry

Trigonometry is absolutely essential in the understanding of how things move and the effect of forces on objects. For instance, if we want to measure how the knee moves and what forces are acting on it while it is moving, we will need trigonometry to find this out. The challenging part of biomechanics is working out what it all means after we have carried out all the calculations, but trigonometry is a vital albeit first step. Most, if not all, movement-analysis systems will work this out for you, but these are useful skills for understanding what is going on, which can in turn help the understanding of patient assessment. The next sections will cover Pythagorean theorem, and tangent, sine and cosine, by considering the positions and orientations of body segments.

Pythagorean Theorem

Pythagoras was alive from approximately 570 to 495 BC. It was Pythagoras who first discovered that in a right-angled triangle, the square of the hypotenuse is equal to the sum of the squares of the other two sides. This only works for right-angled triangles (where one of the internal angles is 90°).

Interestingly, for the majority of problems in biomechanics, this simple property of right-angled triangles is all we need to consider joint movements and forces. This is mostly due to the way in which we divide up the body into three planes. These three body planes are at 90° to one another (or orthogonal, if we wish to use the scientific term). The useful thing from the mathematical point of view is that, whichever anatomical plane we are looking at, we will have a 90° angle present. This is good news, because triangle problems with a 90° angle are a lot easier to solve: mind you everything is relative!

So hopefully you are now convinced that triangles are important for biomechanics. We will now look at Pythagorean theorem and consider the position and angle of the femur. To start we need to know the positions of the distal and proximal ends of the femur. These are often identified by the femoral condyles (A) at the knee and the head of the femur at the hip (C) (Fig. 1.2).

Pythagorean theorem states that the square of the hypotenuse is equal to the sum of the squares of the other two sides. The hypotenuse is the longest side in any right-angled triangle, where the remaining two sides make up the 90° angle. Thus:

$$AC^2 = AB^2 + BC^2$$

where AB is the horizontal distance between the knee and hip joints, BC is the vertical distance between the knee and hip joints, and AC is the hypotenuse or length of the femur.

Movement analysis systems will often tell us the position of the ends of a body segment in x and y coordinates. If we consider that we know the lengths of the horizontal and vertical sides, AB = 20 cm and BC = 50 cm, we can use Pythagorean theorem to find the length of the femur or AC.

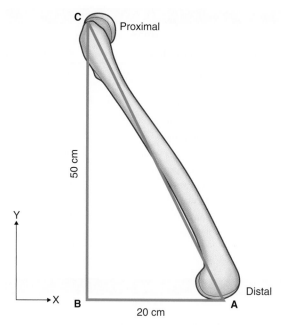

FIGURE 1.2 ■ Pythagorean theorem

So:

$$AC^2 = AB^2 + BC^2$$
$$AC^2 = 20^2 + 50^2$$
$$AC^2 = 400 + 2500$$
$$AC^2 = 2900$$
$$AC = \sqrt{2900} = 53.85\,\text{cm}$$

Therefore, the length of the femur is 53.85 cm.

It is very important to note that the length of the femur, or hypotenuse, is the longest side of the triangle. This will always be the case: if you work out the hypotenuse to be shorter than either of the other two sides, then you have probably got a little mixed up with the equation.

So, if we know any two sides of a right-angled triangle, the third side can be found. Or, to consider the femur, if we know the horizontal and vertical positions of the knee and hip joints, we can work out the length of the femur. Although for most people this is not terribly exciting, without this we would know little about mechanics, and even less about biomechanics.

What Are Tangent, Sine and Cosine?

The best way of thinking about tangent (tan), sine and cosine is as ratios of the different sides of a triangle. In the UK, we often describe the steepness of a hill in terms of how far up we go in relation to how far along we go. For example, a 1 in 4 hill means we go up 1 m for every 4 m we go along: this tells us something about the steepness of slope of the hill.

This is fine until we try to relate this to an angle in degrees as we can't really express a hip flexion angle meaningfully in these terms. At this point tan, sine and cosine come to our rescue; these convert the ratios between the different sides into an angle in degrees. Now at this point I could go into a lot of detail about how tan, sine and cosine work, BUT we want to know how to use tan, sine and cosine, not prove where they come from and why they work! The best way to convert these ratios into angles in degrees is using any scientific calculator. Alternatively, if you are 'electronically challenged' and do not have a scientific calculator, you could use tables that will do the same job.

STUDENTS' NOTE

When solving these with a scientific calculator you will need to use the sin, cos, tan buttons when you know the angle. If you are trying to find the angle from a ratio you will need to use the $\sin^{-1}$, $\cos^{-1}$ and $\tan^{-1}$; you may have to use a second function key to get to these.

The Tangent of an Angle

In a right-angled triangle, the ratios of the sides of the triangle determine the angles within the triangle and vice versa (Fig. 1.3).

$$\text{The tangent of angle } \theta \,(\tan\theta) = \frac{\text{Opposite side}}{\text{Adjacent side}}$$

An important aspect of biomechanics is the calculation of body segment angles in the different planes. We can find these from knowing the location of the proximal and distal ends of a body segment. Now this can get quite complex when we look at all three dimensions (x,y,z), or three planes (sagittal, coronal and transverse), but for the moment we will focus on two dimensions or angles (x,y), or the sagittal plane in anatomical terms. If we consider the femur again with the same measurements as before we know the lengths AB and BC are 20 cm and 50 cm, horizontal and vertical distances, respectively (Fig. 1.3). One of the things

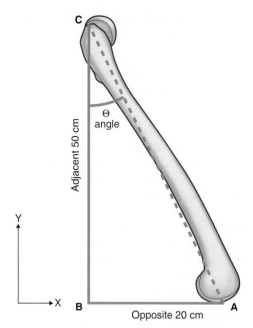

FIGURE 1.3 ■ Tangent of an angle

we may need to find out is the thigh angle θ from the vertical (which in this example shows the femur flexing forwards). The most important thing about working out angles is naming the sides. If a side is opposite the angle we are interested in we call it the 'opposite side', if it is next to the angle we are interested in we call it the 'adjacent side'.

STUDENTS' NOTE

Now at this point you could say there are two sides next to the angle; however, the longest one will always be the hypotenuse, which we are not considering in our angle calculations just yet.

$$\tan\theta = \frac{\text{Opposite side}}{\text{Adjacent side}}$$

$$\tan\theta = \frac{20}{50}$$

$$\tan\theta = 0.4$$

So now we have found tan θ, we need to find the angle θ in degrees. To do this we simply move the tan function over the equals sign, which then becomes $\tan^{-1}$. Then it is a simple matter of putting the number in the calculator.

$$\theta = \tan^{-1} 0.4$$
$$\theta = 21.8°$$

Thus, the thigh flexion angle is 21.8°.

In this way, if we know the length of the side opposite to the angle, and the side adjacent to the angle, we can find the thigh segment angle θ. Likewise, if we know the angle θ and length of the opposite side, we can find the length of the adjacent side.

The Sine ad Cosine of an Angle

Two other ratios exist between the sides of a right-angled triangle and the angles of the triangle. These are sine and cosine, which are commonly written sin and cos. Sine and cosine work in much the same way as tan; however, they use the hypotenuse and the opposite and adjacent sides, respectively (Fig. 1.4).

$$\text{The sine of angle } \theta \; (\sin\theta) = \frac{\text{Opposite}}{\text{Hypotenuse}}$$

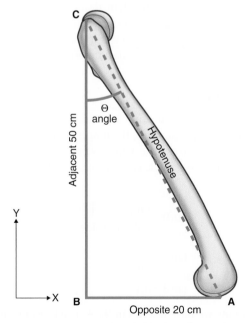

FIGURE 1.4 ■ Sine and cosine of an angle

$$\text{The cosine of angle } \theta \; (\cos\theta) = \frac{\text{Adjacent}}{\text{Hypotenuse}}$$

Sine. If we look at sine first, and consider that we know the thigh angle θ is 21.8°, and we know the opposite side is 20 cm, but we need to find the length of the femur using only this information:

$$\sin\theta = \frac{\text{Opposite}}{\text{Hypotenuse}}$$

$$\sin 21.8 = \frac{20}{\text{Hypotenuse}}$$

$$0.3714 = \frac{20}{\text{Hypotenuse}}$$

We now move the hypotenuse to the other side of the equation, where instead of dividing, it then multiplies:

$$0.3714 \times \text{Hypotenuse} = 20$$

$$\text{Hypotenuse} = \frac{20}{0.3714}$$

$$\text{Hypotenuse} = 53.85 \text{ cm}$$

(which is the same length for the femur as before, not surprisingly).

Cosine. So now for cosine. In this example we will consider the hypotenuse and the adjacent side to find the thigh segment angle. We now know the hypotenuse is 53.85 cm, and the adjacent is 50 cm, how do we find the angle, θ?

$$\cos\theta = \frac{\text{Adjacent}}{\text{Hypotenuse}}$$

$$\cos\theta = \frac{50}{53.85}$$

$$\cos\theta = 0.9285$$

$$\theta = \cos^{-1} 0.9285$$

$$\theta = 21.8°$$

Again, this is the same value of the thigh flexion angle we calculated before. This demonstrates that the different ratios can be used interchangeably depending on what information about a particular triangle you are given. So, there is often more than one way to tackle a particular problem.

A Summary of Sine, Cosine and Tangent

A quick summary of the ratios of the sides of right-angled triangles and a possible memory aid is the word SOHCAHTOA:

$$\mathrm{Sin}\,\theta = \frac{\text{Opposite}}{\text{Hypotenuse}}$$

$$\mathrm{Cos}\,\theta = \frac{\text{Adjacent}}{\text{Hypotenuse}}$$

$$\mathrm{Tan}\,\theta = \frac{\text{Opposite}}{\text{Adjacent}}$$

With this information if we know the length of one side and one angle of a right-angled triangle, we can find the length of all the other sides and their angles.

Within biomechanics it is possible to use only right-angled triangles. With the previously mentioned tools it is possible to solve almost all the trigonometry necessary in biomechanics.

1.2.2 Vectors

What Is a Vector

Vectors have both magnitude (i.e. size) and direction. All vectors can be described in terms of components in the vertical and horizontal directions, or described by a resultant effect acting at a particular angle (Fig. 1.5).

One vector that we will be considering throughout this book is that of the force on the foot from the ground, or 'ground reaction force', which we will consider in much more detail later. Fig. 1.5 shows the horizontal and vertical components of this force and the overall effect, the resultant, of these components. Other examples of vectors include displacement, velocity and acceleration. Vectors may be worked out in exactly the same way as shown in Section 1.1.1 with right-angled triangles, the only difference being the terminology.

The Resultant

This is the combination effect of all the vectors. In the previous example the resultant is the overall force acting from the ground. This in essence is just a hypotenuse and can be found using Pythagoras' theorem, or with sine, cosine and tangent depending on what information is provided.

The Component

The components of the resultant act at 90° from one another; these are equivalent to the opposite and

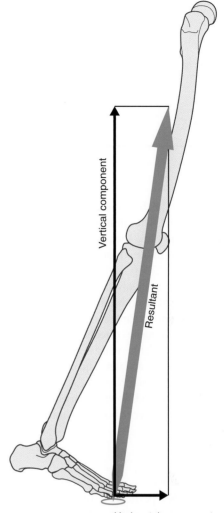

FIGURE 1.5 ■ Vector diagram

adjacent sides of a right-angled triangle. These components act along a coordinate system or frame of reference, which in this case is vertical and horizontal to the ground. So, if we always consider the horizontal and vertical 'effects' we will always create a right-angled triangle.

Adding and Subtracting Vectors

In practical biomechanical problems, the segments or limbs which are analysed will usually be subjected to a number of forces acting in various directions. Often,

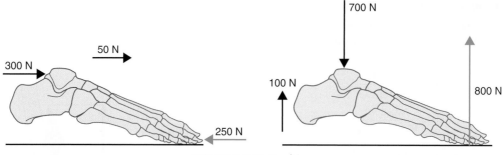

FIGURE 1.6 ■ Resolving

we will wish to 'add' these forces together to determine their overall effect. The simplest example is where the vectors are acting along the chosen frame of reference. Fig. 1.5 shows the different forces acting and the resultant effect.

If all the vectors involved act along the same line they can be added algebraically. That is, forces acting in one direction are regarded as positive, whilst those acting in the opposite direction are regarded as negative—how we define what is positive and what is negative will be covered later. The example in the following sections shows forces on the foot pushing left and right, and up and down, with the overall effect (Fig. 1.6). Do not worry about the units N (newton) just yet!

Resolving

Vectors may act in many different directions as well as magnitudes; this is particularly true when we consider the forces in muscles acting around the joints of the body. When vectors do act in different directions, it is still possible to break these down and find the overall effect if we follow a set of steps no matter how complex the problem looks.

The key to successfully looking at complex systems of vectors is 'resolving'. Resolving is the term used for finding the component vectors from a resultant vector or vice versa, which once again takes us back to right-angled triangles.

Guidelines for Solving Vector Problems

When considering vector problems, we need to first decide on a sensible frame of reference or coordinate system. Frames of reference may be:

1. the vertical and horizontal direction relative to the ground;
2. the planes of the human body, e.g. sagittal, coronal or transverse; or
3. along a body segment and at 90° to it.

To calculate and understand the overall effect of all the vectors we must relate each vector to the same sensible frame of reference. In other words, we are interested in finding the effects along and at 90° to the sensible frame of reference.

We often have problems where we have vectors acting at an angle to the chosen frame of reference. I sometimes refer to this as acting at a 'funny angle', i.e. the vector does not line up with the frame of reference. If this is the case the vector at the 'funny angle' can be 'resolved' along and at 90° to the sensible frame of reference, or, to put it another way, the vector at the funny angle, the hypotenuse, can be split into the opposite and adjacent sides of a right-angled triangle. The opposite and adjacent sides will be the component vectors acting in each of the directions of your reference system. Similarly, if we have the horizontal and vertical components (the opposite and adjacent sides) we can find the resultant (the hypotenuse) using Pythagoras' theorem. Then use sine or cosine we can find the angle at which the resultant acts.

A Simple Vector Problem. This problem deals with the forces we have during push off when walking. At this point we will not concern ourselves with what this means or the nature of the units. The aspect we need to focus on is that we have a vector of magnitude 1000 N acting at an angle of 80°. The question that needs answering is this: what are the magnitudes of the

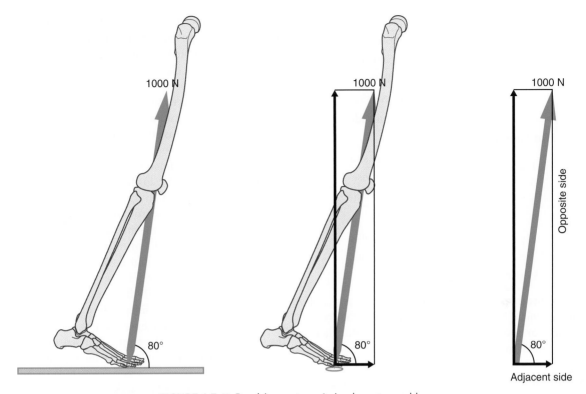

FIGURE 1.7 ◾ Resolving vectors. A simple vector problem

horizontal and vertical components of this resultant force during push off?

Before we do anything, we have to decide on a sensible frame of reference. In this case this is quite straightforward, with the vertical and horizontal to the ground making most sense. We should now draw a box around the ends of the vector at the 'funny angle', i.e. the vector that does not line up with vertical and horizontal to the ground. Do not be distracted by the rest of the anatomy. Although this matters, as the force has an effect on the anatomy, we can ignore it for the moment and just focus on the vector. We now need to consider what the vertical and horizontal components look like, their magnitude and where they originate from.

The magnitudes of the components will be the lengths of the vertical and horizontal sides of the box we have drawn around the vector. All components must originate from the same point, i.e. the tails of the arrows representing the resultant, horizontal and vertical components should all converge on the same point, which in this example is under the metatarsal heads of the foot (Fig. 1.7).

We now have our sensible frame of reference and can visualize the horizontal and vertical components. Now, and only now, are we ready to consider calculating the magnitudes of the horizontal and vertical components; this is what we sometimes refer to as a laboratory or global coordinate system. This may seem like overkill, but without being able to visualize we are likely to make mistakes on more complex problems.

So on to the easy part: the maths! In Fig. 1.7 we now have two identical right-angle triangles. For the lower triangle, we know one of the internal angles is 80°. Now all we need to do is to identify the hypotenuse, and the opposite and adjacent sides of the triangle, and use sine and cosine to find the horizontal and vertical components (Fig. 1.7).

$$\text{The cosine of the angle} = \frac{\text{Adjacent side}}{\text{Hypotenuse}}$$

$$\cos 80 = \frac{\text{Adjacent side}}{1000}$$

$$1000 \cos 80 = \text{Adjacent side}$$

$$173.6 = \text{Adjacent side}$$

Horizontal component = 173.6 N

$$\text{The sine of the angle} = \frac{\text{Opposite side}}{\text{Hypotenuse}}$$

$$\sin 80 = \frac{\text{Opposite side}}{1000}$$

$$1000 \sin 80 = \text{Opposite side}$$

$$984.8 = \text{Opposite side}$$

Vertical component = 984.8 N

To put this in a functional context, this means that a force of 984.8 N is pushing up, whereas 173.6 N is pushing, or propelling the body forwards. But there will be much more on this later.

A More Difficult Vector Problem. This problem deals with the muscle forces acting around the hip joint. The two muscle groups we are considering here are the hip adductors and abductors, the anatomical insertions here being a rough illustration only.

The question is what force is acting along the femur and what force is pushing the femur into the hip joint?

Again, we are first going to consider a sensible frame of reference. However, in this case we want to know what is happening in reference to the femur, which will be different to the horizontal and vertical in relation to the ground we used in the previous example. This is because the femur is not aligned perfectly to the vertical. This is what we sometimes call a local or segment coordinate system (Fig. 1.8). As before we will be using the unit for force, N (newton).

Once we have drawn on this frame of reference we then need to draw boxes around the ends of each of the muscle force vectors, making sure that the components align with this frame of reference. As before we are going to ignore the femur itself and focus on the vectors and their frame of reference only; once we have solved the problem, we will then relate it back to the anatomy. Next, we will consider each muscle separately and work out the components along the long axis of the femur and at 90° to the femur (our sensible frame of reference). We will first consider the hip adductors (Fig. 1.8):

The force along the long axis of the femur is the opposite side to the angle of 40°. Therefore:

$$\text{The sine of the angle} = \frac{\text{Opposite side}}{\text{Hypotenuse}}$$

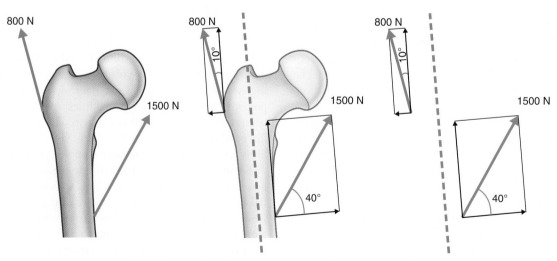

FIGURE 1.8 ■ Resolving vectors. A more difficult vector problem

$$\sin 40 = \frac{\text{Opposite side}}{1500}$$

$$1500 \sin 40 = \text{Opposite side}$$

$$964.2 = \text{Opposite side}$$

Long axis component = 964.2 N

The force along the axis at 90° to the femur is the adjacent side to the angle of 40°. Therefore:

$$\text{The cosine of the angle} = \frac{\text{Adjacent side}}{\text{Hypotenuse}}$$

$$\cos 40 = \frac{\text{Adjacent side}}{1500}$$

$$1500 \cos 40 = \text{Adjacent side}$$

$$1149 = \text{Adjacent side}$$

Component at 90° to the femur = 1149 N

We now consider the hip abductors (Fig. 1.8). The force along the long axis of the femur is the adjacent side to the angle of 10°. Therefore:

$$\text{The cosine of the angle} = \frac{\text{Adjacent side}}{\text{Hypotenuse}}$$

$$\cos 10 = \frac{\text{Adjacent side}}{800}$$

$$800 \cos 10 = \text{Adjacent side}$$

$$787.8 = \text{Adjacent side}$$

Long axis component = 787.8 N

The force along the axis at 90° to the femur is the opposite side to the angle of 10°. Therefore:

$$\text{The sine of the angle} = \frac{\text{Opposite side}}{\text{Hypotenuse}}$$

$$\sin 10 = \frac{\text{Opposite side}}{800}$$

$$800 \sin 10 = \text{Opposite side}$$

$$138.9 = \text{Opposite side}$$

Component at 90° to the femur = 138.9 N

If we now combine what we have found from both muscles.

Total force along the axis of the femur = 964.2 + 787.8

Total force along the axis of the femur = 1752 N

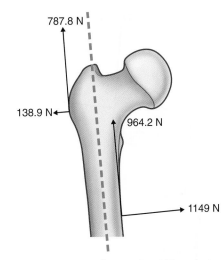

FIGURE 1.9 ■ Forces along and at 90° to the long axis

Total force at 90° to the long axis of the femur
= 1149 − 138.9

Total force at 90° to the long axis of the femur = 1010.1 N

This has now simplified the problem as all the forces are either acting along the axis of the femur or at 90° to the long axis (Fig. 1.9). We can now add the forces acting along the axis and at 90° to the long axis of the femur. To do this we need some simple rules:

- All forces acting up are positive and all forces acting down are negative.
- All forces acting to the right are positive and all forces acting to the left are negative.

The problem can be taken one step further as these forces must be balanced with the joint forces (Fig. 1.10).

This is the beginning of solving quite complex biomechanical problems, which we have achieved by breaking the problem up into more simple parts. We will deal with both muscle and joint forces in far more detail in Chapter 2: Forces, Moments and Muscles.

1.3 MECHANICS

1.3.1 Forces

The fundamental concepts of mechanics and biomechanics include the study of forces, movement and

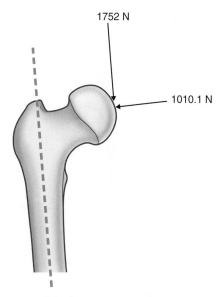

1752 N

1010.1 N

FIGURE 1.10 ■ Joint forces

moments. Forces make things move, stop things moving, or make things change shape. They can either push or pull. In the SI system of units, forces are measured in newton (N). Force is a vector quantity; therefore, all forces have two characteristics, magnitude and direction, which both need to be stated in order to describe the force fully. A good place to start considering the effect of forces is with the laws formulated by Newton.

1.3.2 Newton's Laws of Motion

Isaac Newton (1642–1727) published a three-volume work called 'Philosophiae Naturalis Principia Mathematica' in 1687, which was published first in Latin, revised in 1713 and 1726, and, interestingly, was not translated into English until 1729, after his death. In this epic work, he dealt with many concepts in physics, amongst which were the 'inverse square' law of gravity and the three 'laws of motion', which we will now consider, although somewhat abridged from the original.

Newton's First Law

'Every object in a state of uniform motion tends to remain in that state of motion unless an external force is applied to it.'

This law states that if an object is at rest it will stay at rest; and if it is moving with a constant speed in a straight line it will continue to do so, as long as no external force acts on it, i.e. if an object is not experiencing the action of an external force it will either keep moving or not move at all.

This law expresses the concept of inertia. The inertia of a body can be described as being its reluctance to start moving, or stop moving once it has started. In fact, the state of being at total rest or constant velocity never happens in animals, as there is always some movement, which means a continually changing velocity. Even if we consider a runner at a perceived constant speed, there will in fact be significant changes in vertical and horizontal velocities of the whole body and the individual body segments.

Newton's first law of motion was an important first statement which highlighted the nature of an object travelling in a frictionless environment at a constant speed; this also set the scene for the remaining two laws of motion.

Newton's Second Law

'The relationship between an object's mass (m), its acceleration (a) and the applied force (F) is F = ma. The direction of the force vector is the same as the direction of the acceleration vector.'

This law states that the rate of change of velocity (acceleration) is directly proportional to the applied, external, force acting on the body and takes place in the direction of the force. Therefore, forces can either cause an acceleration or deceleration of an object. **Acceleration** is usually defined as being positive and **deceleration** as being negative.

$$F = ma$$

F = Applied force (N)

m = Mass of the body (kg)

a = Acceleration of the body (m/s^2)

One way to think about this is to consider what would happen if you put a mouse on a skateboard and gave it a push, a single external force. The skateboard and mouse would accelerate off quickly whilst you were pushing, or providing a force (F) as the mass (m) is small. Now consider exchanging the mouse for a large dog and you provided the same force during the push. Clearly the dog would accelerate at a much slower rate; this is due to its larger mass. So, for the same force, two different accelerations would be attained due to the different masses of animal by the relationship of a = F/m.

This law also raises an interesting point about external forces in biomechanics, where we often have many external forces acting at any one time. Therefore, to be able to work out how an object is going to move we need to consider **all** the forces acting. This can make some problems very difficult to solve.

Newton' Third Law

'*For every action there is an equal and opposite reaction.*'

This law states that if a body A exerts a force on a body B, then B exerts an equal and opposite directed force on A. This does not mean the forces cancel each other out because they act on two different bodies. For example, a runner exerts a force on the ground and receives a reaction force that drives him up and forward. This is known as a **ground reaction force** or **GRF,** which we will be considering in greater detail throughout this book.

1.3.3 Mass and Weight

What Is Mass?

Mass is the amount of matter an object contains, or to put it another way, the number of atoms that make up your body. This will not change unless the physical properties of the object are changed, e.g. you change the amount of matter you contain by growing, dieting or losing a body part. One extreme example of demonstrating this is going into orbit or going to the moon; although you may well become weightless, or much reduce your weight, you still contain the same amount of matter. Therefore, the dieting group 'weight watchers' is in fact incorrectly named and 'mass watchers' would be more correct, as the thing which is being changed is the amount of matter, or mass of the body.

What Is Weight?

Weight is an attractive force we have with whichever planet or celestial body we happen to be on or near. In fact, this attractive force is present between all objects;

however, unless the mass of one of the objects is very large, the effects are very hard to observe. This force depends on both the mass of the object and the acceleration acting on it, e.g. gravity. Weight is often interpreted as being the force acting beneath our feet, e.g. bathroom scales measure this force, although they very rarely use the correct units, which *should be* newtons.

So, can we change our weight? A good way to lose weight is to stand in a lift and press the down button. You will lose weight, i.e. the force beneath your feet will reduce as the lift accelerates downwards. Unfortunately, when the lift comes to a stop you will gain weight again as the lift decelerates downwards. What this is doing is temporarily changing the conditions with the addition of the acceleration of the lift as well as the acceleration due to gravity.

Another example of the difference between mass and weight is to consider astronauts. When they are in space they are weightless. This does not mean they have gone on an amazing diet, but it does mean that there is little or no acceleration acting on them, so any force acing on them is zero.

So, weight is a force which is dependent on the mass of the object and the acceleration due to gravity. This brings us back to Newton's second law of motion, F = ma, but with weight as the force and the acceleration being the acceleration due to gravity.

Force = Mass × Acceleration

Weight = Mass × Acceleration due to gravity

Weight = mg

Acceleration Due to Gravity

Wherever you are on planet Earth there is an acceleration due to gravity acting on you. So where does this acceleration due to gravity come from? Once again, we look to Newton, who found what is called the inverse square law:

$$F = \frac{GMm}{r^2}$$

where F = force, G = universal gravitational constant (6.673×10^{-11} Nm^2/kg^2), M = mass of object 1, m = mass of object 2, and r = the distance away from the centre of the objects.

F is the attractive force between any two objects, so the greater the mass the objects contain, the larger the attractive force between the objects. So the larger the body the more attractive it is!

Let's now consider my attractive force with the Earth, or weight. The mass of the Earth is approximately 5.9742×10^{24} kg, the radius of the Earth is approximately 6375 km and my current mass is approximately 75 kg.

$$\text{F or weight} = \frac{6.673 \times 10^{-11} \times 5.9742 \times 10^{24} \times 75}{6\,375\,000^2}$$

This gives a force, $\text{F} = 735.7$ N

So this gives me an attractive force with the Earth of 735.7 N, which is my current weight in newtons. Now if we relate this back to Newton's second law of motion we will find a much easier way of doing this:

$$\text{F} = \text{ma}$$

$$735.7 = 75 \times \text{a}$$

$$\frac{735.7}{75} = \text{a}$$

$$\text{a} = 9.81 \, \text{m/s}^2$$

So the attractive force between each of us and the Earth produces an acceleration due to gravity (g) of 9.81 m/s², which is the accepted value and is only subject to very small geographic variations over the surface of the Earth. For the purposes of rough calculations this is often rounded up to 10 m/s². However, to get the best possible accuracy 9.81 m/s² should be used; therefore, I will be using 9.81 m/s² in this book.

$$\text{Weight} = \text{mass} \times \text{gravity}$$

or

$$\text{Weight} = \text{mg}$$

1.3.4 Static Equilibrium

The concept of static equilibrium is of great importance in biomechanics as it allows us to calculate forces that are unknown. Newton's first law tells us that there is no resultant force acting if the body is at rest, i.e. the forces balance.

Therefore, if an object is at rest, the sum of the forces on the object, in any direction, must be zero. Therefore, when we resolve in a horizontal and vertical direction, the resultant force must also be zero.

If we consider someone standing still, they will have a reaction force from the ground under each foot. This will have a vertical component, but there will also be a small horizontal medial (directed towards the middle) component under each foot, as the feet are wider apart than the width of the pelvis. However, these horizontal forces will in fact act against one another and cancel out in the same way described in the section on vectors, as they act on the same object, in this case a person. The other force we need to consider is the weight acting down: this will be equal and opposite to the vertical component of the reaction forces acting under the feet. So the sum of the forces in the vertical and horizontal will be zero, indicating that the person is indeed standing still (Fig. 1.11).

1.3.5 Free Body Analysis

Free body analysis is a technique of looking at and simplifying a problem by constructing a diagram or sketch showing all the forces acting. We have already seen an example of this in the section on vectors.

If we consider two people having a tug of war, both pulling a rope (Fig. 1.12), we first need to identify all the forces acting. In this example, we have the tension in the rope, which is pulling each person towards the centre, a resultant force pushing up beneath the feet, and the weight of each individual acting down. Once we have drawn these, then consider how they are acting in relation to a sensible frame of reference, and if the forces do not align with this frame of reference, they will need to be resolved vertically and horizontally. This breaking down of a problem and drawing the system of forces is called a free body diagram (Fig. 1.13).

Once we have drawn this diagram we can then start to think about the forces and solving the problem. In this case, we need to find the tension in the rope and the mass of the person. Sometimes we need to make some assumptions—in this case we have to assume that the two people are the same weight and height, that they are in identical positions and that they are in static equilibrium—but we are told the resultant force beneath the feet is 900 N and the angle of the force acting to the vertical is 30° (Fig. 1.14).

Resolving. First, we must resolve the force of 900 N so that it is in the sensible frame of reference, which in this case is horizontal and vertical to the ground.

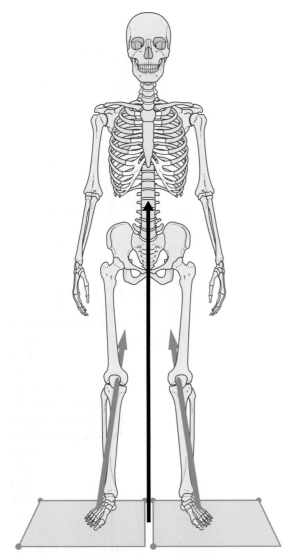

FIGURE 1.11 ■ Static equilibrium

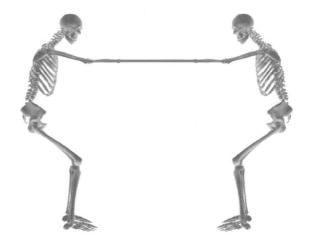

FIGURE 1.12 ■ Tug of war

Horizontally

$$\sin 30 = \frac{\text{Opposite side}}{\text{Hypotenuse}}$$

$$\sin 30 = \frac{\text{Horizontal component}}{\text{Hypotenuse}}$$

$$900 \sin 30 = \text{Horizontal component}$$

$$450 \, \text{N} = \text{Horizontal component}$$

$$\text{Tension in the rope} = 450 \, \text{N}$$

There are no other horizontal forces acting on the person apart from the tension in the rope; therefore, if this is in static equilibrium, this force MUST be equal and opposite to this force in the rope from Newton's third law of motion.

Vertically

$$\cos 30 = \frac{\text{Adjacent side}}{\text{Hypotenuse}}$$

$$\cos 30 = \frac{\text{Vertical component}}{\text{Hypotenuse}}$$

$$900 \cos 30 = \text{Vertical component}$$

$$779.4 \, \text{N} = \text{Vertical component}$$

$$\text{Weight of person} = 779.4 \, \text{N}$$

There are no other vertical forces apart from the weight of the person; therefore, if this is in static equilibrium, the weight MUST also be equal and opposite to this force from Newton's third law of motion.

But what is the person's mass? If we now consider the concept of mass and weight again:

$$\text{Weight} = \text{Mass} \times \text{Acceleration due to gravity}$$

$$779.4 \, \text{N} = \text{Mass} \times 9.81 \, \text{m/s}^2$$

$$\frac{779.4}{9.81} = \text{Mass}$$

Therefore, the mass of the person is 79.45 kg.

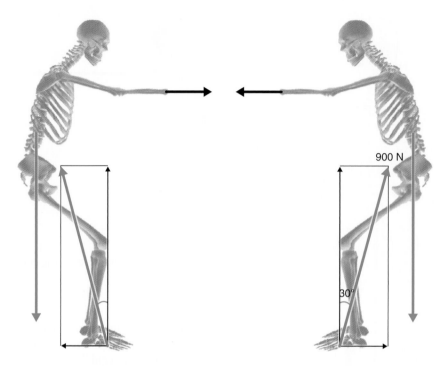

FIGURE 1.13 ■ Free body analysis

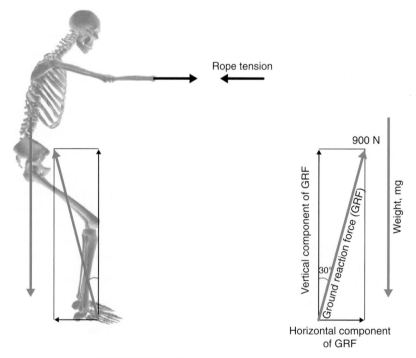

FIGURE 1.14 ■ Forces during a tug of war

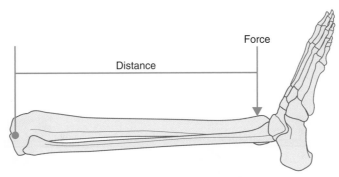

FIGURE 1.15 ■ Moments

1.3.6 Moments and Forces

When a force acts on a body away from its pivoting point a turning effect is set up. Consider opening and closing a door. You are in fact creating sufficient force to turn the door on its hinges: the force required to do this multiplied by the distance away from the hinges you are pushing is the moment (Fig. 1.15).

A turning moment is defined as:

$$M = F \times d$$

where M = turning moment, F = magnitude of force (how hard you push) and d = distance from the pivot.

Balancing Moments

This is very much like working out the unknown forces using static equilibrium. As with static equilibrium we can balance the moments by making the overall effect zero, i.e. the effect of one moment cancels out the effect of another moment. The best way of thinking about this is considering the turning forces on a seesaw (Fig. 1.16).

In Fig. 1.16A it is clear that this seesaw will not balance. In fact, a better way of describing this would be that the seesaw would rotate in the clockwise direction. This would have the effect of moving the heavier person down until their feet touched the ground, at which point some of the force would be removed from the seesaw. So, it is clear to balance the seesaw we need to move the pivot point closer to the heavier person (Fig. 1.16B).

The Mathematics Behind Balancing Moments

To solve problems with moments we have to consider what the action of each force would be in turn. To do

this we consider if each force will try to rotate the object (in this case a seesaw) in a clockwise or anticlockwise direction. If it is in a clockwise direction, it is considered to be in a positive direction, and if anticlockwise, it is considered to be in a negative direction. Thus, if we consider the mathematics of the example in Fig. 1.16B:

The 500 N weight will try to turn the seesaw anticlockwise.
The 1000 N weight will try to turn the seesaw clockwise.

If the seesaw balances, then the sum of the clockwise turning effects and anticlockwise turning effects must be zero. To do this we are going to break the problems down by considering the effect of each force separately, and then consider the overall effect.

If we consider anticlockwise moments as negative and clockwise as positive:

$$\text{Moment} = \text{Force} \times \text{Distance to pivot}$$
$$\text{Moment} = -(500 \times 2) + (1000 \times 1)$$
$$\text{Moment} = -1000 + 1000$$

i.e. the moments cancel out and the seesaw is balanced.

Although there seems to be little effect, if we now consider Newton's third law of motion then we do in fact have a third force acting. If we consider the seesaw vertically, we have two forces (weights) on either side acting down, giving a total force of 1500 N. From Newton's third law of motion there must be an equal and opposite reaction acting up. The only place for this to act is at the pivot; therefore, there must be a force

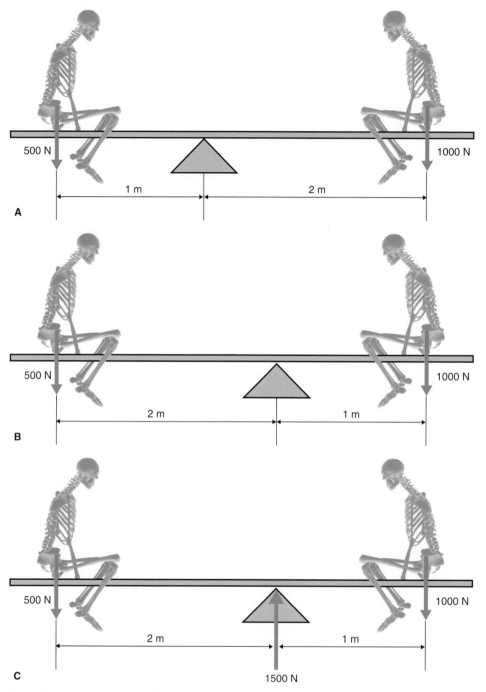

FIGURE 1.16 ■ (A) Unbalanced seesaw. (B) Balanced seesaw. (C) Pivot reaction force *(From Tidy's Physiotherapy 15e, Porter, 2013.)*

acting up of 1500 N as it is static equilibrium (Fig. 1.16C).

These techniques of finding forces and moments are identical to the ones we use to find muscle and joint forces in the body, which will be dealt with in more depth in Chapter 2: Forces, Moments and Muscles.

1.3.7 Pressure

What Is Pressure?

Pressure is best thought about as the force acting over an area. If we consider a large force distributed over a very large area the pressure will be relatively small; alternatively if we consider the same force acting over a very small area the pressure will be very high. Therefore, pressure is dependent on the force applied and the area over which the force acts.

$$\text{Pressure} = \frac{\text{Force}}{\text{Area}}$$

Pressure has been measured using a variety of units over the course of history, which unfortunately still plague us today. These include millibars, bars, pascals, kilopascals, megapascals, newtons per square centimetre, newtons per square metre, atmospheres, inches of water, feet of water, mm of water, inches of mercury, mm of mercury, kilograms per cm squared, pounds per square inch (psi), pounds per square foot, and tonnes per metre squared, to mention just some of them!

Within biomechanics, thankfully, we are a little more consistent, although there is still a variety that are used, including newton per millimetre square (N/mm^2), newton per centimetre square (N/cm^2), kilograms per millimetre square (kg/mm^2), kilograms per centimetre square (kg/cm^2), and sometimes pounds per square inch (psi) and millimetres of mercury (mmHg).

Any measurement of pressure involving kilograms is strictly speaking wrong, as this is a measure of mass and NOT a measure of force. However, more people are likely to have a conceptual idea of what a kilogram is, which has contributed to its use. The same argument could be made for the use of pounds, even though the Conférence Générale des Poids et Mesures (CGPM) agreed over three decades ago in 1971 to maintain the metric standards of the Convention du Mètre of nearly 100 years earlier in 1875! The use of different units makes the comparison of clinical data difficult and only makes the calculations harder.

So, what units should we use? The SI unit for pressure is pascal (Pa), although in pressure, because of the size of the measurements, kPa makes more sense, i.e. 1 kPa = 1000 Pa. The unit pascal is named after Blaise Pascal (1623–1662), a French mathematician and physicist. A pascal is the pressure produced when a force of 1 N is distributed over an area of 1 m^2.

Finding Pressures When Standing

If a subject of weight (force) 700 N stands on a block with a side of 4 cm, as in Fig. 1.17A, then we can find

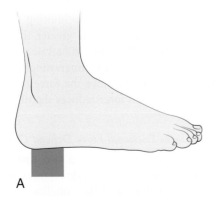
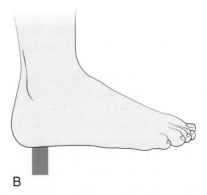

FIGURE 1.17 ■ Standing on a block with a side of (A) 4 cm, (B) 2 cm

the average pressure beneath the area of the foot in contact with the block:

$$Area = 0.04 \times 0.04 = 0.0016 \, m^2$$

The force beneath the foot will be equal to the weight from Newton's third law. The average pressure between the foot and the block = force / area:

$$Pressure = \frac{Force}{Area}$$

$$Pressure = \frac{700}{0.0016}$$

$$Pressure = 437\,500 \, N/m^2 \, (Pa) \, or \, 437.5 \, kN/m^2 \, (kPa)$$

If the subject of weight 700 N now stands on a block with a side of 2 cm, the average pressure beneath the foot will be greater. Again, this depends on the area of the foot in contact with the block and the subject's weight (Fig. 1.17B).

The area beneath the foot in contact with the block is calculated by:

$$Area = 0.02 \times 0.02 = 0.0004 \, m^2$$

The average pressure between the foot and the block = force / area:

$$Pressure = \frac{Force}{Area}$$

$$Pressure = \frac{700}{0.0004}$$

$$Pressure = 1750\,000 \, N/m^2 \, (Pa) \, or \, 1750 \, kN/m^2 \, (kPa)$$

For example, if an individual stands with flat feet, they will have lower pressures on the foot than an individual with arched feet as the area over which the force is distributed is larger.

This only considers the average pressure over the whole foot. However, the load will not be evenly distributed over the whole of the base of the foot, but will be concentrated at various points on the foot. The distributions of pressure beneath the foot are extremely important in both pain relief and prevention of tissue breakdown; this will be dealt with in much more detail in Chapter 7: Measurement of Force and Pressure.

1.3.8 Friction

What Causes Friction?

We have already talked briefly about friction, or at least the absence of it, when considering Newton's first law of motion. But what is friction and what causes it?

When the two surfaces meet, what happens? The surfaces in contact weld together microscopically. Then, when the objects continue to move against each other, the peaks break off. This process creates a force that tries to resist the motion and this force is what we call the frictional force.

This process leaves fragments that also resist the movement and wear down the surfaces further; this is known as three body wear, the three bodies being the two opposing surfaces and the fragments themselves. So, where there is friction, there is wear, and you need look no further than the soles of your shoes.

Static Friction

When a body moves, or tries to move, over a surface it experiences a frictional force. The frictional surfaces act along a common surface, and are in a direction to oppose the movement. If we consider pulling an object along the ground, an unwilling dog or child perhaps, initially the force is small and the object does not move. As the pulling force is increased there reaches a point where the object starts to slide (note: this should not be considered necessarily as good parenting or dog ownership practice).

This indicates that for small values of pulling force the frictional force is equal and opposite, but there is a maximal frictional force that can be brought into play; this is known as the limiting frictional force. When the pulling force is greater than the limiting frictional force, the object will accelerate in accordance to Newton's second law. Interestingly, however, once the object starts to move the force required to overcome the friction force reduces slightly.

What Does Frictional Force Depend On?

Frictional force depends on two main factors; how hard the object is pressing down onto the surface (the normal reaction) and the roughness of the contact between the two surfaces (coefficient of friction).

Coefficient of Friction

The frictional force can be shown to be proportional to the normal reaction force by placing increasing loads on a platform and then applying a horizontal force. The force required to move the platform will increase with the loading on the platform. Therefore, we can say as the reaction force increases so does the limiting frictional force. Therefore, as the normal reaction (R) increases, the limiting frictional force (F) would also increase. The limiting frictional force will vary depending on the materials the two surfaces are made of. Each combination of materials will have its own coefficient of friction, given the symbol μ.

Maximum (Limiting) Frictional Force Available

A good way of considering the limiting frictional force is by thinking of two objects of the same shape but different weights. An example would be a mouse and a dog standing on a block. If we try to push the block the coefficient of friction will in fact be the same for both conditions; however, the dog's larger weight will produce a larger reaction force, which will mean the amount of frictional force available is much greater than the same block with the mouse.

Maximum frictional force available may be found using the following equation:

Maximum frictional force available
= coefficient of friction (μ) × normal reaction

or

$$F = \mu R$$

where F = maximum frictional force available, μ = coefficient of friction, and R = normal reaction.

Fig. 1.18 shows the frictional force increasing in proportion to the applied vertical force on a shoe. If the shoe is then pulled along a flat surface at a constant velocity, the frictional force equals the pulling force as the shoe is neither accelerating nor decelerating.

Limiting Frictional Forces During Walking

Limiting frictional forces are extremely important for our stability during walking. If we consider the effect of walking on two different surfaces with different coefficients of friction, we can predict whether an individual is likely to slip. So, if a person walks first on a carpet, $\mu = 0.55$, and then on a tiled floor, $\mu = 0.18$, as in Fig. 1.19, will they slip?

First, we have to find the vertical and horizontal components of the GRF.

$$\text{The cosine of the angle} = \frac{\text{Adjacent side}}{\text{Hypotenuse}}$$

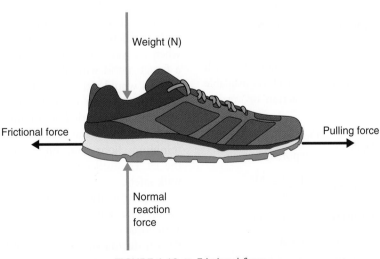

FIGURE 1.18 ■ Frictional force

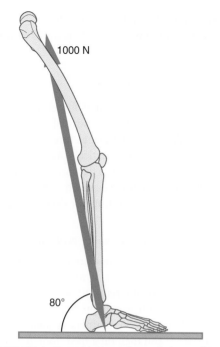

1000 N

80°

FIGURE 1.19 ■ Frictional forces during walking

$$\cos 80 = \frac{\text{Adjacent side}}{1000}$$

$$1000 \cos 80 = \text{Adjacent side}$$

$$173.6 = \text{Adjacent side}$$

Horizontal component = 173.6 N

This horizontal force will have to be equal to or less than the maximum friction force available for each surface. To find the maximum frictional force available we first need to find the 'normal reaction' or 'vertical component' of the GRF.

$$\text{The sine of the angle} = \frac{\text{Opposite side}}{\text{Hypotenuse}}$$

$$\sin 80 = \frac{\text{Opposite side}}{1000}$$

$$1000 \sin 80 = \text{Opposite side}$$

$$984.8 = \text{Opposite side}$$

Vertical component = 984.8 N

Maximum Frictional Force Available for Carpet. The size of the vertical component will determine the amount of frictional force available with each coefficient of friction.

Maximum frictional force available
= normal reaction × μ

$$F = \mu R$$

Maximum frictional force available for carpet
= 0.55 × 984.8 = 541.6 N

The maximum frictional force available when walking on carpet is 541.6 N, which is much greater than that of the horizontal force, 173.6 N; therefore the person is quite safe.

Maximum Frictional Force Available for Tiled Floor

Maximum frictional force available for tiled floor
= 0.18 × 984.8 = 177.3 N

The maximum friction force available when walking on the tiled floor is much reduced, 177.3 N, but there is just sufficient frictional force available to support the 173.6 N horizontal force to stop the person slipping. However, any greater horizontal force or any slight reduction in the coefficient of friction will mean the person's foot will slip, as the floor will no longer be able to maintain the anterior frictional force to stop it.

The Clinical Relevance of Friction

Friction affects many aspects of biomechanics—without friction there would be no horizontal forces when we walk. You only have to consider walking on a frozen pond with shoes with no grip; the lack of friction would either not allow us to move at all or make us slip over if we ever did get moving. Much work has been carried out on non-slip surfaces which allow, in theory, the best coefficient of friction with minimum floor wear under all conditions to yield the largest available frictional force to enable us to walk safely.

Friction is also used in many external prostheses; for instance some external prosthetic knee units have a frictional brake, which allows an amputee some knee flexion during stance phase whilst remaining stable. Conversely, internal prosthetic devices (knee and hip replacements) aim to reduce the friction as much as possible, which increases the life of the prosthesis.

SUMMARY: MATHS AND MECHANICS

- Vectors are quantities that have magnitude and direction. We can describe these in terms of a value and an angle from a frame of reference. From this, their effects can be found along anatomical axes.
- When a body moves, it is in a continuous state of acceleration and deceleration. Internal and external forces are continually at work to drive these movements.
- Weight and mass are not the same. Mass is the amount of matter a body contains, whereas weight is a force due to the body mass and the accelerations acting on it.
- Moments describe turning effects about pivot points. Understanding moments is essential when considering the action of muscles and how they control joint movement.
- Pressure is a measure of the distribution of the force, or pressure, beneath the foot or contact area. Pressure can be used to determine variations in the loading or pressure patterns. High pressures can cause tissue breakdown and injury.
- Frictional forces are always present. Perhaps the most important frictional force is between the foot and the ground. Without this force, we would not be able to propel or stop ourselves when walking.

FORCES, MOMENTS AND MUSCLES

JIM RICHARDS

This chapter considers the use of the techniques covered in Chapter 1, in relation to the musculoskeletal system, in more detail. It also considers the properties of the body segments, the joint moments, muscle forces, and joint reaction forces in upper and lower limbs.

AIM

To relate forces and moments to muscle forces and joint forces.

OBJECTIVES

- To explain the nature of centre of mass and how to find body segment information
- To understand the calculation of moments about joints in static problems
- To understand the calculation of muscle forces in static problems
- To understand the calculation of joint forces in static problems.

2.1 CENTRE OF MASS

What is the centre of mass? The centre of mass is a point on an object where all the mass can be considered to act. The centre of mass of an object does not always coincide with its geometric centre as it is affected by not just the shape of the object, but also the distributions of the densities of material throughout the object. The concept of 'centre of gravity', a term often used interchangeably with centre of mass, was first introduced by the Ancient Greek mathematician, physicist and engineer Archimedes. The defining of a

single point where all the mass can be considered to act is very useful in mechanics and biomechanics as it allows us to study the force due to an object's position and its response to external forces.

2.1.1 The Centre of Mass by Calculation

The calculation of an object's centre of mass can be done by finding the moments around a given point on an object. One way to show this is by considering the centre of mass of two people on a seesaw. The centre of mass will be the horizontal position directly under the balancing point, although in the following example, we will ignore the weight of the seesaw (Fig. 2.1).

The centre of mass of an object is the point at which its weight acts, and in this case the object is the entire system of the seesaw and the two people. If the object is pivoted about its centre of gravity, there would be no turning moment about that point and therefore the seesaw would balance. However, in the example in Fig. 2.1, I have neglected to tell you the horizontal position of the balance point, as this is the very thing we are trying to find out. So, the balance point will be X metres from one end: all we have to do now is find out what X is.

As in Chapter 1: Moments and Forces, we are going to consider the moment from each end of the seesaw, but now we are going to do this in terms of our unknown value X. As before, clockwise moments are

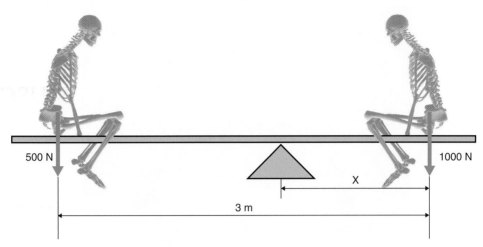

FIGURE 2.1 ■ Centre of mass by calculation

considered as positive and anticlockwise moments are considered as negative. So:

The moment due to the 1000 N

– The moment due to 500 N = 0

1000 N × its distance to balance point

– 500 N × its distance to balance point = 0

$$1000\,X - 500\,(3-X) = 0$$

$$1000\,X - 1500 + 500\,X = 0$$

$$1500\,X - 1500 = 0$$

$$1500\,X = 1500$$

$$X = \frac{1500}{1500}$$

Therefore, X = 1 m

This tells us the exact location of the centre of mass of the object, which is 1 m from the heavier person.

2.1.2 Finding the Centre of Mass by Experiment

The calculation method can be very useful; however, if we have to find the two-dimensional or three-dimensional location of the centre of mass of irregular objects, such as body parts, this technique becomes harder. In these situations, we should consider another technique for finding the centre of mass by experiment.

Ankle foot orthoses (AFOs) are ankle supports frequently used in the management of cerebral palsy and stroke. We could find the centre of mass using a computer-aided design (CAD) package if we have a precise three-dimensional representation of the device. However AFOs are often handmade from casts on the subject's foot and ankle; therefore the best way of finding the centre of mass is to find it experimentally. To do this we need to suspend the object from a location near an edge and drop a plumb-line and mark this on the object. We then suspend the object from another location not too close to the first and drop a second plumb-line and mark again. The intersection of the two lines will be at the centre of mass. This may then be checked by choosing a third suspension location, which should coincide with the intersection already marked (Fig. 2.2 A, B, C).

2.1.3 Centre of Mass and Stability of the Body in Different Positions

For an object of variable shape, such as the human body, the precise position of the centre of mass of the whole body will clearly change with the position of the limbs. Different body positions may even result in the centre of mass falling inside or outside of the body. During standing, for instance, the centre of mass of an adult lies within the pelvis in front of the upper part of the sacrum, its exact location depending on the build, sex and age of the individual. However, as the person moves the relative position of the body segments move. Fig. 2.3 shows how the

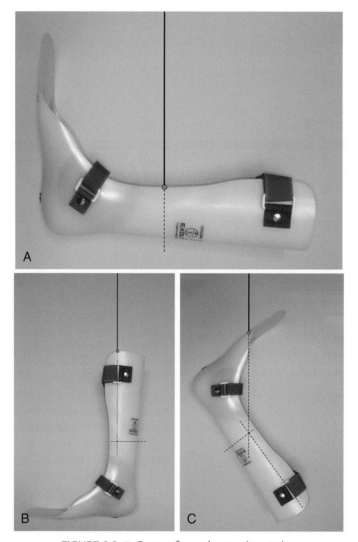

FIGURE 2.2 ■ Centre of mass by experimentation

position of the centre of mass moves during a sit-to-stand task.

The first diagram shows the sagittal plane view just after the person has left the chair. Notice that the force falls behind the foot. If the person were to stop at this point they would fall back onto the chair. The second diagram shows the force is nearer the pelvis and the line of force now falls in the base of support of the feet. If the person were to stop at this point they would now be in a stable crouched position. The third and fourth diagrams show the sagittal and coronal views at the end standing position.

2.2 ANTHROPOMETRY

2.2.1 Background to Anthropometry

The study of muscle and joint forces requires data regarding the length of body segments, mass distributions, centre of mass and the radius of gyration of body segments. Information about human body dimensions was first collected in the late 19th century by Braune and Fischer (1889). Since then more comprehensive studies have been undertaken, most notably by Dempster (1955), Dempster and colleagues (1959), Drillis and Contini (1966), Clauser and colleagues

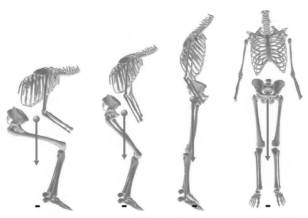

FIGURE 2.3 ■ Centre of mass during a sit-to-stand task

(1969), Chandler et al. (1975), Zatsiorsky and Seluy-anov (1983) and de Leva (1996). Some of this information has been obtained from cadavers and some used measured segment volumes in conjunction with density tables. Kingma and colleagues (1995) discussed the errors associated with using stereotyped anthropometric data and suggested procedures to optimize the calculation of segment centres of mass and centre of mass of the whole body. A greater degree of accuracy may be achieved by taking measurements of segment lengths and therefore removing some of the errors due to natural variation. To remove all errors associated with these calculations, full anthropometric measurements need to be taken; however, this takes a considerable amount of time and is also open to numerous measurement errors itself.

The work by Dempster in 1955 is still considered by many as the best to work with and much of the subsequent research into anthropometry has been on adjustments to Dempster's values, rather than new independent work. However, these must be viewed as estimates of segment values as Dempster's report in 1955 only included data from eight cadavers, and Dempster saw some variations between the cadavers. Dempster's report covered many aspects of anthropometry including the design of a mannequin, but the most quoted are the values of the mass of body segments, the position of the centres of mass, and the moments of inertia. Drillis and Contini (1966) reported the relative segment lengths based on the

overall body height. From these data, we are able to find an estimate of all the physical properties of all body segments using measures of only the subject's mass and height. However, this is open to error due to the natural variation in anatomical proportions. Although we should go to reasonable lengths to reduce any errors, most sources of anthropometric data provide similar results and it is debatable whether the different models produce clinically significant different results.

2.2.2 Common Anthropometric Parameters

The most common anthropometric measurements used in biomechanics are: segment lengths (L) with respect to body height (H), segment mass with respect to total body mass (m), and the position of centres of mass (r) and radius of gyration (k) with respect to segment length (Fig. 2.4). The relationship between the location of the centre of mass and radius of gyration with segment length varies for different body segments; therefore, we have to use specific values for each body part. The radius of gyration of the body segments only needs to be used when the acceleration and deceleration of the body segments are being considered. For the moment, we will consider the body segments to be static, although we will need to consider the radius of gyration later when calculating dynamic moments. Table 2.1 shows a summary of the data found by Dempster (1955), and Table 2.1 and Fig. 2.5 show a summary of the data from Drillis and Contini (1966).

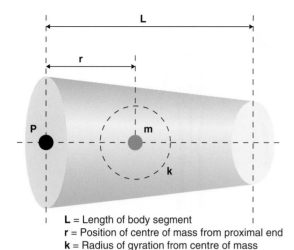

L = Length of body segment
r = Position of centre of mass from proximal end
k = Radius of gyration from centre of mass
m = Body segment mass
P = Point of rotation at proximal end

FIGURE 2.4 ■ Anthropometric parameters

2.2.3 Anthropometric Calculations

Anthropometric calculations are quite simple as we work from percentages or ratios of the body segment parameters in relation to height, mass or segment length. Therefore, using Table 2.1 the length, mass, centre of mass and radius of gyration may be found from the person's overall height and weight.

The following examples show calculations of body segment parameters in a person of mass 80 kg and height 1.8 m.

The mass of a foot from Table 2.1 is 0.014 of the total body mass. Therefore, the mass of a foot may be found by simply multiplying this by the total body mass:

$$\text{The mass of a foot} = 0.014 \times 80$$

$$\text{The mass of a foot} = 1.12 \, \text{kg}$$

TABLE 2.1				
Anthropometrics of Body Segments				
Body Segments	**Length of Segment Body Height**	**Mass of Segment Body Mass**	**Centre of Mass (CoM) Segment Length (Measured from proximal end)**	**Radius of Gyration Segment Length (Measured about CoM)**
Foot	0.152 (length) 0.055 (width) 0.039 (height)	0.014	0.429	0.475
Shank	0.246	0.045	0.433	0.302
Shank and foot	0.285	0.0595	0.434	0.416
Thigh	0.245	0.096	0.433	0.323
Entire lower extremity (including pelvis)	0.530	0.157	0.434	0.326
Upper arm	0.186	0.0265	0.436	0.322
Forearm	0.146	0.0155	0.430	0.303
Hand	0.108	0.006	0.506	0.297
Forearm and hand	0.254	0.0215	0.677*	0.468
Entire upper extremity	0.441	0.0485	0.512	0.368
Head, trunk and pelvis minus limbs	0.52	0.565	0.604 (from top of head)	0.503
Head and neck	0.182	0.079		0.495

*The position of the centre of mass of the forearm and hand is 0.677 of forearm length only.

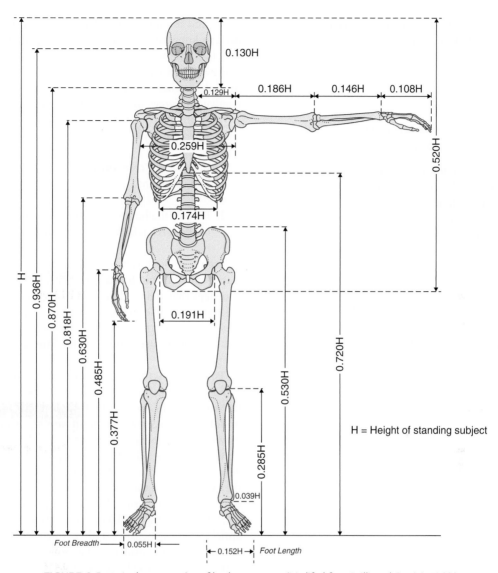

FIGURE 2.5 ■ Anthropometrics of body segments *(Modified from Drillis and Contini, 1966.)*

The length of a body segment can also be found in a similar way. The length of the thigh is 0.245 of the total body height. Therefore, the length of the thigh may be found by simply multiplying this by the height of the person:

The length of the thigh = 0.245 × 1.8
The length of the thigh = 0.441 m

The position of the centre of mass may also be found; however, this depends on the length of the segment and not just the height of the person. Therefore, to find the location of the centre of mass of the forearm from the proximal joint we first have to find the segment length and then the centre of mass:

The length of the forearm = 0.146 × 1.8
The length of the forearm = 0.2628 m

The position of the centre of mass is $0.430 \times$ segment length from the proximal joint. Therefore, the position of the centre of mass of the forearm may be found:

The position of the centre of mass of the forearm
$$= 0.2628 \times 0.430$$

The position of the centre of mass
$$= 0.113 \text{ m from the proximal joint}$$

So far we have not considered the forces when we move; this will be covered later in this book. One parameter we will need to consider to work out rotational forces is the 'radius of gyration'. This can be found in a similar way to the position of the segment centre of mass. Therefore, to find the radius of gyration we again use the segment length:

The length of the forearm $= 0.2628$ m

The radius of gyration is $0.303 \times$ segment length from the proximal joint.

The radius of gyration $= 0.303 \times 0.2628$

The radius of gyration $= 0.0796$
about the centre of mass

STUDENTS' NOTE

Many movement analysis software packages do not use the person's height to find the centre of mass as they automatically find the length of the body segments by knowing the joint centres and then use this value to determine the position of the centre of mass and radius of gyration. This is considered more accurate as it removes some of the error due to natural variation in relationship between body segment lengths and height.

2.3 METHODS OF FINDING MOMENTS, MUSCLE AND JOINT FORCES

2.3.1 How to Find Forces and Moments Acting on the Musculoskeletal System

We are now going to use the concepts of mechanics and anthropometry to consider the forces and moments acting on the musculoskeletal system in exactly the same way as we did on a seesaw (see Section 1.3.6 Moments and Forces).

To find the mechanics of forces and moments we need to know the force or forces acting about a joint and the distance at which they act from the joint. However, in biomechanics the forces seldom act at 90° to body segments. Therefore, we invariably need to **resolve the forces** to find what we need (see Chapter 1: Vectors). There are two ways of doing this and both are mathematically correct. Either:

1. Resolve the component of force at 90° to the body segment
   or
2. Resolve the horizontal and vertical components of the force relative to the ground.

Each technique can be described in a number of ways but to the uninitiated I break this down into main two steps that can be described in very simple terms: (1) spot where the triangle is, and (2) spot where the seesaw is. In fact, the difficulty in biomechanics is often not the maths itself, but spotting these two 'more simple' aspects.

This allows us to consider muscle forces in the same way as balancing forces on a seesaw and joint forces in the same way as finding the force at the pivot in the middle of the seesaw. The joint and segment angles may be considered as triangles.

When dealing with the seesaw problem earlier we also considered the effect of the forces, and whether they would try to turn in a clockwise or anticlockwise direction. For the examples covered in this chapter I will be considering clockwise moments as positive and anticlockwise as negative. Other conventions can also be used, such as considering if a moment about a joint is trying to flex or extend a joint. Although anatomically more meaningful, this does not actually change the mathematics when considering the effects of moments on the muscles forces.

Any students who have not got a strong maths background should not become too worried as this section uses exactly the same principles dealt with in Chapter 1. Look back at this section and see how the balancing force and the force on the pivot are found, and how vectors can be resolved.

The following techniques are a simplification; however, they are useful to get an idea of the effect forces have in and around joints. For a more advanced model of joint moments and forces during dynamic activities see Chapter 6: Inverse Dynamics Theory.

2.3.2 How to Find Muscle Force

To find the forces in muscles we will consider the forces acting around the elbow joint. The biceps muscle acts at an inclined (funny) angle to the forearm. Therefore to work out its effect at the elbow, we have to break the force up into two perpendicular components, one horizontal and one vertical to the frame of reference. In this case a sensible frame of reference is along the forearm and at right angles to the forearm; this involves us resolving the force by using sine and cosine. These two components of a muscle are sometimes referred to as rotary and stabilizing components (Fig. 2.6). These can be found using the following equations, where the angle A is the angle between the biceps muscle and the forearm:

Rotary component = Muscle force × sin A

and

Stabilizing component = Muscle force × cos A

The rotary component is the force that tries to turn the body segment around the proximal joint (e.g. flexing or extending the joint). The stabilizing component is the force that acts along the body segment (e.g. forearm shown in Fig. 2.6) forcing into, or pulling out of the joint.

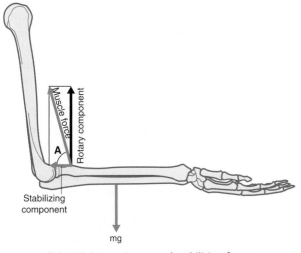

Muscle force

Rotary component

A

Stabilizing component

mg

FIGURE 2.6 ■ Rotary and stabilizing forces

The stabilizing component acts through the joint, and therefore has no effect on the joint moments, whereas the rotary component will produce a moment about the proximal joint. Although only the rotary component has an effect on the muscle forces, both the rotary and stabilizing components have an effect on the joint forces.

2.3.3 How to Find the Joint Force

To find the joint force we need to think back to the seesaw problem, where the force at the pivot was equal to the sum of the two forces acting down. This is the same technique as we will adopt here. However, for joint forces we have to consider the forces acting horizontally and vertically (Fig. 2.7). If we want a more complete picture we also need to consider the moments in the coronal, or frontal plane and the transverse planes, and not just the sagittal plane (but more of that later).

There are two main ways we can consider joint forces; either looking at the forces acting horizontally and vertically at the joint, or working out the overall resultant force acting at the joint. There are good points to both methods. The resultant, for instance, gives us the total force acting on the joint, but this does not necessarily relate the forces in an anatomical frame, whereas considering the forces in their horizontal and vertical components on a body segment can give information about compressive forces and shear forces acting on a joint.

To find the vertical component of the joint force we need to consider all the forces acting in a vertical direction or at 90° to the body segment. These consist of the vertical component of the muscle force, the weight of the body segment and the vertical joint force. The sum of all these forces must equal zero, which means if we know two of these forces we can find the third.

As with the vertical forces, the sum of all the forces acting in the horizontal direction must equal zero. In this example, we only have two forces, the horizontal component of the muscle force and the horizontal joint force.

We can also find the resultant joint force by using the Pythagorean theorem. So, to find the resultant joint force we need to use the following equation:

Resultant joint force² = Vertical joint force² +
Horizontal joint force²

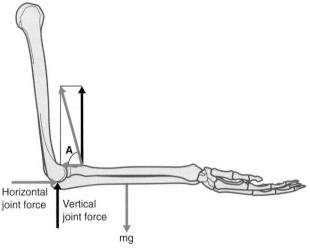

Horizontal
joint force   Vertical
joint force

mg

FIGURE 2.7 ■ Joint forces

Dynamic Joint Forces. In more dynamic situations this is a simplification as the true forces on the joints are going to also depend on the accelerations and decelerations of the body segments, but perhaps even more significant is the muscle action around the joints. This last point is quite an important one. Although we have a way of calculating the muscle forces in our simplified model in 'How to find the muscle force', this does not consider all the muscles that may be acting, or even the way they might be acting. For example, if we were to stand upright and relax, and were then to tense our muscles in the lower limb, the mechanics may not show any difference between the two situations using the previously mentioned methods. However, the co-contractions around the joints will, in fact, pull the body segments together into the joints, which will increase the forces experienced by the joint surfaces. One way of improving our modelling of joint forces is by considering muscle tensions and forces in more detail, and using what is sometimes referred to as an electromyography (EMG) assisted model. These factors aside, the techniques described in the following sections are very useful to get an idea of the nature of the forces acting at joints.

2.4 JOINT MOMENTS, MUSCLE FORCES AND JOINT FORCES IN THE LOWER LIMB

We will now consider some examples of how we can find the joint moments, muscle forces and joint forces during different lower-limb tasks. The two tasks we consider here are a lower-limb squat exercise and walking. In this section, we will consider the moments by considering the force and the distance to the joint; for a more advanced method see Chapter 6: Inverse Dynamics Theory.

2.4.1 Joint Moments During a Squat Exercise

The squat exercise aims to work the quadriceps; however, we can determine any additional effects this may have about the ankle and hip joints also. To determine the effect, we need to know the point of application of the ground reaction force (GRF) and the position of the ankle, knee and hip joints. If we assume the squat exercise is performed slowly, then there will be little acceleration and deceleration of the body segments; therefore we can consider that the problem is not dynamic. If we also assume that the descent is well-controlled, then the GRF will be acting straight up.

Fig. 2.8 shows that the GRF falls in front of the ankle joint, behind the knee joint and through the hip joint. Before we consider the effects on the muscles around the ankle, knee and hip joints we are first going to consider the moments acting about the ankle knee and hip joints created by the GRF. If we know the magnitude of the GRF and the horizontal distances from the GRF to the ankle, knee and hip joints, we can simply consider the effect of the GRF for each joint separately. So, for each joint we focus only on the position of the joint and the vertical force acting.

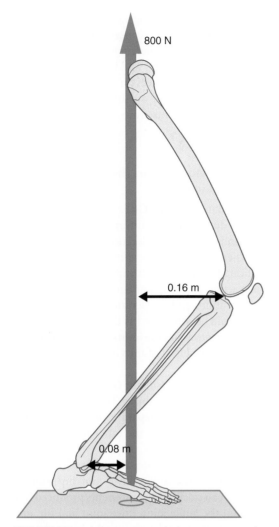

FIGURE 2.8 ■ Joint moments during a squat exercise

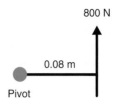

So:

Moment = Force × Distance

Moment about the ankle = −(800 × 0.08)

Moment about the ankle = −64 Newton meter (Nm)

The negative sign indicating the effect is to try to turn the distal (foot) segment anticlockwise in relation to the proximal (tibial) segment of the joint. This could also be described anatomically as a 64 Nm dorsiflexing moment, as the moment will be trying to dorsiflex the ankle. Describing this anatomically often makes more sense as the clockwise/anticlockwise will depend of which direction the person is facing.

Moments About the Knee Joint. The knee joint is at a horizontal distance of 0.16 m away from the GRF of 800 N. Again, if we consider the pivot is fixed in space, then the 800 N force will try to push up and try to turn the distal segment of the joint in a clockwise direction, as the force is now on the left-hand side of the joint; therefore, the moment will be considered as a positive moment.

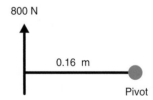

So:

Moment = Force × Distance

Moment about the knee = +(800 × 0.16)

Moment about the knee = 128 Nm

The positive sign indicating the effect is to try to turn the distal (tibial) segment clockwise in relation to the proximal (femoral) segment of the joint. Again, this could also be described anatomically as a 128 Nm knee flexing moment.

Moments About the Ankle Joint. If the ankle joint is a horizontal distance of 0.08 m away from a GRF of 800 N, all we have to do is consider the force times the distance, and whether the force will try to turn clockwise or anticlockwise. So, all we are in fact considering is what effect an 800 N force will have about a pivot shown hereafter. If we consider the pivot is fixed in space, then the 800 N force will try to push up, which will try to turn the distal segment of the joint in an anticlockwise direction; therefore the moment will be considered as a negative moment.

Moments About the Hip Joint. As with the ankle and knee joints, we can consider the hip joint. However, in this case the 800 N force acts straight through the hip joint. This has the same effect as sitting in the middle of the seesaw, directly over the pivot. In this case, no moment is produced as the distance the force acts away from the pivot is in fact zero.

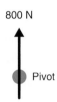

800 N

Pivot

So:

$$\text{Moment} = \text{Force} \times \text{Distance}$$
$$\text{Moment about the hip} = +(800 \times 0)$$
$$\text{Moment about the knee} = 0 \, \text{Nm}$$

What Are the Effects of These Moments on the Muscles? About the ankle joint the GRF had an anticlockwise turning effect, which would try to dorsiflex the ankle joint. This would be commonly referred to as a dorsiflexing moment of 64 Nm. If the subject is holding this position in a stable manner, then there must be an equal and opposite balancing clockwise moment of 64 Nm, which is provided internally by the muscles posterior to the ankle joint (calf group) or plantarflexors. Therefore, this is sometimes referred to as a plantarflexor moment, as the posterior muscles are responsible for plantarflexing the ankle joint.

Similarly, around the knee joint the GRF had a clockwise turning effect, which would try to flex the knee joint. This would be referred to as a flexing moment of 128 Nm. The muscles active holding this position would be the knee extensors (quadriceps), so this is also sometimes referred to as the extensor moment.

With the hip joint, there was no moment as the force passed through the hip joint, although in reality there will be stabilizing muscle contractions around the hip joint. These will not be as a result of the GRF but will be more due to the position of the joint.

The concept of flexing/extensor moments and extending/flexor moments can get confusing. Throughout this book, I will be referring to the external effects from the GRF on the joint and whether these produce a flexing or extending effect on the joints and describe the corresponding effect on the muscles, rather than describing the moments from the muscles the 'point of view'.

2.4.2 Joint Moments in the Lower Limb During Walking

During walking, external GRFs act on the lower limb. These are due to the foot hitting or pushing off from the ground and the deceleration/acceleration of the body. Therefore, the GRFs during walking are more complicated than those during the squat as they are not just acting vertically. For a more complete analysis, the accelerations and deceleration of all the body segments should be considered as these will have an effect on the moments about the joints; however, first we will only consider the effect of the GRF on the joints of the lower limb.

Fig. 2.9 shows the resultant GRF (thick arrow) seen at foot flat in a normal subject. The resultant GRF can be broken up into two separate components: one in the vertical direction and the other in the horizontal direction.

If the point of application and angle of the GRF and the position of the ankle, knee and hip joints are known, we can calculate the moments produced by the GRF about the ankle joint, knee joint and hip joints. From this we can then determine which muscle groups must be acting to support these moments. To do this we are going to use exactly the same steps as with the squat exercise, except the GRF we have to deal with here is acting at an angle. Therefore, the first thing we have to do is to resolve it into its vertical and horizontal components (Fig. 2.9A, B).

The GRF, 1000 N, is acting at 80° to the horizontal.

Resolving

$$\text{Horizontal GRF} = 1000 \times \cos 80$$
$$\text{Vertical GRF} = 1000 \times \sin 80$$
$$\text{Horizontal GRF} = 173.6 \, \text{N}$$
$$\text{Vertical GRF} = 984.8 \, \text{N}$$

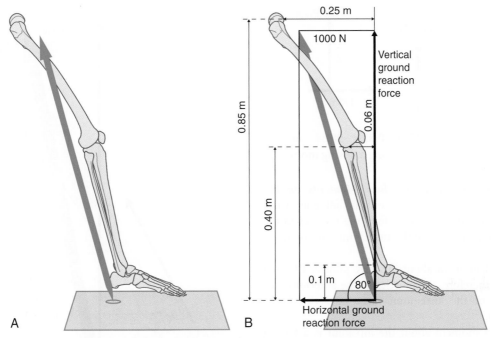

FIGURE 2.9 ■ (A and B) Joint moments in the lower limb during walking

Once the vertical and horizontal forces have been found, we can consider the action of each component about each joint separately.

Moments About the Ankle. The vertical component of the GRF acts straight through the joint; therefore, this will not produce a moment. The horizontal component of the GRF acts to the right and below the ankle, creating a clockwise moment or plantarflexing moment:

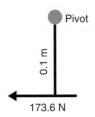

Moment about the ankle $= -984.8 \times 0 + 173.6 \times 0.1$

Moment about the ankle $= 17.36$ Nm

(positive in this case meaning a plantarflexing moment)

Moments About the Knee. The vertical component of the GRF acts in front of the knee joint creating an anticlockwise or extending moment. The horizontal component of the GRF acts to the right and below the knee, creating a clockwise or flexing moment:

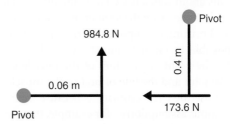

Moment about the knee $= -984.8 \times 0.06 + 173.6 \times 0.4$

Moment about the knee $= 10.35$ Nm (positive in this case meaning a flexing moment)

Moments About the Hip. The vertical component of the GRF acts in front of the hip joint, creating an anti-clockwise or flexing moment. The horizontal component acts to the right and below the hip, creating a clockwise moment or extending moment:

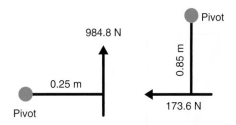

$$M_{hip} = -984.8 \times 0.25 + 173.6 \times 0.85 = -98.64 \text{ Nm}$$
(negative in this case meaning a flexing moment)

What Are the Effects of These Moments on the Muscles? The net moment about the ankle joint is a plantarflexing moment; therefore, the muscles in the anterior compartment of the ankle joint must be active (dorsiflexors).

The net moment about the knee joint is a flexing moment; therefore, the muscles in the anterior compartment of the knee joint must be active (knee extensors).

The net moment about the hip joint is a flexing moment; therefore, the muscles in the posterior compartment of the hip joint must be active (hip extensors).

2.4.3 Muscle Forces in Lower Limb

As in the previous example, we first need to resolve the GRF into its vertical and horizontal components. These component vectors can then be used to find the moments about joints if we know the horizontal and vertical distances from the joint of interest to the vertical and horizontal component forces. However, it is also possible to estimate muscle forces by knowing how far the point of insertion of the muscle is away from the joint and the line of pull of the muscle. The reason I use the term 'estimate' is because we have to make various assumptions—for example, that we only have one muscle active and that there is no co-contraction. These are BIG assumptions; however, it is still possible to a gain a useful estimate of muscle forces.

The example considers the forces in the Achilles tendon during running in a person of 80 kg. During running it is common to have a GRF of 2.5 times bodyweight. We know the angle of the GRF from the horizontal is at an angle to the horizontal of 75° (Fig. 2.10).

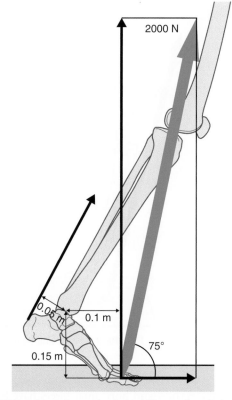

FIGURE 2.10 ■ Forces in the Achilles tendon during push off when running

Resolving Forces

As in the previous example the first step is to resolve the GRF into its horizontal and vertical components.

Vertical component of the GRF = 2000 sin 75°
Vertical component of the GRF = 1931.9 N
Horizontal component of the GRF = 2000 cos 75°
Horizontal component of the GRF = 517.6 N

Taking Moments

We can now find the moments about the joints. In this case, we will just consider the ankle joint:

Moment about the ankle = −1931.9 × 0.1 − 517.6 × 0.15
Moment about the ankle = −270.8 Nm

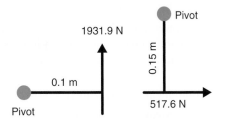

Muscle Forces

The net moment about the ankle is the external moment that must be supported by an internal moment. This internal moment is produced by the muscles and must be equal and opposite to the external moment. The external moment due to the GRF is, in this case, acting anticlockwise trying to dorsiflex the foot; therefore, the internal muscle moment will be acting in a clockwise direction, produced by the plantarflexors which travels through the Achilles tendon. Therefore the muscles force can be found by balancing the equation:

$$\text{Muscle force} \times \text{Distance to joint centre} - 270.8 = 0$$

$$\text{Muscle force} \times 0.05 = 270.8 \, \text{Nm}$$

$$\text{Muscle force} = \frac{270.8}{0.05}$$

$$\text{Muscle force} = 5416 \, \text{N}$$

This equates to an Achilles tendon force of approximately 6.9 times bodyweight in the 80-kg person, which is within the range values reported by Scott and Winter (1990) of 6.1–8.2 times bodyweight.

2.4.4 Joint Forces in Lower Limb

Joint forces depend on all the forces acting horizontally and vertically around a joint. However, there are variations in the way we can do this. The two techniques relate to whether the muscle forces are included in the calculations or not. Muscle forces can contribute quite significantly to the magnitude of the joint force. In the following section, I have run through the same example, firstly considering the joint force with no muscle force and secondly considering the joint force with the action of a single muscle.

Joint Force Without Muscle Forces

We will start by considering the vertical and horizontal components of the GRF. These forces must be equal and opposite to the joint forces, or to put it another way, the sum of all the forces in the vertical and horizontal direction must be zero (Fig. 2.11).

Vertical component of the GRF = 2000 sin 75°

Vertical component of the GRF = 1931.9 N

Horizontal component of the GRF = 2000 cos 75°

Horizontal component of the GRF = 517.6 N

Vertical joint force = −1931.9

(the negative referring that this acts down, i.e. the force from the tibia onto the ankle joint)

Horizontal joint force + 517.6 = 0

Horizontal joint force = −517.6 N (the negative referring that this acts to the left or posterior)

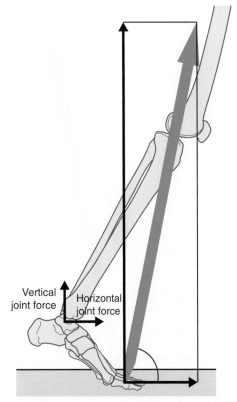

FIGURE 2.11 ■ Forces in the Achilles tendon during push off when running

From the horizontal and vertical joint forces, we can find the resultant joint force:

$$\text{Resultant joint force}^2 = 517.6^2 + 1931.9^2$$

Resultant joint force = 2000 N, which is the same as the resultant GRF we started with.

Joint Force With Muscle Forces

In this example, we will consider the force acting in the muscles posterior to the ankle joint (calf muscles) found previously. As with the GRF the muscle forces must be resolved vertically and horizontally with respect to the frame of reference. To find the joint force all the vertical and horizontal forces need to be considered separately and balanced with the vertical and horizontal joint forces (Fig. 2.12).

$$\text{Vertical component of the GRF} = 1931.9\,\text{N}$$
$$\text{Horizontal component of the GRF} = 517.6\,\text{N}$$

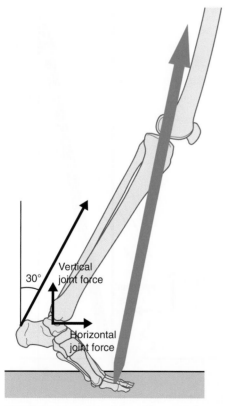

FIGURE 2.12 ■ Joint force with muscle forces

The force in the muscle is 5416 N from the earlier calculation. However, this is acting at an angle of 30° to the vertical and so has to be resolved into its horizontal and vertical components:

$$\text{Vertical component of muscle force} = 5416 \cos 30$$
$$\text{Vertical component of muscle force} = 4690.4\,\text{N}$$
$$\text{Horizontal component of muscle force} = 5416 \sin 30$$
$$\text{Horizontal component of muscle force} = 2708\,\text{N}$$

Again, if we consider up as positive and down as negative and right as positive and left as negative we get the equations:

Vertical joint force + Vertical component of the GRF + Vertical component of muscle force = 0

Vertical joint force + 1931.9 + 4690.4 = 0

Vertical joint force = −1931.9 − 4690.4

Vertical joint force = −6622.3 N

(the negative referring that this acts down, i.e. the force from the tibia onto the ankle joint)

Horizontal joint force + Horizontal component of the GRF + Horizontal component of muscle force = 0

Horizontal joint force + 517.6 + 2708 = 0

Horizontal joint force = −3225.6 N

(the negative referring that this acts to the left, i.e. a posterior joint force)

Again, from the horizontal and vertical joint forces we can find the resultant joint force:

$$\text{Resultant joint force}^2 = 3225.6^2 + 6622.3^2$$
$$\text{Resultant joint force} = 7366.1\,\text{N}$$

This equates to a compressive force on the tibia of approximately 9.4 times bodyweight in the 80-kg person, similar to the values reported by Scott and Winter (1990) of 10.3–14.1 times bodyweight.

Please note that in these examples we could also include the weight of the body segments distal to the joint of interest, in this case the foot. Although strictly speaking this should be included, its effect in this particular example is negligible; however, as we consider the effects further up the lower limb these become more important.

*So Why the Difference in Methods of
Finding Joint Forces?*

Clearly the introduction of an additional large muscle force will affect the forces at the joint itself, but why the two methods? In most biomechanical models of the lower limb we use the first method where muscle forces are not considered. This is mainly due to the fact that internal forces in muscles and ligaments are VERY hard to find or estimate accurately. At this point we should consider the work by John Paul in 1967, who calculated the forces acting at the hip joint and then, in 1970, found the effect of walking speed on the force transmitted at the hip and knee joints. Paul calculated the inter-segment forces, i.e. the joint forces, dynamically, but he also considered the forces in both tendinous and ligamentous structures to infer a joint force. This in essence is what was found in the previous example 'Joint force with muscle forces'; except Paul did not just consider one internal force from one muscle, but many forces from many structures, a feat which few have tried to replicate today, let alone with the computing power of the 1960s! Around the same time Rydell (1966) took a more 'direct' approach to finding joint forces during walking. Rydell reported two clinical cases of patients 6 months post total-hip-replacement surgery, where the hip prostheses fitted contained electrical strain gauges. Paul (1967) found that the pattern of forces acting at the hip joint varied between 1.7 to 9.2 times body weight during the gait cycle. To put this into perspective a 1000 N person would have a force of more than 9000 N acting on the hip during level walking, or to convert crudely to non-SI units nearly a tonne of force! Rydell found smaller forces of up to 3 times body weight; however, Rydell's subjects had undergone hip-replacement surgery and had significantly reduced stride lengths compared to Paul's subjects.

So, if Paul could do this why don't we do it now? The simple answer is that in most biomechanical studies the forces at the joints are not necessarily the primary research outcome measures; therefore, we do not consider the internal forces in the tendinous and ligamentous structures. More often we consider the external moments that are balanced with the internal moments the muscles provide, rather than estimating the muscle forces themselves. Interestingly though, by not including the muscle forces, the moment calculations are still correct, as all the internal forces and moments of the muscles are influenced directly from the external forces and moments, i.e. the GRF and the weight of the body segments are balanced by the internal structures.

It should be noted, though, that the action of the forces at joints is still extremely important in many musculoskeletal conditions, and models do exist that consider the forces in muscles and their effects on joint forces (Fig. 2.13). These have been very important in assisting the design of internal prosthetics such as hip joint replacements.

2.4.5 The Effect of the Weight of the Segments on Moment Calculations

In the lower-limb examples covered here we have only considered the effect of the GRFs as an external force. However, the weight of the segments themselves will also have an effect. In simple calculations, this is often ignored in lower-limb problems as the GRF will have the largest effect due to its magnitude. The weight of the segments is also frequently ignored as it does make the problems a little harder to solve!

By including the weight of the segments, the accuracy of the calculations is improved; however, it should be noted that even if we do include the segment weights this still does not consider all the forces involved when the body starts to accelerate and decelerate. To assess the effect of all the 'dynamic' forces we need to consider the topic of inertia (from Newton's second law of motion) in a lot more detail, the effect of the inclusion and exclusion of the different 'dynamic' forces is covered in Chapter 6: Inverse Dynamics Theory.

2.5 CALCULATION OF MOMENTS, MUSCLE AND JOINT FORCES IN THE UPPER LIMB

One only has to do a quick search of the biomechanics literature to realize the majority of the attention has been on the lower limb and pelvis, with less attention to the spine and upper limbs. Although the upper limb is not often in contact with the ground, this does not mean it is not subjected to significant forces. The examples covered in this section relate to normal

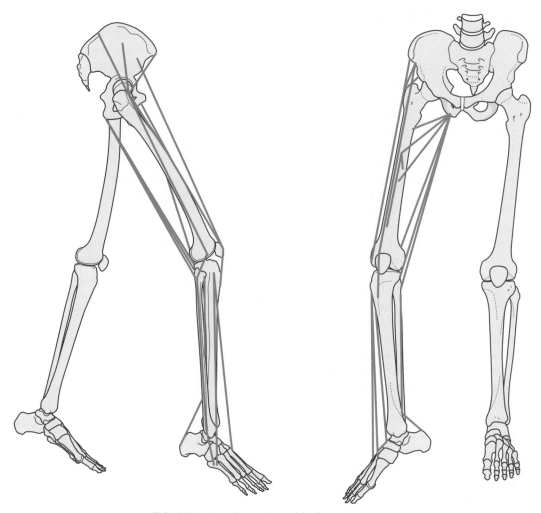

FIGURE 2.13 ■ Dynamic model of muscle actions

situations for the upper limb. In upper-limb problems, we cannot ignore the weight of the body segments, as these have a far more substantial effect as the other external forces are much smaller compared with the effect of the GRF on the lower limb, unless we are doing a handstand of course! In each of the examples the weight of the body segments, muscle forces and joint forces are considered.

2.5.1 Moments, Muscle and Joint Forces While Holding a Pint of Beer

The lower-limb examples covered so far have all involved the effect of the GRFs. The biomechanics of

drinking beer may not be the most clinically important topic in this book; however, the same concepts used here are transferable to many upper-limb activities. In the example of drinking a pint of beer (we will deal with SI units shortly), we have to consider the weight of a pint of beer, the weight of the forearm and hand, the length and position of the centre of mass of the forearm and hand, any inclination of the forearm, the point of insertion of the muscle and the line of action of the muscle. To make this slightly easier, at first we will consider that the forearm is held in a static horizontal position, not much good for drinking, but one step at a time.

The Weight of a Pint of Beer. If we assume a pint is roughly 0.568 litres and specific gravity of well-brewed beer is 1000 kg/m³, then the mass of a pint of beer is 0.568 kg. If we also add on the mass of the glass, about 0.25 kg, then the mass comes to 0.818 kg, a weight of 8.02 N.

The Anthropometry. If the person holding the glass is 1.7 m tall with a mass of 71 kg, then the length of the forearm and hand, the weight of the forearm and hand and the position of centre of mass of the forearm and hand may be found using anthropometry. The mass of the forearm and hand is 0.0215 × body mass, the length of the forearm and hand is 0.254 × height, and the centre of mass of the forearm and hand is 0.677 × forearm length. Therefore:

Mass of the forearm and hand $= 0.0215 \times 71$

Mass of the forearm and hand $= 1.5265$ kg

or weight $= mg = 14.975$ N

The length of the forearm and hand
$= 0.254 \times 1.7 = 0.4318$ m

The length of the forearm $= 0.146 \times 1.7 = 0.2482$ m

The centre of mass of the forearm and hand
$= 0.677 \times 0.2482 = 0.168$ m

These calculations assume the hand is fully extended and not gripping the glass; however, this will serve as an estimate of the forces involved (Fig. 2.14).

Moments About the Elbow. The moments about the elbow joint = (Weight of forearm × centre of mass) + (Weight of beer × Length of forearm)

Moment about the elbow joint
$= (14.975 \times 0.168) + (8.02 \times 0.4318)$

Moment about the elbow joint $= 2.526 + 3.463$

Moment about the elbow joint $= 5.989$ Nm

2.5.2 Finding the Force in the Muscle

If the muscle is inclined to the forearm at 80° and the muscle insertion point is 0.06 m away from the elbow joint, how can we find the muscle force?

The muscle must provide an equal and opposite turning moment to support the weight of the arm and the weight of the beer. However, the muscle is inclined to the forearm so the muscle force needs to be resolved so that it is perpendicular to the forearm.

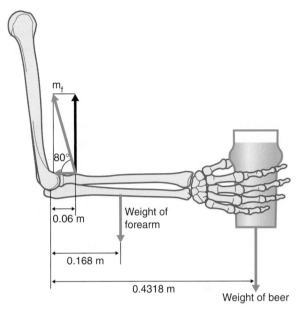

FIGURE 2.14 ■ Moments while holding a pint of beer

If the muscle force is given the symbol m_f, the vertical component (or rotary component) will be m_f sin 80.

As the muscle produces an equal and opposite turning moment, the clockwise component must equal the anticlockwise component, i.e. the muscle must provide an equal and opposite moment to the moment about the elbow joint.

Therefore:

$$m_f \sin 80 \times 0.06 = 5.989$$

$$m_f = \frac{5.989}{0.06 \times \sin 80}$$

$$m_f = 99.6 \text{ Nm}$$

It should be noted that the muscle force is significantly larger than both the weight of the forearm (14.975 N) and the weight of the beer (8.02 N). The close proximity of the muscle insertion point to the elbow joint requires a comparably large muscle force to balance the external moment.

2.5.3 Finding the Joint Force

To find the joint force we once again have to consider the forces in the vertical and horizontal directions to a frame of reference. The best frame of reference for this example would be along and at 90° to the forearm.

The forces at 90° to the forearm consist of the weight of the beer, the weight of the forearm, the vertical component of the muscle force and the joint force (Fig. 2.15):

Weight of beer + Weight of forearm − $m_f \sin 80$
    + Vertical joint force = 0

$8.02 + 14.975 − 101.36 \sin 80 + \text{Vertical joint force} = 0$

Vertical joint force $= −8.02 − 14.975 + 99.82$

Vertical joint force $= 76.83$ N

The forces along the forearm consist of the horizontal component of the muscle force and the joint force:

Horizontal joint force $− 101.36 \cos 80 = 0$

Horizontal joint force $= 17.60$ N

These horizontal and vertical components are along and at 90° to the segment. Using this frame of reference, the horizontal force will relate to the stabilizing or compressive force on the joint and the vertical force will relate to the rotary component. However, as in previous examples, we could also find the resultant joint force using the Pythagorean theorem:

$\text{Resultant joint force}^2 = 76.3^2 + 17.60^2$

$\text{Resultant joint force} = 78.8$ N

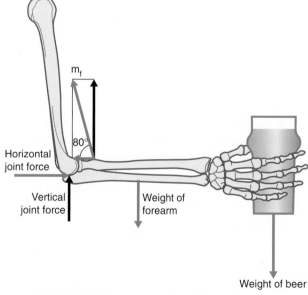

FIGURE 2.15 ■ Joint force while holding a pint of beer

2.5.4 Moments and Forces About the Elbow Joint While Holding a 20 kg Weight

Now we will use the same method to find the forces acting when holding a much larger object of mass (20 kg), with the forearm inclined down by 30° to the horizontal (Fig. 2.16).

External Moments

To find the moments about the elbow we need to find out the length and the centre of mass of the forearm and hand, and the position of the centre of mass, all of which may be found from anthropometry:

Mass of the forearm and hand $= 0.0215 \times 71 = 1.5265$ kg

or

Weight $= mg = 14.975$ N

The length of the forearm and hand
    $= 0.254 \times 1.7 = 0.4318$ m

The length of the forearm $= 0.146 \times 1.7 = 0.2482$ m

The centre of mass of the forearm
    $= 0.677 \times 0.2482 = 0.168$ m

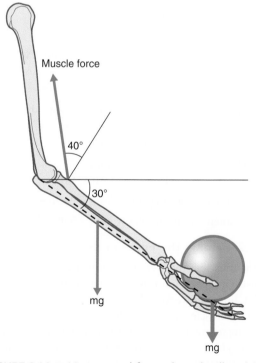

FIGURE 2.16 ■ Moments and forces about the elbow joint while holding a weight

The forces causing external moments are the weight of the body segment and the weight of the 20 kg mass (Fig. 2.17):

The moments about the elbow joint = (weight of forearm cos 30 × centre of mass) + (weight cos 30 × length of forearm)

From the previous example, the weight of the body segment was 14.975 N. The weight of the 20 kg mass = mg = 20 × 9.81 = 196.2 N:

Moment about the elbow joint
   $= (14.975 \cos 30 \times 0.168) + (196.2 \cos 30 \times 0.4318)$

Moment about the elbow joint $= 2.179 + 73.369$

Moment about the elbow joint $= 75.548 \, \text{Nm}$

Muscle Forces

If the biceps were angled at 50° to the forearm and had the same muscle insertion point as before, 0.06 m, the muscle force may be found. The external moments must be balanced with the internal moments provided by the internal structures, in this case the biceps (Fig. 2.18).

$$\text{Muscle force} \sin 50 \times 0.06 = 75.548$$

$$\text{Muscle force} = \frac{75.548}{0.06 \times \sin 50}$$

$$\text{Muscle force} = 1643.68 \, \text{N}$$

Joint Forces

Because the forearm is inclined at an angle, we need to resolve all the forces so that there is a component of the force acting along the segment and a component at 90° to the segment. This is our frame of reference or segment coordinate system. From this we can consider the forces acting at the joint (Fig. 2.19).

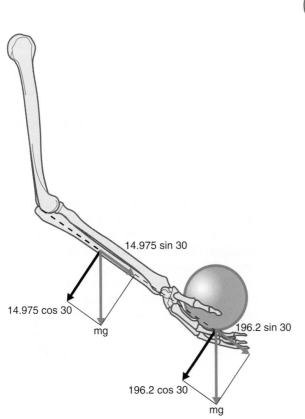

FIGURE 2.17 ■ External forces about the elbow joint while holding a weight

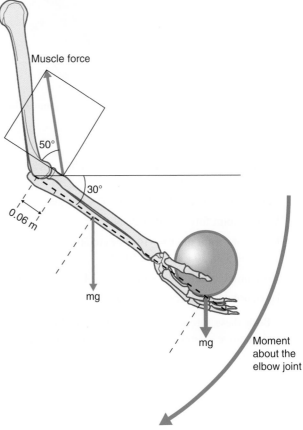

FIGURE 2.18 ■ Muscle forces about the elbow joint while holding a weight

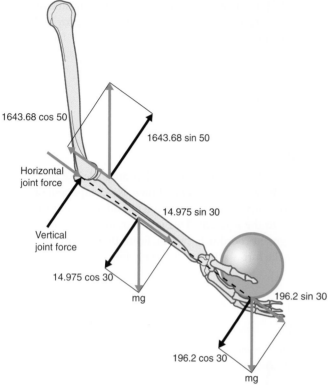

FIGURE 2.19 ■ Joint forces about the elbow joint while holding a weight

Vertical joint force $+ 1643.68 \sin 50 - 14.975 \cos 30$
$\quad - 196.2 \cos 30 = 0$

Vertical joint force $+ 1259.13 - 12.97 - 169.9 = 0$

Vertical joint force $= -1076.26$ N

Horizontal joint force $= -1643.68 \cos 50$
$\quad + 14.975 \sin 30 + 196.2 \sin 30 = 0$

Horizontal joint force $= -1056.54 + 7.4875 + 98.1 = 0$

Horizontal joint force $= -950.95$ N

Both these forces can be very useful to know. The vertical force relates to the shearing force across the joint and the horizontal force relates to the compressive joint force or axial force along the forearm. We can also find the overall resultant force acting on the joint, although this will not have any anatomical reference:

$$\text{Resultant joint force}^2 = 1076.26^2 + 950.95^2$$

$$\text{Resultant joint force} = 1436.19 \text{ N}$$

The resultant force on the elbow joint in this example is significantly greater than the person's body weight. Although 20 kg is a fairly large weight, it is far from the largest weight that could be held in this manner.

SUMMARY: FORCES, MOMENTS AND MUSCLES

- Anthropometry allows us to find important information about the proportions of body segments, including the mass, weight and the position of the centre of mass.
- The net or overall moments about joints may be found by considering all the forces acting on the distal body segments to a particular joint. Moments can also be used to describe muscle action during different movement tasks.
- By knowing joint moments and the approximate muscle insertion points, useful estimates of the muscle forces may be obtained. It is very important to consider as many of the forces present as possible to ensure a good estimate.
- Joint forces may be found by considering all the forces acting vertically and horizontally, or along and at right angles to the distal body segment. Sometimes these are found by including the estimates of the muscle forces that provide the most useful estimate of the joint forces, although these are often not included when using the inverse dynamics approach, which is considered in Chapter 6.

3

GROUND REACTION FORCES AND PLANTAR PRESSURE

JIM RICHARDS ■ AOIFE HEALY ■ NACHIAPPAN CHOCKALINGAM

This chapter considers the use of ground reaction forces (GRFs) as functional measures and the consideration of the nature of various measures that may be drawn from them. This covers GRFs during postural sway, walking and different running styles.

AIM

To consider measurement derived from GRFs in individuals who are pain and pathology free during different functional tasks.

OBJECTIVES

■ To describe the nature and use of centre of pressure measurements by considering standing balance

■ To interpret vertical, anterior–posterior and medial–lateral forces during walking in relation to function

■ To explain how to construct Pedotti or vector diagrams

■ To calculate impulse and momentum from GRF data

■ To apply the use of impulse and momentum to the GRF during running

■ To interpret vertical, anterior–posterior and medial–lateral forces during different running styles.

3.1 GROUND REACTION FORCES DURING STANDING

Borrelli in 1680 was the first to measure the centre of mass of the body and described how balance is maintained during gait. Borrelli was also the first to consider the effect of GRFs acting about joints.

A GRF is the force that acts on a body as a result of the body resting on the ground or hitting the ground. This relates to Newton's third law of motion, relating to an equal and opposite reaction force, and to the second law of motion relating to deceleration during an impact or acceleration during propulsion. If we first consider a person standing on the floor without moving, the person will be exerting a force on the floor, but the floor will be exerting an equal and opposite reaction force on the person. This reaction force is known as the GRF (Fig. 3.1).

If the person does not move, the forces under each foot will remain in exactly the same position and will not move. The position on the floor of these GRF vectors is known as the centre of pressure. The term 'centre of pressure' can be misleading as it is not a measure of pressure but a measure of position and refers to the average pressure point beneath the foot or feet. Fig. 3.1 shows the two GRF vectors under each foot. Each of these will have a relative position (centre of pressure) under each foot. I have also included a combined force that is a summation of the two forces. The centre of pressure of the combined force is often used to study postural sway, which we will consider next.

In reality, we are never completely static and, although we may well be in this position from time to time, we will also be prone to postural sway both in

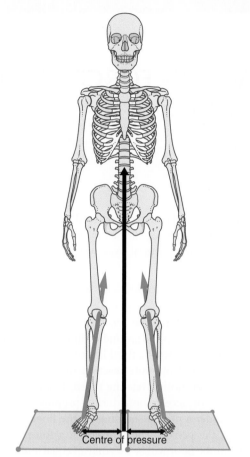

Centre of pressure

FIGURE 3.1 ■ Force and centre pressure during standing

the medial–lateral direction (Fig. 3.2), and the anterior–posterior direction (Fig. 3.3). These figures show an exaggerated postural sway test to illustrate how the forces beneath the feet vary. During postural sway the GRF will always try to point towards the approximate position of the centre of mass; so, the force vector will point inwards in relation to the base of support. If the force vectors point outwards this will produce an acceleration vector away from the centre of mass and the person will become unstable and is, therefore, likely to fall.

The position the force is acting on (centre of pressure) will also move as the person sways from side to side and back to front (Fig. 3.4). The amount of movement of the centre of pressure is often used to quantify dynamic stability during postural sway tests.

A common experiment is to look at the amount of movement in the medial–lateral and the anterior–posterior directions with eyes open and eyes shut.

3.2 GROUND REACTION FORCES DURING WALKING

3.2.1 General Description of Graph Shapes

It is possible to measure vertical, anterior–posterior and medial–lateral force graphs, centre of pressure (or point of application) graphs, and Pedotti or butterfly diagrams. For each of these graphs we can observe the shape by eye and look for differences between the plots obtained for a particular subject and a non-pathological gait pattern. Key points in the gait cycle can then be identified on the traces and measurement taken. For each of these measurements, the percentage difference can be studied between the left and right side, and between the subject tested and individuals who are pain and pathology free. This will not only identify what differences are present in the walking patterns but also how big these differences are.

3.2.2 Vertical Force Measurements

Vertical force measurements are by far the most quoted in the literature. In both this case and in the previous section, a variety of useful measurements may be taken. These include: the first peak or maximum vertical loading force; the dip trough, which can give useful information about the movement of the body over the stance limb; and the second peak or maximum vertical thrusting force.

All these measures may be reported in newtons, but are more usually reported with respect to body weight. This is where the value in newtons is simply divided by the person's body weight. Normal values for the two peaks are in the order of 1.2 the value of body weight, with the trough usually being in the order of 0.7 the value of body weight, although these are dependent on walking speed.

The timing at which these peaks and troughs occur can also be a very useful measure. So, corresponding timing measures are: the time to the first and second peaks (the time to maximum vertical loading and maximum vertical push-off force), the time to the trough (which is sometimes referred to as midstance, although we will consider this again when looking at

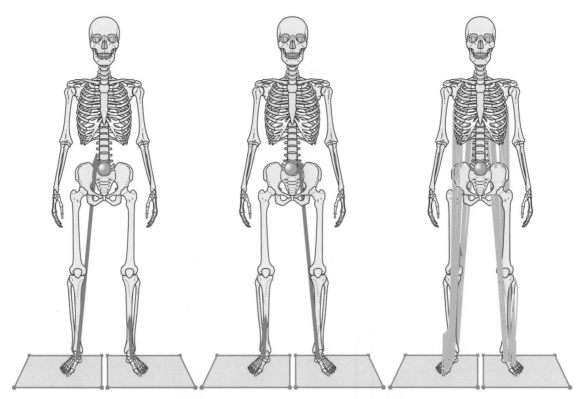

FIGURE 3.2 ■ Postural sway in the medial–lateral direction

the anterior–posterior forces) and the total contact time. The usefulness of time in seconds is debatable as this will relate directly with walking speed; however, the time to each of these events in relation to the total stance time can be very useful (Fig. 3.5A, B). The vertical component of the GRF can be split into four sections. Each section may be related to functional events during foot contact and can give us important information about the overall functioning of the lower limb.

Heel Strike to First Peak

This is where the foot strikes the ground and the body decelerates downwards and transfers the loading from the back foot to the front foot during initial double support. The first peak should be in the order of 1.2 times the person's body weight.

The first peak relates to the amount of loading the person is putting onto the front foot. In amputee gait, for example, this can relate to the person's confidence in the prosthetic limb, a reduction in the loading relating to poor confidence. A reduced loading could also relate to the presence of any pain and discomfort, poor functional movement of the joints of the lower limb or a slow walking speed.

First Peak (F1) to Trough (F2)

As the body starts to progress the knee extends, raising the centre of mass. As the centre of mass approaches its highest point it is slowing down or decelerating its upwards motion. This has the same effect as going over a humped-backed bridge in a car: as you reach the top of the hump you feel very light, i.e. the contact force is reduced. This deceleration of the body upwards produces a dip or trough in the vertical force pattern, with the normal value being in the order of 0.7 times the person's body weight.

The depth of the trough, therefore, relates to how well the person moves over their stance limb, which again could be affected by pain and/or lower-limb dysfunction. A high trough value, or shallow trough,

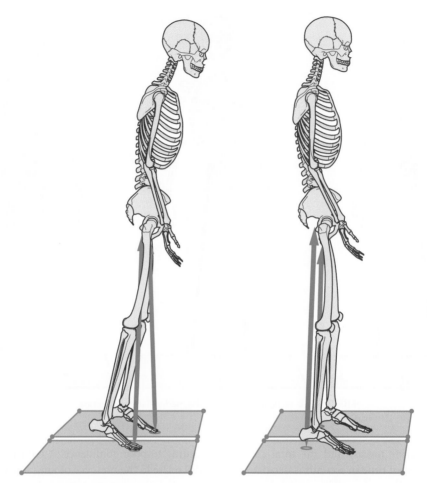

FIGURE 3.3 ■ Postural sway in the anterior–posterior direction

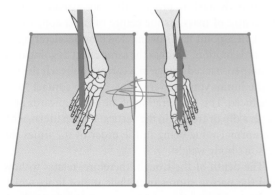

FIGURE 3.4 ■ Movement of the centre of pressure during postural sway

can relate to a poor movement of the body over the stance limb or a slow walking speed. A low or deep trough may be produced by a fast walking speed or large vertical translations of the body during walking. The trough usually occurs at 50% of the stance phase, although this does vary between individuals, and should occur at approximately the same time as the crossover point of the anterior–posterior force (Fig. 3.6).

Trough (F2) to Second Peak (F3)

The centre of mass now falls as the heel lifts and the foot is pushed down and back into the ground by the action of the muscles in the posterior compartment

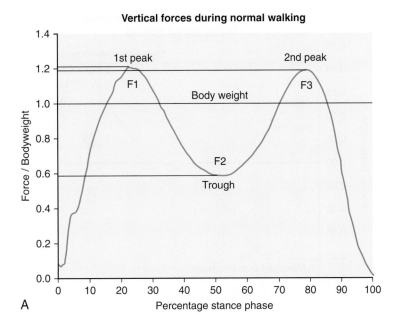

Vertical forces during normal walking

Time to vertical forces during normal walking

FIGURE 3.5 ▪ (A and B) Timings and force in the vertical direction during normal walking

of the ankle joint. Both the deceleration downward and propulsion from the foot and ankle complex cause the second peak. The second peak should be in the order of 1.2 times the person's body weight.

The second peak relates to the amount of vertical propulsive force, which drives the person upwards. A low peak relates to a poor ability to push off, whereas a high peak could relate to the person accelerating. Although this is important, this does

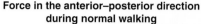

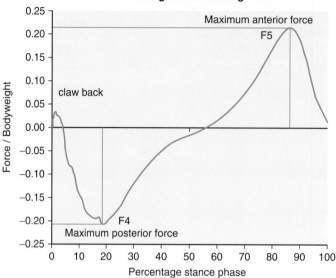

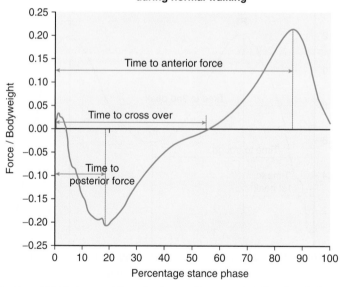

FIGURE 3.6 ■ (A and B) Timings and force in the anterior–posterior direction during normal walking

not tell us about the force that drives the person forwards.

Second Peak (F3) to Toe Off

The foot is unloaded as the load is transferred to the opposite foot. The time taken to offload from the back foot will relate to the speed of transfer of the weight to the front foot; therefore, the longer the offloading period from the back foot, the lower the first peak during loading on the front foot. When investigating force patterns, great care should be taken in considering the forces under both feet as a poor push off and offloading phase may cause changes in the initial loading on the opposite side.

Anterior–Posterior Force Measurements

As with the vertical force measurements, the anterior–posterior forces may also be studied. These include: the negative peak or maximum posterior loading force, and the positive peak or maximum anterior thrusting force. Again, both of these measures may be reported in newtons or with respect to body weight. Normal values for the two peaks are in the order of 0.2 the value of body weight, but again these are dependent on walking speed.

The negative and positive aspect of these peaks can cause some confusion, as this depends on the direction of walking over the force platforms; for example, if an individual walked in exactly the same way to the right and then to the left, then the results would show mirror images if the mirror were placed along the time axis. Therefore, the value of the maximum loading force and maximum thrusting forces would be the same, apart from in one direction, they would be negative and positive, and in the other direction they would be positive and negative. There is no clear right and wrong to this, other than NEVER refer to them as positive and negative forces and ALWAYS refer to them as posterior and anterior forces.

As with the vertical forces, the timing at which these peaks occur can also be a very useful measure. So, the corresponding timing measures are the time to the maximum posterior loading force and maximum anterior thrusting force. A third timing measure may also be taken, which is the time to crossover; this is where there is no anterior or posterior force acting. If this is the case, then the body must be directly over the stance limb. Therefore, this would be a better measure of the point in time mid-stance occurs, rather than the trough on the vertical force graph. It is interesting to note that in most people who are pain and pathology free these can be subtly different; however, during many pathological movement patterns these can be very different (Fig. 3.6A, B).

One final measurement that can be taken is the impulse, the area under the force–time graphs. These may be considered separately as posterior and anterior impulse in the same way as earlier (Fig. 3.7).

As with the vertical forces, the anterior–posterior GRFs can also give us important information about the overall functioning of the lower limb. The

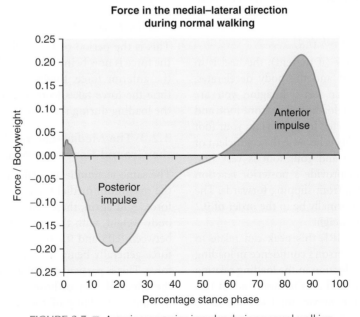

FIGURE 3.7 ■ Anterior-posterior impulse during normal walking

anterior–posterior component of the GRF during walking may also be split into four sections (Fig. 3.6A).

Claw Back and Heel Strike Transients

Claw back is an initial anterior force that is not always present during walking. This is caused by the swinging limb hitting the ground with a backwards velocity, thus causing an anterior force as the leg decelerates. Claw back is often exaggerated during marching as the swing limb is driven back to meet the ground. A heel strike transient is a rapidly increasing force due to an impact. Consider walking with and without shoes: with shoes we will have better shock absorption due to the properties of the shoes; however, when shoes are removed a rapid 'transient' force can often be seen as the unprotected heel strikes the ground.

The following example shows a subject with a large heel strike transient force with claw back. In this example the transient force at impact is greater than that of both the loading and propulsive forces. The claw back has the effect of sending the force anterior to (in front of) both the knee and hip joint. This person had problems with knee and hip pain; however, with the introduction of heel cushioning, this effect was significantly reduced and the person's pain was alleviated. Although patterns as extreme as this are rare, this pattern demonstrates the nature of transient GRFs and claw back (Fig. 3.8A, B).

Heel Strike to Posterior Peak (F4)

After the initial claw back (if present), the heel is in contact with the ground and the body decelerates, causing a posterior shear force. Imagine you are walking on a thick carpet, loading your front foot, and suddenly you are transported to an ice rink; your foot would slide forwards. This is because the coefficient of friction between the ice and your foot is very low, whereas the carpet can provide a posterior reaction force that stops your leg from slipping forwards. The posterior peak should normally be in the order of 0.2 times the person's body weight.

As with the vertical GRF, this peak can relate to speed of walking or the person's confidence in loading the front foot, with a reduction in the loading relating to poor confidence. We will also naturally adapt this value depending on the maximum frictional force available, which will depend on the coefficient of friction between our shoes/feet and the surface we are walking on.

Posterior Peak to Crossover

The posterior component reduces as the body begins to move over the stance limb, reducing the horizontal component of the resultant GRF. At the crossover point, the horizontal force is zero; therefore, the only force acting is that of the vertical GRF. At this point the body is directly positioned above the foot. This is sometimes defined as the point of midstance, which is usually at 55% of stance phase, and should also approximately correspond with the trough in the vertical force pattern.

Crossover to Anterior Peak (F5)

The heel lifts and the foot is pushed down and back into the ground by the action of muscles in the posterior compartment of the ankle joint. This has the effect of producing an anterior component of the GRF that propels the body forwards. As with the other force measurements, this is dependent on walking speed; however, the anterior peak should be in the order of 0.2 times the person's body weight. A reduced peak would tell us the person is not propelling the body forwards well no matter what the vertical force pattern may show.

Anterior Peak to Toe Off

This is the period of terminal double support where the force is now being transferred to the front foot and the anterior force, therefore, reduces. The length of time the force takes to reduce and offload can affect the loading during the next foot contact.

3.2.3 The Medial–Lateral Component of the Ground Reaction Force

The same technique as earlier may also be applied to the maximum medial force and the maximum lateral force. And again, these can be related to the person's body weight, with the maximum medial force being between 0.05 and 0.1 of body weight and the lateral force generally being less than the maximum medial force. This is particularly interesting when we consider the effect of foot orthoses: in particular, the use of posting or wedging of the rearfoot, which can have substantial effects on these forces.

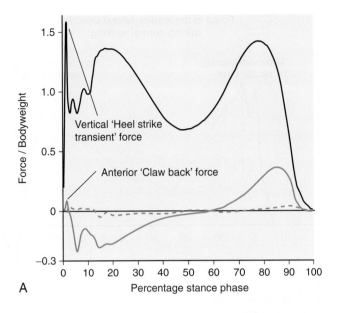

A

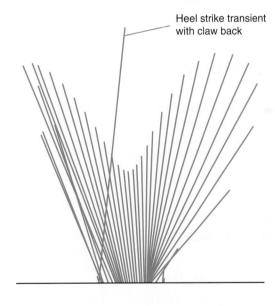

B

FIGURE 3.8 ■ (A and B) Claw back and heel strike transients

The medial–lateral component may be split into two main sections. Initially, at heel strike there is a lateral thrust during loading, during which time the foot is working as a mobile adaptor and generally moving from a supinated position into pronation. After the initial loading, the forces push in a medial direction as the body moves over the stance limb. Small lateral forces are often seen during the final push-off stage (Fig. 3.9).

The medial–lateral forces are the most variable of the three components, and can be easily affected by footwear and foot orthoses. Normally the maximum medial force is between 0.05 and 0.1 of body weight. The maximum lateral force should generally be less

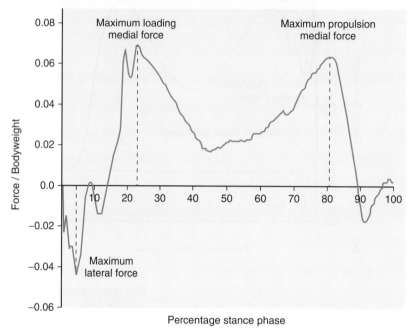

FIGURE 3.9 ■ Medial-lateral component of the GRF

than the maximum medial force. Although the medial–lateral forces can be variable, they can have a substantial effect on the loading of the ankle and knee in the coronal plane and should not be ignored when considering the effects of shoe modifications and orthotic management.

3.3 CENTRE OF PRESSURE AND FORCE VECTORS DURING NORMAL WALKING

3.3.1 Centre of Pressure During Walking

The term 'centre of pressure' is the position of the force coming out of the floor, as mentioned earlier. The centre of pressure can move forward and backward (anterior and posterior) and side to side (medial and lateral) under each foot. Fig. 3.10A, B shows the force appearing under the heel during loading and under the toes during push off.

The centre of pressure during walking is often presented as (A) the anterior–posterior against time, (B) medial–lateral centre of pressure against time, or (C)

a combined medial–lateral and anterior–posterior (Fig. 3.11). The centre of pressure against time in the different directions allows the progression and speed of progression to be investigated in more detail. The combined anterior–posterior and medial–lateral graph gives a useful picture of how the force moves from heel to toe and any variations in the medial–lateral movement of the centre of pressure.

One example where we can see the use of centre of pressure is in the examination of a subject with an early heel lift during walking. Fig. 3.12A shows the anterior–posterior centre of pressure against time and Fig. 3.12B shows the vector (Pedotti) diagram. This subject showed a pronounced loading pattern (1), coupled with a fast movement of the centre of pressure forwards until it reaches the metatarsal heads when the heel begins an early lift off the ground (2). This dwelling of the force under the metatarsal heads can be seen by a close grouping of the force vectors. The subject then shows a faster final movement of the centre of pressure forwards from the metatarsal heads to the toes during push off (3).

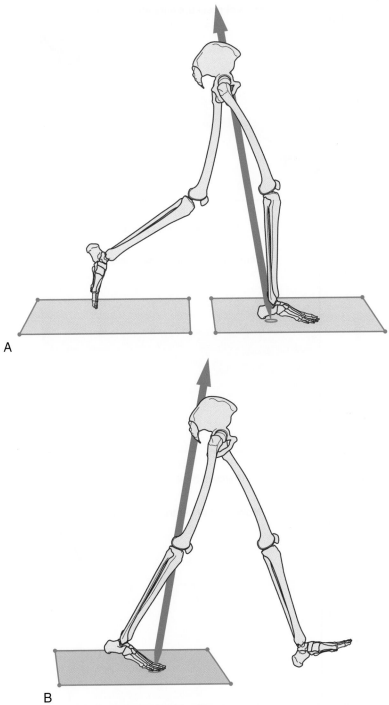

FIGURE 3.10 ■ GRF during walking: (A) loading force and (B) push-off force

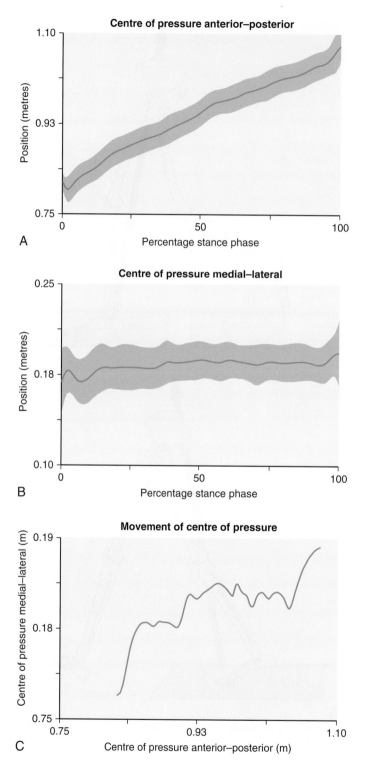

FIGURE 3.11 ■ (A) The anterior–posterior against time, (B) medial–lateral centre of pressure against time, and (C) a combined medial–lateral and anterior–posterior

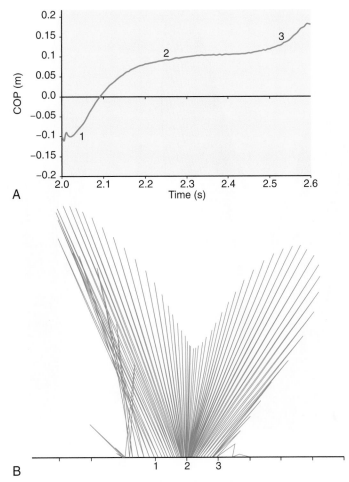

FIGURE 3.12 ■ Forces and centre of pressure (COP) with an early heel lift: (A) anterior–posterior COP and (B) vector (Pedotti) diagram

3.3.2 Resultant Ground Reaction Forces and Pedotti Diagrams?

The resultant GRF is made up from three components: vertical, anterior–posterior and medial-lateral. The interaction of the vertical, anterior–posterior and medial–lateral force may be shown with a Pedotti diagram. Figs 3.13A and B show a Pedotti diagram; note how the magnitude and direction of the force changes and how the centre of pressure moves from heel to toe. The Pedotti diagram shows the magnitude of the resultant GRF.

The GRF points posterior for the first part of stance phase; the magnitude of the force increases during loading response (deceleration phase), then decreases as the body moves over the stance limb during mid-stance. After midstance, the magnitude of the force increases again, but now the force is pointing in an anterior direction during propulsion (acceleration phase).

The Pedotti diagram shows the magnitude of the resultant GRF and is a good way of visualizing the interaction of the forces in different directions with the centre of pressure. However, to determine the magnitude and function of the different aspects of the GRF it is easier to consider each component separately.

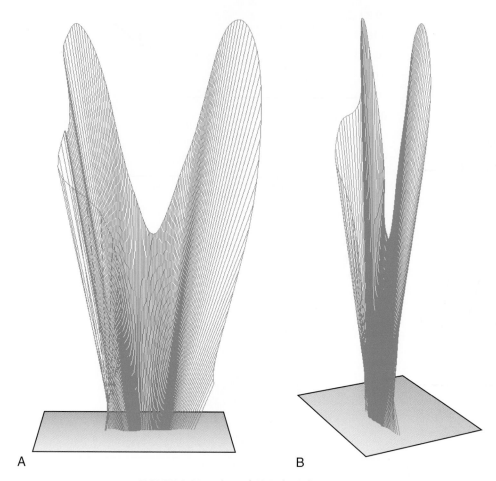

A   B

FIGURE 3.13 ■ (A and B) Pedotti diagrams

3.3.3 Construction of Pedotti Diagrams

Pedotti diagrams rely on the information provided by force platforms. To construct a Pedotti diagram we need to know the vertical and horizontal forces, and the position of the centre of pressure in the plane of interest for each moment in time.

During stance phase the forces move forward from under the heel to the under the toes. So, the centre of pressure is described as moving from posterior to anterior. As we have seen in the previous sections, the vertical and horizontal GRFs are continually changing during stance phase, therefore, changing the direction and magnitude of the resultant GRF.

Fig. 3.14A–F shows the vertical, horizontal and resultant GRF components being drawn at heel strike. The centre of pressure then moves forwards and the new vertical, horizontal and resultant GRF components are drawn from the new position. This is repeated throughout stance phase, giving a butterfly-like diagram.

3.3.4 How Force Vectors Relate to Muscle Activity

To estimate the muscle activity, consider which side of the joint the GRF passes, i.e. in front or behind. If the GRF passes in front of the knee joint this will try to extend the knee; therefore, the knee flexors need to be

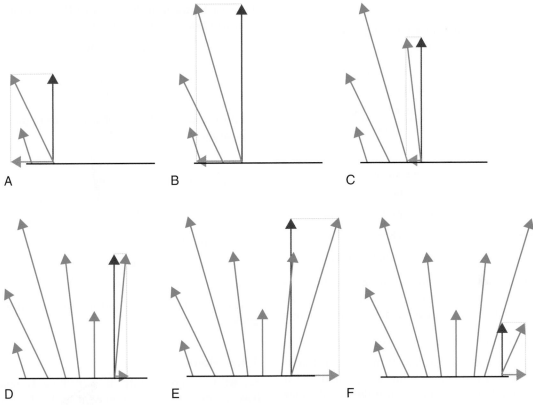

FIGURE 3.14 ■ (A–F) Construction of Pedotti diagrams

working to support the turning moment produced. This is sometimes referred to as a flexor moment (meaning the external moment will try to extend but the body needs to produce an internal flexion moment from the knee flexors). If the GRF passes behind the knee joint this will try to flex the knee; therefore, the knee extensors need to be working to support the turning moment produced. This is sometimes referred to as an extensor moment (meaning the external moment will try to flex but the body needs to produce an internal extension moment from the knee extensors). It must be emphasized that, although very useful in observation gait analysis, these can only be used for estimation and do not give precise numerical data.

During swing phase, there is no effect due to the GRF. However, muscle activity is still required to overcome the inertial forces due to the acceleration and deceleration of body segments. Moments during the accelerations and decelerations during swing phase

can be found using a technique called inverse dynamics (Chapter 6: Inverse dynamics theory).

3.4 IMPULSE AND MOMENTUM

3.4.1 Impulse

Although force alone is extremely useful, as it tells us how big propulsion forces are, it does not tell us all we may need to know about the entire propulsive stage. Impulse and momentum can give us important additional information if we have a force acting over a given time. This can be useful when considering the effect of an impact where we may have rapid changes in velocity over very short periods of time or when considering changes in velocity during propulsion. When considering the concepts of impulse and moments we first need to go back to Newton's second law of motion:

$$F = ma$$

where:

F = force applied, m = mass of the object,
a = acceleration of the object

If we now consider what we mean by acceleration, i.e. a change in velocity over time:

$$\text{Acceleration} = \frac{\text{change in velocity}}{\text{time}}$$

So, we could say that:

$$\text{Force} = \frac{\text{m} \times \text{change in velocity}}{\text{time}}$$

From this we can find the equation for force × time:

$$\text{Force} \times \text{time} = \text{Ft} = \text{mass} \times \text{change in velocity}$$

$$\text{Ft} = \text{m (final velocity} - \text{initial velocity)}$$

This product of both force and time is known as the impulse force or impulse. This tells us that an increase in force or a longer time will give a larger impulse.

Consider the impact or impulse of an object of mass 2 kg as it hits the ground. If the object is travelling at 20 m/s just before it hits the ground and it comes to rest after the impact, then the impulse may be found (Fig. 3.15).

$$\text{Ft} = \text{m (final velocity} - \text{initial velocity)}$$

$$\text{Ft} = 2\,(0 - 20)$$

$$\text{Ft} = -40$$

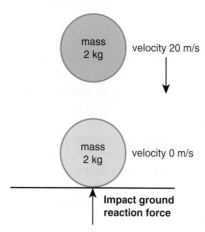

FIGURE 3.15 ■ Impulse during impact

But what are the units for impulse? If we consider the right-hand side of the equation:

$$\text{mass (kg)} \times \text{velocity (m/s)}$$

the units may be written:

$$\text{Kg m/s}$$

But if we now consider the left-hand side of the equation we get:

$$\text{Ns}$$

These are in fact the same thing:

$$F(N) = m(kg) \times a(m/s^2) \quad \text{or} \quad F(N) = \frac{kg \times m}{s^2}$$

So, the units of impulse can be written:

$$F(N)\,t(s) = \frac{kg \times m}{s} \text{ or } kg \cdot m/s$$

However, impulse is usually written with the units Ns.

3.4.2 Momentum

So, what is momentum? If something is said to have momentum, this implies it is hard to stop. A heavy object travelling fast will have a much greater momentum than a light object travelling slowly. This is due to both the mass of the object and the object's velocity. So, momentum may be considered as simply the mass of the object multiplied by its velocity.

$$\text{Momentum} = \text{mass} \times \text{velocity}$$

If we consider impulse again, the impulse is the mass multiplied by the change in velocity. Therefore, the impulse tells us about the change in momentum.

$$\text{Change in momentum} = \text{mass} \times \text{change in velocity}$$

$$\text{Force} \times \text{time} = \text{Ft} = \text{mass} \times \text{change in velocity}$$

3.4.3 Impulse and Change in Momentum During a Sprint Start

We have seen earlier how impulse relates to Newton's second law of motion, but how can we relate this to GRF patterns? A good example is to consider the impulse and change of momentum during a sprint start. Fig. 3.16 shows graphs of the vertical and anterior–posterior forces of the back and front feet during a sprint start in an individual who is able to run 100 m in approximately 11 seconds.

Impulse may be written as force × time. This means that the area under a force–time graph must relate to the impulse and, therefore, the change in momentum and velocity. So how large is the change in velocity during the sprint start and what are the contributions of the front and back foot to the total change in momentum? It should be stressed at this point that the data presented were collected without the use of starting blocks (Fig. 3.16). The net anterior–posterior impulse from the back foot and the front foot may be found by finding the **area under the force–time graph** or '**integrating**' the force–time graph between the start of the movement to the point each foot leaves the ground:

The net anterior–posterior impulse for the back foot
= 50.6 Ns
The net anterior–posterior impulse for the front foot
= 69.5 Ns

Therefore, the back foot is responsible for 42% of the horizontal acceleration and the front foot is responsible for 58%. The total net anterior–posterior impulse for the sprint start may be found by simply adding these values together:

$$\text{Total impulse} = 50.6 + 69.5 = 120.1 \text{ Ns}$$

If the mass of the individual is 62 kg, then the change in velocity just during the push-off phase of the sprint start may be found.

$$\text{Impulse (Ft)} = \text{mass} \times \text{change in velocity}$$
$$120.1 = 62 \times \text{change in velocity}$$
$$\text{Change in velocity} = 1.94 \text{ m/s}$$

This is a significant proportion of the average velocity of the total 100 m sprint:

$$\text{Average velocity of sprint} = \frac{100}{11}$$
$$\text{Average velocity of sprint} = 9.09 \text{ m/s}$$

From the change in velocity occurring over 0.45 s, i.e. the time taken from the start of the movement to the moment the front foot leaves the ground, the acceleration over the start may also be found.

$$\text{Average acceleration} = \frac{\text{change in velocity}}{\text{time}}$$
$$\text{Average acceleration} = \frac{1.94}{0.45}$$
$$\text{Average acceleration} = 4.3 \text{ m/s}^2$$

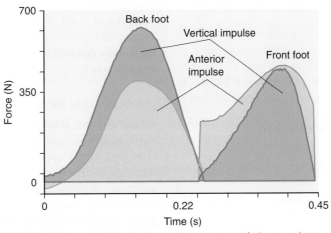

FIGURE 3.16 ■ Impulse and change in momentum during a sprint start

3.4.4 Protection Against the Force of Impacts

The relationship between the impulse, velocity and forces is very important when considering the mechanics of impacts. If a person jumps or falls, if they can lengthen the time of impact, the magnitude of the force is reduced in proportion. One example of this is bending the knees when landing from a jump. Increasing the length of time of the impact using joint movement is not always possible. One solution is to use different materials to act as a 'shock absorber'. Effective shock absorbers lengthen the time over which the impact is occurs. This may be seen in many types of running shoe which incorporate many different devices and materials that act as shock absorbers, but more about this when we consider impact loading forces during running in Section 3.6.

3.5 INTEGRATION AND THE AREA BENEATH DATA CURVES

In the previous section, we considered the concept of impulse and momentum, which may be found by calculating the area under the force–time graph. Integration, or finding the area beneath data curves, can tell us important biomechanical information: the area under a velocity–time graph can tell us the distance travelled; the area under force–distance graphs can tell us about the work done; and the area under force–time graphs can tell us the impulse or change in momentum. However, all these patterns can be extremely complex, so how can we integrate complex data curves?

3.5.1 Integration

Integration is a way of finding the sum of the area under the graph. We can integrate between limits, i.e. choose the starting point and the finishing point between which we can find the area.

The general formula for integration is:

$$\text{Integral of } X^n = \frac{X^{n+1} + c}{n+1}$$

So if a curve had the relationship:

$$y = x$$

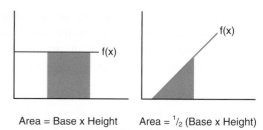

Area = Base x Height   Area = $\frac{1}{2}$ (Base x Height)

FIGURE 3.17 ■ Integration of simple shapes

which can be written x^1, the integral would be:

$$\frac{1}{2}x^2$$

3.5.2 Integration of Simple Shapes

If we first consider the area, or integral, of simple data curves, these can be found by considering the height and width of simple shapes, such as triangles and rectangles (Fig. 3.17).

Although this is a useful technique in biomechanics, it is usual that the relationships between parameters are far more complex. To use this technique, we also need to know the equation for a data line. In biomechanics this is seldom the case (Fig. 3.18). In cases such as the force–time graph of the sprint start, we need a simple and accurate method of finding the area under the graph that does not require knowing the equation to describe the line.

3.5.3 Counting the Squares

One method is to divide the pattern into a series of squares and count them (Fig. 3.19). This is time consuming and is open to errors such as miscounting, and inaccuracy due to the size of the squares, and can give you a headache!

3.5.4 Bounds for the Area

When it is not possible to find the area under the curve exactly we need to approximate. One method is to look at the upper bounds of the area, where a rectangle is drawn around the data. When using the technique of drawing the rectangle it is clear that in the previous case the rectangle is going to be much greater than the two areas, and will give a considerable error (Fig. 3.20A, B).

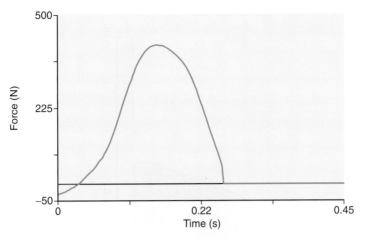

FIGURE 3.18 ■ Force during a sprint start

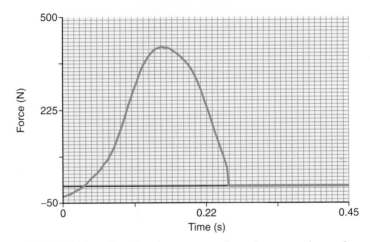

FIGURE 3.19 ■ Counting the squares to determine area under graph

3.5.5 The Rectangular Rule

If we play around with the height of the rectangle we will find one with a height that gives us a completely accurate estimate of the area, but this is not always easy to find. What we need in practice is to obtain a fairly good answer easily.

3.5.6 Trapezium Rule

If we consider the graph in Fig. 3.18 and draw vertical lines of set width from the horizontal axis until they meet with the data curve, and then draw a line between these two meeting points, we create a series of trapeziums. The area of each trapezium may then be found separately (Fig. 3.21A, B, C).

The area of each trapezium may be found by the following equation:

Area = ½(height at 1 + height at 2) × width of base

What we are doing is finding the average height of a trapezium and multiplying by the width of the base:

Area = average height of strip × width of strip

The larger the widths of strip the less accurate this method is. The more strips we divide the curve into the greater the accuracy of the estimation of the area under the graph. Fig. 3.21C shows 15 strips, whereas

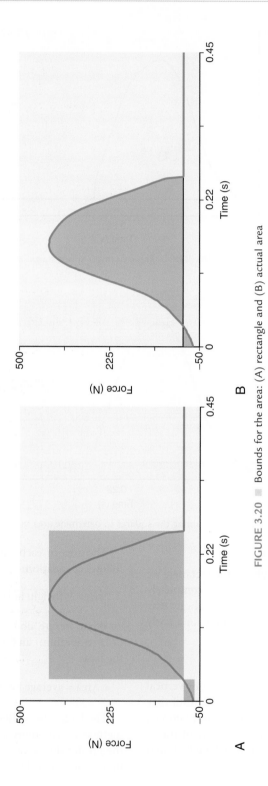

FIGURE 3.20 ■ Bounds for the area: (A) rectangle and (B) actual area

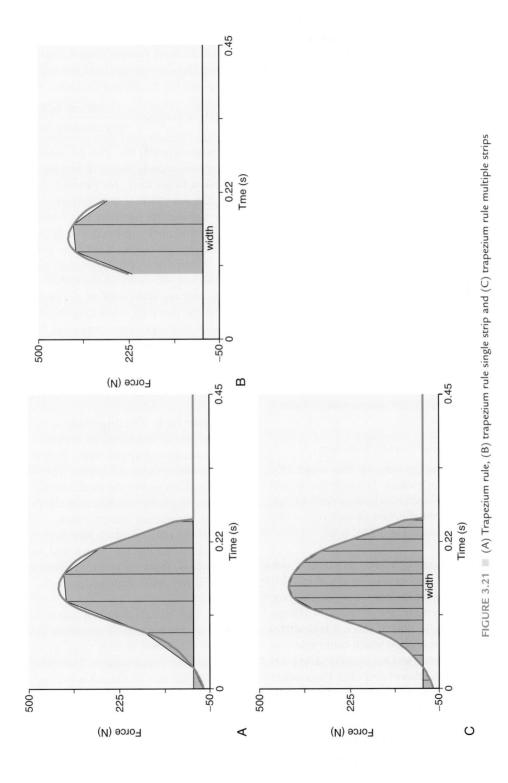

FIGURE 3.21 ■ (A) Trapezium rule, (B) trapezium rule single strip and (C) trapezium rule multiple strips

in Fig. 3.21A we considered only 6. This shows that the top of the trapeziums fits much better to the line of the data, although we are still in essence fitting a straight line to a line that is, in fact, curved.

When we use this technique in biomechanics the widths of the strips are the individual frame rates. These particular data were collected at 400 readings per second, with the total time of the data pattern being 0.25 second; therefore, we would normally divide up this pattern into 100 trapeziums. Although this is still 'an estimate' of the area under the curve, it is clear that the error becomes very small indeed.

3.6 GROUND REACTION FORCE PATTERNS DURING RUNNING

As with walking, we can study the forces during any activity where there is either contact or impact with the ground. One activity which demonstrates the concept of impulse and momentum is running, as the force acts over a very short period of time. During walking the contact phase is in the order of 0.6 second or 60% of the gait cycle; during running the force acts for approximately 0.25 second or 30% of the gait cycle, although this is open to considerable variation depending on speed.

3.6.1 Vertical Forces During Running

The vertical forces during running show very little similarity with those for walking. Although the pattern may still loosely be divided into a loading peak, trough and propulsive peak, the function of each of the different sections is different (Fig. 3.22A, B, C).

Rate of Loading. The rate of loading tells us how well the shock from initial contact is being absorbed. This is found by taking two measurements of vertical force and the time that they occur and dividing the change in force by the change in time. If this is a high value, then shock absorption is poor, which could relate to poor function of the ankle and knee joint and/or poor shock absorbency of the shoes being used. High values of loading rate occur if the force is reached too quickly. In the graphs in Fig. 3.22 we can see that rearfoot strikers have the greatest impact loading rate, followed by midfoot strikers, with forefoot strikers having the lowest rate of loading. This parameter is particularly important for footwear design and any shock-related problems and some overuse injuries. The data from the three different types of impact would indicate that different footwear designs and orthotic management are required for the different running styles.

$$\text{Rate of loading} = \frac{\text{change in force}}{\text{time taken for that change}}$$

The measurements that can be taken to calculate loading rate include the peak impact force and the time taken to get there. For instance, the values from Fig. 3.22A were 1800 N over 0.03 second, which gives an average loading rate over this time of 60 000 N/s. This is sometimes expressed as rate of loading per body weight (BW/s). So, if we know the body weight is 750 N then this would equate to 80 BW/s. However, this does not tell us about the instantaneous rate of loading and can underestimate the peak loading rate. This can be found by calculating the change in force over each successive time period, in this case 1/400 seconds or 0.0025 second (Fig. 3.23). It is also possible to calculate the loading rate in the anterior–posterior and medial–lateral directions which could also be relevant to some overuse injuries in running.

The Impact Peak. The magnitude of this shows how hard the person is hitting the ground, i.e. the force which is attenuated up the stance limb in the vertical direction during the initial impact. This initial loading peak should not exceed the maximum vertical propulsive force. Again, the magnitude of this peak relates to the shock absorbency characteristics, in particular the nature of the initial contact, with heel strikers (rearfoot strikers) having the largest and the most discernible peak, followed by the midfoot strikers, with forefoot strikers often with no discernible impact peak.

Trough. This trough does not correspond with the zero crossing in the anterior–posterior forces as it does in walking; however, it does relate to a reduction in force after the initial impact. The reduction of this force approximates to the ankle moving rapidly into plantarflexion to the foot flat position. This trough is barely discernible in midfoot strikers, and not present during forefoot running due to the fact that the ankle moves less and the movement is into dorsiflexion and not plantarflexion.

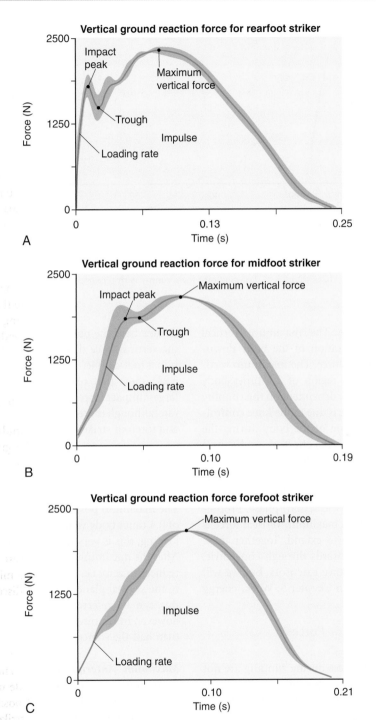

FIGURE 3.22 ■ Vertical forces during running: (A) typical rearfoot striker, (B) typical midfoot striker and (C) typical forefoot striker

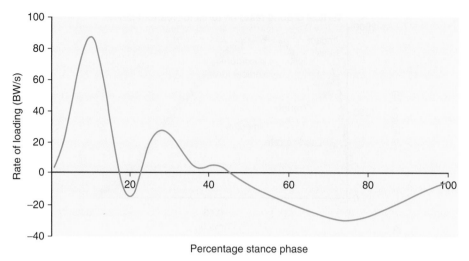

FIGURE 3.23 ■ Loading rate for a runner with a rearfoot strike

Maximum Vertical Force. The maximum vertical force relates to the deceleration of the body downwards or vertical breaking force. The maximum vertical force is usually in the region of 2.5 times body weight; however, this is very dependent on the running speed. The increase in force is due to the knee controlling the vertical deceleration of the body during the loading phase. This also has the effect of producing a stretching/lengthening effect (eccentric muscle action), on the knee extensors, which initiates a stretch shortening cycle, which in turn aids propulsion during the shortening (concentric muscle action) phase. Propulsion starts shortly after the maximum vertical force is reached as the knee starts to extend. Together these act to propel the body forwards through concentric, power production of the knee extensors, but we will cover this in more detail in Chapter 5: Work, energy and power.

3.6.2 Anterior–Posterior Forces During Running

The anterior–posterior forces during running are not dissimilar to those for walking; as with walking, the pattern can be divided up into a period of loading and a period of propulsion (Fig. 3.24A, B, C).

The Posterior Impact Peak. As with the vertical impact peak, this is the magnitude of the force telling us how hard the person is hitting the ground. As with the vertical force this is dictated by the nature of the initial foot contact and the footwear. The graphs (Fig. 3.24) show that rearfoot strikers again produce the largest impact forces, with the largest posterior loading rate, although there is some variation between midfoot and forefoot strikers.

Maximum Posterior Breaking Force. This is the maximum posterior force that occurs during the loading or breaking as the body decelerates at impact. The maximum posterior force is usually in the region of 0.4 times body weight; however, as with the vertical loading, this is very dependent on the running speed. After the maximum posterior breaking force, the force reduces to zero, i.e. the crossover point when the force in the sagittal plane acts straight up. As with walking, this can be referred to as the point of midstance; however, in running this usually occurs at slightly less than half the stance time.

Maximum Anterior Thrusting Force. This is the maximum anterior force that occurs during propulsion as the body accelerates forward. During this horizontal propulsion phase, the centre of pressure moves forward under the forefoot, which puts the force a greater distance away from the ankle joint. This maximizes the moment about the ankle and the power

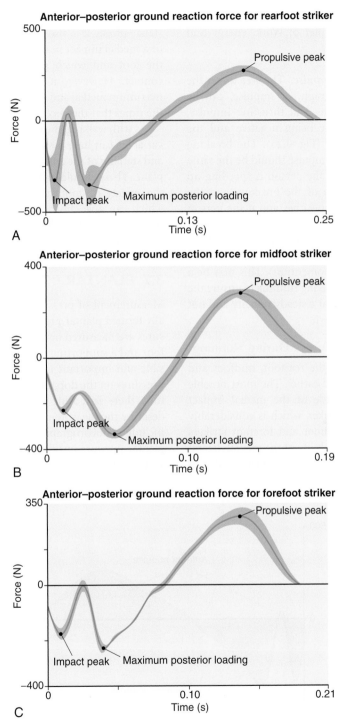

FIGURE 3.24 ■ The anterior–posterior forces during running: (A) typical rearfoot striker, (B) typical midfoot striker and (C) typical forefoot striker

production and is responsible for the driving of the person forwards (see Chapter 5: Work, energy and power).

Breaking and Thrusting Impulse. The area under the anterior–posterior force graph, or impulse, can be divided easily into breaking and thrusting impulse, with the breaking impulse being negative and the thrusting impulse positive (Fig. 3.25). The breaking impulse and the thrusting impulse should be the same in magnitude. However, if the person is speeding up the **net impulse** (the sum of the breaking and the thrusting impulse) will be positive, and if the person is slowing down the net impulse will be negative. From the breaking and thrusting impulse, the exact deceleration or acceleration of the body may be found (see Section 3.4 Impulse and momentum). This may be a very useful check when studying running to ensure the subject is, in fact, running at a steady velocity and not speeding up or slowing down.

3.6.3 Medial–Lateral Forces During Running

The following data show the rearfoot, midfoot and forefoot strikers considered earlier. The most notable difference is the magnitude of the medial impact force with the rearfoot striker, which is considerably greater than with the midfoot and forefoot strikers (Fig. 3.26A, B, C). The variability of the medial–lateral forces between individuals can be considerable due to

varying amounts of pronation and supination during stance phase. It is also possible to have either a lateral or a medial impact peak, depending on the position of the foot and which part of the foot makes initial contact. However, it is still possible to find the maximum medial and the maximum lateral forces, and the times that they occur from the force–time graphs.

As with walking, the medial–lateral forces (although variable) can have a substantial effect on the loading and stability of the ankle and knee joints in the coronal plane. These should not be ignored when considering the effects of footwear and orthotic management, which can give clinically significant changes to both the medial and lateral force patterns for an individual (Fig. 3.27A, B).

3.7 PLANTAR PRESSURE

Measurement of pressure related to the foot is generally termed plantar pressure analysis as typically pressures are measured between the plantar surface of the foot and a supporting surface. However, it is also possible and important in certain situations to measure pressures on the dorsum of the foot. Within the literature, there is a wide range of terminology used to describe the measurement of plantar pressures. These include pedobarography, foot pressure measurement, plantar pressure imaging and load distribution analysis (see Chapter 7).

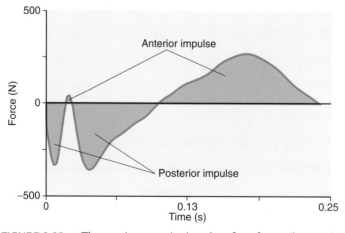

FIGURE 3.25 ■ The anterior–posterior impulse of rearfoot striker running

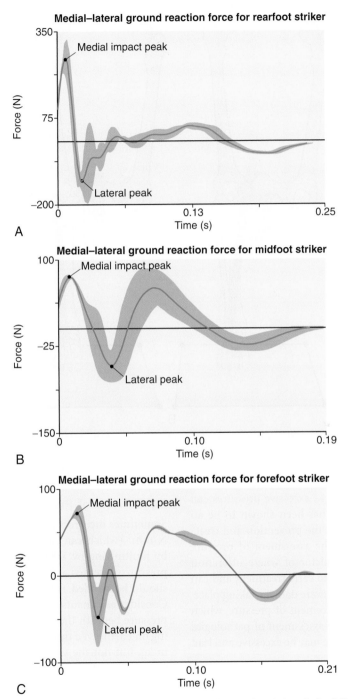

FIGURE 3.26 ■ Medial–lateral forces during running: (A) typical rearfoot striker, (B) typical midfoot striker and (C) typical forefoot striker

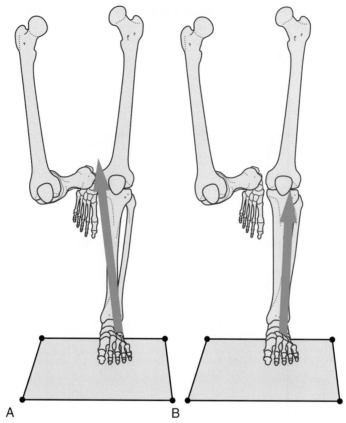

FIGURE 3.27 ■ (A) Medial loading peak and (B) lateral peak

3.7.1 Why Is Foot Pressure Measurement Important?

Pressure is very important as excessive pressures can cause tissue damage. This has been shown to be an important consideration in the prevention and treatment of ulceration, with the assessment of pressure being a possible early predictor of where ulceration might occur. This can lead to early treatment and can allow clinicians to prevent severe damage taking place. This section reviews measurement of pressure, which is of particular value in the assessment of pathologies where pressure on soft tissue may be excessive and lead to injury.

3.7.2 Definition

Pressure is the ratio of force to the area over which that force is being applied. This force is normally considered to be in a direction perpendicular to that of the surface of the area that it is applied.

$$\text{Pressure} = \frac{\text{Force}}{\text{Area}}$$

3.7.3 Units of Pressure

Sometimes incorrect units are presented: for example, N/mm^2 (which can easily be converted back to N/m by multiplying by $1\,000\,000$). Worse still are units of pressure presented as kg/m^2 or kg/mm^2. These last two should be avoided as kg is the unit of mass not force. Consider astronauts on the moon; they will have less pressure beneath their feet than when they are standing on Earth, even though they will have the same mass in kg, but not the same weight (force on the feet) in newtons (Table 3.1).

3.7.4 Display and Presentation of Data

There are many different ways to display the data recorded from pressure measurement devices. A colour scale similar to a weather map is one of the most used

TABLE 3.1

Conversion of Commonly Used Units of Pressure Into SI Units

Units Commonly Used	Corrected Values	How to Report Results
N/mm^2	$1\ N/mm^2 = 1\,000\,000\ N/m^2$	$1\ MN/m^2$ or $1\ MPa$
N/cm^2	$1\ N/cm^2 = 10\,000\ N/m^2$	$10\ kN/m^2$ or $10\ kPa$
kg/mm^2	$1\ kg/mm^2 = 9\,810\,000\ N/m^2$	$9.81\ MN/m^2$ or $9.81\ MPa$
kg/cm^2	$1\ kg/cm^2 = 98\,100\ N/m^2$	$9.81\ kN/m^2$ or $9.81\ kPa$
mmHg	$1\ mmHg = 133.3\ N/m^2$	$133.3\ N/m^2$ or $133.3\ Pa$
psi (pounds per square inch)	$1\ psi = 6867\ N/m^2$	$6.867\ kN/m^2$ or $6.687\ kPa$

methods; typically, this uses shades of blue to represent low pressure areas and shades of red to represent high pressure areas of the foot (Fig. 3.28A). In many of these systems the scale for the colour range is set by the user and it is always essential to analyse the actual pressure values and not just the colours; it is incorrect to immediately see an area coloured red and assume it is bad! For example, if the scale is set to show a red colour for pressures greater than 200 kPa, then if the pressure is 200 or 500 kPa the area will be the same red colour even though there is a large difference in the actual pressure values. Three-dimensional plots are also used where increases in height are used to display increases in pressures (Fig. 3.28B) and dynamic bar charts.

3.7.5 Interpretation of Data

Commercial companies offer different levels of software depending on the level of detail required from the pressure data. The entry-level software allows for simple and quick analysis procedures aimed for use by clinicians with limited time who only need a visual representation of the pressures, whereas the higher levels of software allow for a more detailed analysis of the data with exportation options to allow further analysis in other software packages.

Regions-of-Interest. For many applications, it is necessary to assess pressures in specific areas of the foot,

termed regions-of-interest. Within the software, users can manually select the region-of-interest or use an algorithm within the software to automatically segment the foot into a number of regions (Fig. 3.29A and B). Different manufacturers use different terminology within their software to refer to the selection of regions-of-interest; these include masks, boxes and zones.

Whereas the assessment of regions-of-interest is valuable, it is always important to consider the effect of an intervention on the whole foot. For example, in the high-risk diabetic foot with high pressures in the forefoot, a footwear intervention will focus on reducing the pressure in this area to reduce the risk of ulceration. It will be important to not only consider the effect of the footwear intervention on the forefoot though, as what may occur is a reduction in pressure in this area but an increase in pressure in an adjacent area. The intervention may prevent ulceration in the forefoot but may increase the risk of ulceration in another area of the foot.

Average Force. A measure that can sometimes be found is the average load (force N) under different areas of the foot. However, much care needs to be taken with such measures as they are often influenced by the size of the area chosen (Fig. 3.30). Often the most useful aspect of this is to replicate the vertical force patterns seen on force platforms.

Contact Area. The contact area is the area of the sensor that is loaded by the foot, which is usually reported in mm^2 or cm^2.

Average Pressure. The average pressure over the whole of the base of the foot tells us very little about what is going on under the structures of the foot. A better measure is the average pressure under specific areas of the foot (regions-of-interest) (Fig. 3.31A, B).

Maximum/Peak Pressure. Maximum pressure, also termed peak pressure, is the maximum instantaneous pressure experienced over time. It is of far more interest when looking for signs of excessive pressure leading to tissue damage. Although this still uses the same area defined on the foot as in average pressures, the largest pressure reading in each of the areas is now plotted

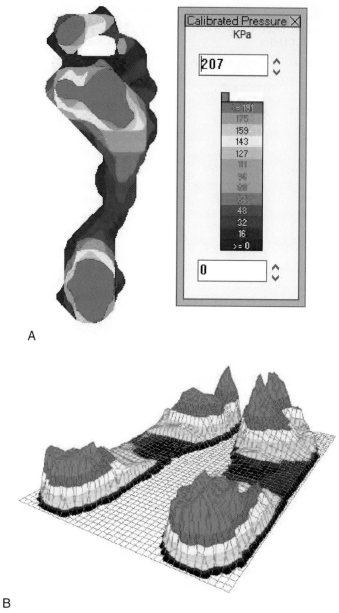

A

B

FIGURE 3.28 ■ Pressure measurement display (A) Peak stance display; showing the maximum pressure reached by each sensor during the stance phase (Tekscan Inc., Boston, MA, USA); (B) 3D display of peak stance (Novel GmbH, Munich, Germany)

rather than the averaged value, which may contain small areas of high pressure.

For normal subjects, typical peak pressures beneath the foot are 80–100 kPa in standing and 200–500 kPa in walking. In diabetic neuropathology, pressures can be as high as 1000–3000 kPa. To put this into perspective, 3000 kPa is the same pressure as 30.5 kg acting on 1 cm², or half a 61-kg person's weight (610 N) acting on 1 cm² (*Ouch!*). In general, the area around the second and third metatarsal heads experiences the

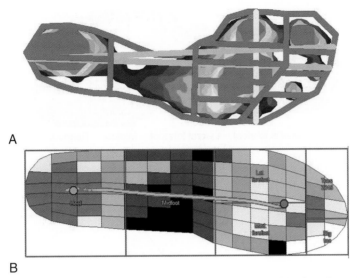

A

B

FIGURE 3.29 ■ (A) Twelve regions-of-interest (Tekscan Inc., Boston, MA, USA); (B) Six regions-of-interest (Novel GmbH, Munich, Germany)

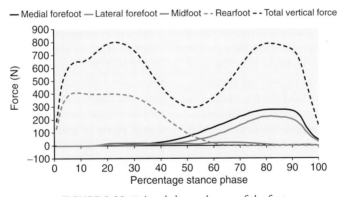

FIGURE 3.30 ■ Loads beneath area of the foot

highest maximum pressure for the foot during walking in healthy adults. Currently it is not possible to predict the location or value of peak pressures during walking; however, factors such as age, linear kinematics, arch structure, plantar soft tissue thickness, radiographic measurements and gastrocnemius activity have been

identified as factors that affect peak pressures (Morag & Cavanagh, 1999).

Pressure–Time Integral. How areas of peak pressure change over time at points on the foot is extremely important. If the pressure produced is only acting for a

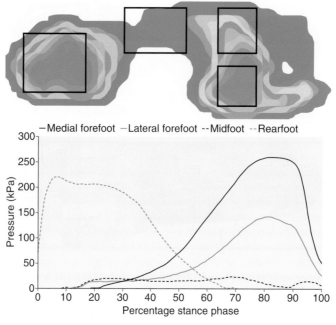

FIGURE 3.31 ■ Average pressures on a normal foot

very short period of time, then it has less time to cause tissue damage; if it acts for longer then there is a greater chance of tissue damage. To measure this, the area under the peak pressure–time or average pressure–time curve is taken for a particular area under the foot, usually under a particular anatomical landmark. This gives a single value for the pressure–time effect on the selected area. This measure is called as pressure–time integral. The integral is the sum of all pressure values in the discrete time series. The units are pascal seconds (pascal second or newton second), or newton seconds per square metre (Ns/m^2 or kNs/m^2). Impulse force is also sometimes reported from the force–time curves, although its usefulness is perhaps questionable for pressure analysis. Fig. 3.32A and B provides an example of where, if only peak pressures are examined, important information can be missed. Here the peak pressures recorded over the time period are equal (200 kPa) for both pressure measurements but the pressure–time integral is much greater for (A) than (B) as the peak pressure was maintained for a longer period in (A).

Centre of Pressure (COP). The COP of the foot is the location of where the GRF acts on the foot; it is represented by x and y coordinates related to the origin of the sensor. Its location changes during walking, moving from the rearfoot at foot contact to the forefoot as the foot leaves the ground. If the COP is plotted over the stance phase of gait, what is displayed is termed the path of the COP (Fig. 3.33). The typical path of the COP for walking starts slightly laterally to the midline of the heel at foot contact, progresses along the midline of the foot up to the metatarsal heads, after which it moves medially and, as the foot prepares to leave the ground, it lies under the first or second toe. Deviations from what is considered the normal path of the COP can be used to identify pathology.

COP can also be utilized to examine balance. Patients can stand on a pressure platform while completing the Romberg test and the movement of the COP in the anteroposterior and mediolateral directions can be quantified. Please note it is important to remember that COP and centre of gravity (COG) are not interchangeable terms (Winter, 1995a,b).

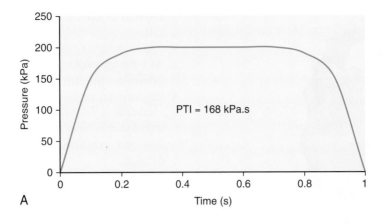

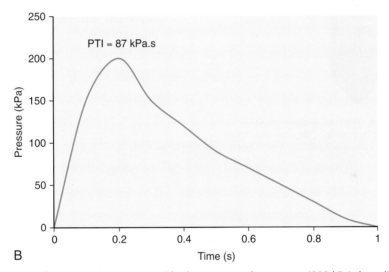

FIGURE 3.32 ■ Example of pressure–time curves with the same peak pressure (200 kPa) but different pressure–time integrals (PTI)

SUMMARY: GROUND REACTION FORCES, IMPULSE AND MOMENTUM

- The centre of pressure is a single point where the GRF can be considered to act. Centre of pressure can be very useful in assessing the movement of the GRF beneath each foot during walking and the amount of postural sway during standing balance tests.
- The overall effect or resultant of the GRF may be shown using a Pedotti diagram. The compo-

nents of the GRF may be considered in the vertical, anterior–posterior and medial–lateral planes of the body.

- Separating the GRF into the different body planes yields important information about the loading, propulsion and the stability of the body during different movement tasks.
- Impulse, the area under a force–time graph, corresponds with the change in momentum of an object. This can be useful in determining change in velocity during walking and running

FIGURE 3.33 ■ Path of centre of pressure (COP) during stance phase of walking (Tekscan Inc., Boston, MA, USA)

and may be used as an assessment of steady-state gait.

■ GRFs can be very useful in determining dysfunction during walking and may also be used to determine different running styles.

■ Pressure can be described as the force acting on a small area of the foot. This is not necessarily linked to the size of the GRFs acting on the whole foot and excessive pressures can cause tissue damage.

■ Pressure analysis can tell us about the forces on regions-of-interest; these can give important information about the load on specific anatomical structures.

■ Pressure analysis of the whole foot can give important information on the effect of an intervention.

4 MOTION AND JOINT MOTION

JIM RICHARDS ■ DOMINIC THEWLIS ■ ROBERT NEEDHAM ■
NACHIAPPAN CHOCKALINGAM

T his chapter covers the basic methods of gait assessment through to the description and discussion of the involvement of the three-dimensional movement of the foot, ankle, knee, hip and pelvis during walking in individuals who are pain and pathology free.

AIM

To consider the function of the movement of the lower limb in individuals who are pain and pathology free.

OBJECTIVES

■ To relate simple methods of temporal and spatial parameters of gait to functional assessment of gait

■ To recognize and draw the movement patterns of the lower limb and pelvis

■ To interpret the movement patterns of the lower limb and pelvis in relation to their functional contribution to walking

■ To interpret the different methods of graphing the interaction of the movement of the lower limb and pelvis.

4.1 MOVEMENT ANALYSIS IN CLINICAL RESEARCH

4.1.1 The Early Pioneers

The study of human movement can be dated back to Borrelli in 1680. However, it wasn't until the late 19th century with the invention of photography that the analysis of movement was possible. Muybridge and Marey both collected movement data from humans. Marey (1873) was the first to produce a stick figure of

human movement, whereas Muybridge produced a photographic investigation of human movement during many activities between 1872 and 1885, which was published in 1907.

However, the majority of the work has been carried out in the second half of the 20th century. Bresler and Frankel (1950) were the first to carry out a mechanical analysis of walking. Their work included studying joint motion and inertial forces involved during gait. Saunders and colleagues (1953) referred to the major determinants in normal gait and applied these to the assessment of pathological gait. Inman (1966, 1967) and Murray (1967) both published detailed analyses on the kinematics and conservation of energy during human locomotion, which are frequently referred to today. Inman and colleagues (1981) later published *Human Walking*, a comprehensive textbook on human locomotion. Many of the techniques of collection and analysing human locomotion have been applied to clinical practice and this has led to more detailed clinical assessment of therapeutic and surgical intervention; one example of which is the assessment of the treatment of cerebral palsy (Sutherland & Cooper, 1978; Davids et al., 1993; Gage, 1994).

4.1.2 Clinical Gait Analysis

The walking cycle or gait cycle is often studied with respect to foot contact times. One complete gait cycle is defined as the period from initial contact of the foot to the next initial contact of the same foot. The gait cycle may be simply divided into stance and swing phases, where the foot is either in contact with the ground or not. Stance may then be divided into periods of single and double support, i.e. when either one or both feet are in contact with the ground (Murray et al., 1964).

The gait cycle can also be studied in more detail, by studying the periods of foot contact and the action and motion of the different body segments separately during the gait cycle. This is commonly known as the study of kinematics. The variables involved with kinematics are foot contact times and distances, linear and angular displacements, velocities, and accelerations of body segments. Kinematics is not concerned with the internal and external forces, but with the movement itself.

Brand and Crowninshield (1981) highlighted the distinction between the use of biomechanical techniques to 'diagnose' or 'evaluate' clinical problems. They stated: 'Evaluate, in contrast to diagnose, means to place a value on something. Many medical tests are of this variety and instead of distinguishing diseases, help determine the severity of the disease or evaluate one parameter of the disease. Biomechanical tests at present are of this variety.' Brand and Crowninshield (1981) gave a guide of six criteria for tools used in patient evaluation:

1. The measured parameter(s) must correlate well with the patient's functional capacity.
2. The measured parameter must not be directly observable and semi-quantifiable by the physician or therapist.
3. The measured parameters must clearly distinguish between normal and abnormal.
4. The measurement technique must not significantly alter the performance of the evaluated activity.
5. The measurement must be accurate and reproducible.
6. The results must be communicated in a form which is readily identifiable.

Brand and Crowninshield stated: 'It is clear to us that most methods of assessing gait do not meet all of these criteria. We believe that it is for this reason that they are not widely used.'

Advances in biomechanical assessment in the last 35 years have been considerable. The description of normal gait in terms of movement and forces about joints is now commonplace. The relationship between normal gait patterns and normal function is also well supported in both peer-reviewed papers and textbooks (Bruckner, 1998; Perry, 1992; Rose & Gamble, 1994). This allows deviations in gait patterns to be studied in relation to changes in function in subjects with particular pathologies. It is possible for a clinician or physician to subjectively study gait; however, the value and repeatability of this type of assessment is questionable due to poor inter- and intra-tester reliability. For instance, it is impossible for one individual to study, by observation alone, the movement pattern of all the main joints involved during an activity like walking simultaneously. Studying movement patterns requires objective motion analysis which allows information to be gathered simultaneously with known accuracy and reliability. In this way changes in movement patterns due to intervention by physical therapists and surgeons and their effect on function may be assessed unequivocally. Most motion analysis systems now report on the joint kinematics for the individual recorded, and also contain information for the mean for normal on the same graph, allowing a direct comparison of the individual's movement pattern in relation to a predefined normal.

Patrick (1991) reviewed the use of movement analysis laboratory investigations in assisting decision making for the physician and clinician. Patrick concluded that the reasons for the use of such facilities not being widespread was due to: the time of analysis being considerable, bioengineers designing systems and presenting results for researchers and not clinicians, and a lack of understanding by physicians and clinicians of applied mechanics and its relevance to assessment of treatment outcome. Since 1991 the movement analysis laboratory has become more widely accepted by physicians, and the time needed for analysis is ever decreasing, resulting in new laboratories appearing in the clinical setting.

Winter (1993) reviewed techniques of gait analysis under the title *Knowledge base for diagnostic gait assessments*. This was a reply to the criticisms from Brand

and Crowninshield. Winter gave evidence to show that clinical gait assessments can give a valuable contribution to diagnostic information to assist surgeons in planning orthopaedic procedures, planning of rehabilitation, and in the assessment of prosthetic devices. Winter also demonstrated the use of a generalized strategy and diagnostic checklist developed for all pathologies. This checklist did not focus on a particular pathology, but rather targeted gait problems that may be common to many pathologies. Winter demonstrated the use of such a checklist using five case studies: knee arthroplasty, below knee amputee, cerebral palsy hemiplegia, above knee amputee and patellectomy. The paper concluded by stating that assessment of pathological gait is not an easy task, and can require considerable expense in equipment, software and specialized personnel. Winter also stressed the need for a database of normal data for children, adult and elderly subjects. A common argument against movement analysis laboratories has been cost. The cost of movement analysis equipment and its potential use in the clinical setting has been reported (Bell et al., 1996). A broader question, indeed, could be put to any clinical assessment or treatment that requires the use of technology. One example of this is the relative cost of radiography to movement analysis equipment, which in comparison is modest (Bell et al., 1996). Gage (1994) claimed that gait analysis costs are comparable with MRI or CAT scans. Gage also stated that the use of movement analysis, as a detailed form of assessment, may have wider cost benefits and improve clinical services more than first realized. Bell and colleagues (1995) highlighted the use of a holistic approach to motion analysis, including muscle performance and joint range of motion, as well as kinematic and kinetic parameters of gait. This holistic approach may be applied to many pathologies to give a detailed assessment of pathology and the subsequent effects of treatment.

4.2 THE GAIT CYCLE

For normal walking the obvious division is the duration when the foot is in contact with the ground and the period when it is not. These are known as stance phase (approximately 60% of the gait cycle) and swing phase (approximately 40% of the gait cycle),

respectively. The stance phase can be subdivided by specific events (Fig. 4.1A–E): (A) heel strike, (B) foot flat, (C) midstance, (D) heel off and (E) toe off. The swing phase can be subdivided into three phases (Fig. 4.2A, B, C): (A) early swing, (B) mid swing and (C) late swing. The simplest way in which we can look at walking patterns is by studying distances and times while the foot is in contact with the ground.

4.2.1 Spatial Parameters

To define a subject's walk, the spatial parameters of foot contact during gait should be considered (Fig. 4.3). The spatial parameters of foot contact during gait are step length, stride length, foot angle, and base width (Murray et al., 1964, 1970; Rigas, 1984).

Step length and stride length are defined as the distance between two consecutive initial contacts by different feet and the distance between two consecutive initial contacts by the same foot, respectively. Foot angle is defined as the angle of foot orientation away from the line of progression. Base width is defined as the medial–lateral distance between the centre of each heel during gait (Murray et al., 1964, 1970; Rigas, 1984). Two other parameters may easily be calculated using this information; these are cadence and average velocity. The spatial parameters of foot contact during gait are:

Step length – This is the distance between two consecutive heel strikes.
Stride length – This is the distance between two consecutive heel strikes by the same leg.
Foot angle or angle of gait – This is the angle of foot orientation away from the line of progression.
Base width or base of gait – This is the medial–lateral distance between the centre of each heel during gait.

4.2.2 Temporal Parameters

Foot contact times are important temporal parameters of gait. If the time of each consecutive heel strike and toe off is recorded, then the step and stride time may be calculated. Step time and stride time are defined as the time between two consecutive initial contacts by different legs and the time between two consecutive initial contacts by the same leg; therefore, one complete gait cycle is the same as one stride

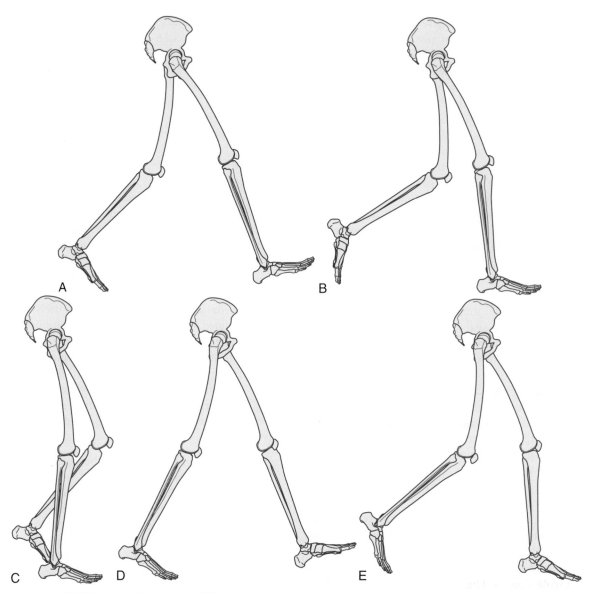

FIGURE 4.1 ■ Stance phase: (A) heel strike, (B) foot flat, (C) midstance, (D) heel off and (E) toe off

(Murray et al., 1964, 1970; Rigas, 1984). However, this is not the only information that can be derived; the single support time and double support time can be found for each leg. Single support time and double support time may be defined as the time when one foot is in contact with the ground and the time when both feet are in contact with the ground, respectively. Swing time is the same as single support time on the opposite leg. From this information, the symmetry of single support time, double support time and step time can also be found. Foot contact timing and the parameters that can be derived from them are represented in Fig. 4.4.

Step time – This is the time between two consecutive heel strikes. This is 50% of the stride time if the person is walking with perfect symmetry between the left and right sides.

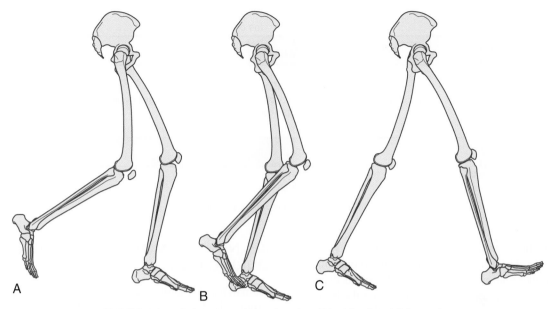

FIGURE 4.2 ■ Swing phase: (A) early swing, (B) mid swing, (C) late swing

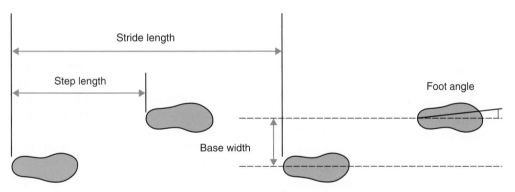

FIGURE 4.3 ■ Spatial parameters

Stride time – This can be defined as the time between two consecutive heel strikes by the same leg, one complete gait cycle or 100% of the gait cycle for that limb.

Single support – This is the time over which the body is supported by only one leg, which is approximately 40% of the gait cycle.

Double support – This is the time over which the body is supported by both legs. This comprises two periods, each lasting 10% of the gait cycle.

Swing time – This is the time taken for the leg to swing through while the body is in single support on the other leg. Therefore, this is the same proportion as single support, i.e. 40% of the gait cycle.

Total support – This is the total time the body is supported by one leg during one complete gait cycle. This time is the single support time and the two double support times, giving 60% of the gait cycle.

Although the percentages of time spent in single and double support time are useful, they need to be used carefully for clinical assessment as the

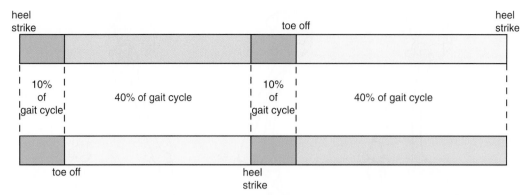

FIGURE 4.4 ■ Temporal parameters

proportions are very dependent on the speed of walking. The previously mentioned proportions are for walking at 'normal' speed; as the speed decreases the greater the double support time and the lower the single support time, until we get to the point of zero speed when double support is 100%, i.e. both feet are in contact with the ground all the time and the person is no longer walking! As the speed increases, the double support period decreases until the stance phase and swing phase are nearly equal, with very little double support time; this is the transition between race walking to running, where there is no longer any double support time and swing time is greater than stance time.

Other parameters may easily be calculated using this information: these are cadence and velocity. The cadence is the number of steps taken in a given time, usually steps per minute. Velocity may be calculated by the following formula:

$$\text{Cadence} = \text{The number of steps per min}$$

$$\text{Velocity} = \frac{\text{Step length (m)} \times \text{cadence (steps/min)}}{60 \,(\text{number of seconds in 1 minute})}$$

Measures of symmetry between the left and right side can also be easily found by dividing the value of a parameter found for the left over that of the right. This can be carried out for all the measures mentioned earlier:

$$\text{Symmetry of step length} = \frac{\text{Step length for the left}}{\text{Step length for the right}}$$

$$\text{Symmetry of step time} = \frac{\text{Step time for the left}}{\text{Step time for the right}}$$

4.3 NORMAL MOVEMENT PATTERNS DURING GAIT

Human walking allows a smooth and efficient progression of the body's centre of mass (Inman, 1967). To achieve this there are a number of different movements of the joints in the lower limb. The correct functioning of the movement patterns of these joints allows a smooth and energy-efficient progression of the body. The relationship between the movements of the joints of the lower limb is critical: if there is any deviation in the coordination of these patterns, the energy cost of walking may increase and also the shock absorption at impact and propulsion may not be as effective.

Joint motion patterns commonly reported include: ankle plantar–dorsiflexion, foot rotation, knee flexion–extension, knee abduction–adduction (valgus–varus), knee internal–external rotation, hip flexion extension, hip abduction–adduction, hip internal–external rotation, pelvic tilt, pelvic obliquity, and pelvic rotation. The movement patterns considered in this chapter include the movement of all the following joints and segments in the sagittal, coronal and transverse planes:

- Ankle joint
- Rearfoot, midfoot and forefoot motion
- Tibial segment
- Knee joint
- Hip joint
- Pelvis.

Additional graphing techniques to show coordination between joints are also reviewed, these include:

- Angle–angle diagrams
- Angle versus angular velocity diagrams.

It is also possible to find the energy involved in moving each body segment. More on this can be found in Section 5.7: Body Segment Energy.

4.3.1 Plantarflexion and Dorsiflexion of the Ankle Joint

The movement of the foot as a whole about the tibia is referred to as ankle joint motion and is the most commonly reported movement pattern of the foot and ankle complex. This overall movement of the foot to the tibia in the sagittal plane is of great importance as it allows shock absorption at heel strike and during stance phase, as well as being vital in the 'push off' or propulsive stage immediately before the toe leaves the ground. During swing phase, the motion of the ankle joint allows foot clearance, which can be lacking in some pathological gait patterns and is generally known as drop foot. The range of motion that occurs in walking varies between 20° and 40°, with an average range of motion of 30°. However, this does not tell us how the motion of the ankle varies throughout gait. During gait the ankle has four phases of motion (Fig. 4.5).

Phase 1. At initial contact, or heel strike, the ankle joint is in a neutral position; it then plantarflexes to between 3° and 5° until foot flat has been achieved. This is sometimes referred to as 'first rocker' or 'first segment', which refers to the foot pivoting about the heel or calcaneus. During this period, the dorsiflexor muscles in the anterior compartment of the foot and

ankle are acting eccentrically, controlling the plantarflexion of the foot. This gives the effect of a shock absorber and aids smooth weight acceptance to the lower limb.

Phase 2. At the position of foot flat the ankle then begins to dorsiflex. The foot becomes stationary and the tibia becomes the moving segment, with dorsiflexion reaching a maximum of 10° as the tibia moves over the ankle joint. The time from foot flat to heel lift is referred to as 'second rocker' or 'second segment', which refers to the pivot of the motion now being at the ankle joint with the foot firmly planted on the ground. During this time the plantarflexor muscles are acting eccentrically to control the movement of the tibia forwards.

Phase 3. The heel then begins to lift at the beginning of double support, causing a rapid ankle plantarflexion, reaching an average value of 20° at the end of the stance phase at toe off. This is referred to as 'third rocker' or 'third segment', as the pivot point is now under the metatarsal heads. During this time the ankle reaches an angular velocity of 250°/s plantarflexion, which can be associated with power production. This is the propulsive phase of the gait cycle during which the plantarflexor muscles in the posterior compartment of the foot and ankle concentrically contract, pushing the foot into plantarflexion and propelling the body forwards. This rapid plantarflexion is responsible for the majority of the power production to propel the body forwards (see Section 5.4: The relationship between moments, angular velocity and joint power during normal gait).

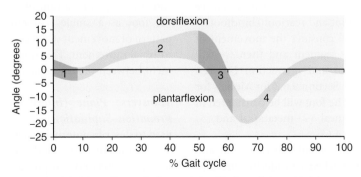

FIGURE 4.5 ■ Plantarflexion and dorsiflexion of the ankle joint

TABLE 4.1			
Rocker/Segments			
Phase 1	**Phase 2**	**Phase 3**	**Phase 4**
1st rocker	2nd rocker	3rd rocker	Swing phase
Heel rocker	Ankle rocker	Forefoot rocker	Swing phase
1st segment	2nd segment	3rd segment	4th segment
Contact	Midstance	Propulsion	Swing phase

Phase 4. During the swing phase the ankle rapidly dorsiflexes (150°/s) to allow the clearance of the foot from the ground. A neutral position (0°) is reached by mid swing, which is maintained during the rest of the swing phase until the next heel strike. This is referred to as the 'fourth segment'. It has been recorded that there is sometimes 3° to 5° of dorsiflexion during the swing phase. During this phase the ankle dorsiflexors concentrically contract to provide foot clearance from the ground and prepare for the next foot strike.

Different Terms Commonly Used to Describe Ankle Motion. There are a number of terms commonly used by clinicians and bioengineers alike. In the previous text I have used **first, second and third rocker** and **first, second, third and fourth segments**. These terms are also commonly used interchangeably (Table 4.1).

4.3.2 Movement of the Ankle, Rearfoot, Midfoot and Forefoot

The vast majority of studies have considered the foot as a single segment, often defined by the malleoli and the lateral aspect of the metatarsal heads, or the heel and the medial and lateral aspects of the metatarsal heads. However, for decades, podiatrists and many other clinicians have been considering the foot in three parts: the forefoot, midfoot and rearfoot (hindfoot).

In this section, we will consider the movement of the foot first as a single segment and then we will consider the contribution and interaction between each of the segments (see Section 9.6.2 for Models for multiple segment foot). The foot will be considered in three segments: (1) calcaneal, (2) metatarsal and (3) phalangeal segment (Fig. 4.6).

To demonstrate the movement that can occur and the differences that exist when considering the foot either as a **single segment** or as **multiple segments**,

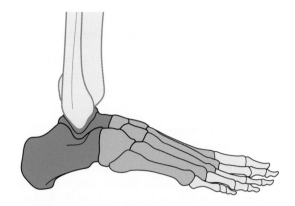

FIGURE 4.6 ■ The foot considered in three segments

the data presented show the differences that may be seen in a single individual walking barefoot. The data show the mean and standard deviations for five trials. These data should not be used as a normative data set, but should act as a guide to relative movement patterns of the different foot segments.

Tibia-to-Foot Movement

Sagittal Plane (Plantarflexion–Dorsiflexion). Tibia-to-foot movement in the sagittal plane is the same as the pattern described in Section 4.3.1 Plantar and dorsiflexion of the ankle joint, where the whole foot is defined as a single segment (Fig. 4.7A).

Coronal Plane (Inversion–Eversion). The coronal plane may be used to describe the inversion–eversion pattern of the foot as a whole. At heel strike, the foot lands in an inverted position and moves into eversion during loading. Rapid inversion then takes place just prior to 50% of the gait cycle (Fig. 4.7B). This pattern shows a range of motion of 7°; however, considering the foot as a single segment produces a different pattern of movement when compared to the calcaneal-to-tibial movement. This discrepancy is due to movement between the calcaneal and metatarsal segments.

Transverse Plane (Internal–External Rotation or Pronation–Supination). The transverse plane can be used to describe internal–external rotation. The transverse plane movement of the foot to tibia has also been used as a descriptor for pronation–supination (Nester et al., 2003). This pattern shows the foot landing in a

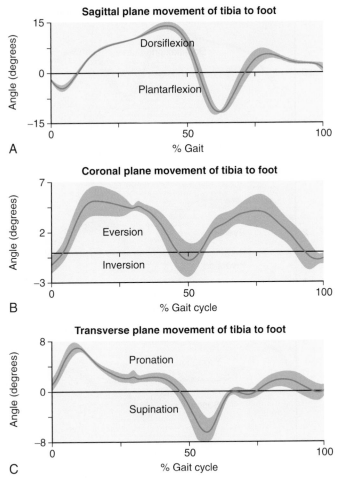

FIGURE 4.7 ■ (A) Tibia-to-foot movement in the sagittal plane, (B) tibia-to-foot movement in the coronal plane and (C) tibia-to-foot movement in the transverse plane

slightly pronated position and moving quickly into further pronation: this shows slight differences with the work by Nester who showed a slightly supinated position at heel strike. The pronation then reduces and levels off before moving into a supinated position in late stance phase (Fig. 4.7C). It is interesting to note that this pattern, which considers the foot as a single segment, is similar to that of the calcaneal-to-tibial movement, although the range of movement is greater.

Calcaneus-to-Tibia Movement

When describing the ankle joint we can use the definition of tibia-to-foot movement; however, this uses a combination of the calcaneal and metatarsal segments

to define the distal segment. However, the calcaneal-to-tibia segment movement would be more meaningful, in particular when considering the coronal and transverse planes.

Sagittal Plane (Plantarflexion–Dorsiflexion). The sagittal plane movement shows much the same functional movements during early-to-midstance as the foot to the tibia. However, the amount of plantarflexion occurring between the calcaneal-to-tibia segments is noticeably less than when the metatarsal segment is included (Fig. 4.8A). This is due to a clear pattern of dorsiflexion and plantarflexion of the metatarsal segment in relation to the calcaneal segment.

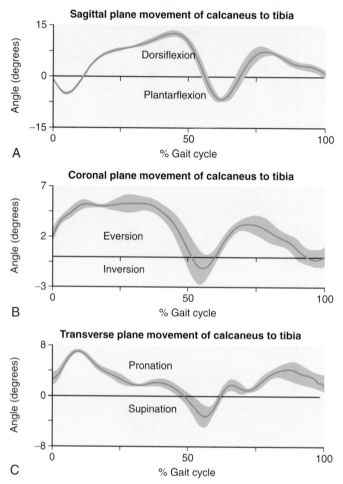

FIGURE 4.8 ■ (A) Calcaneus-to-tibia movement in the sagittal plane, (B) calcaneus-to-tibia movement in the coronal plane and (C) calcaneus-to-tibia movement in the transverse plane

Therefore, including the metatarsal segment may well give an overall function of the foot; however, it gives a distorted view of 'ankle' plantarflexion.

Coronal Plane (Inversion–Eversion). At heel strike, the foot lands in an everted position and moves into further eversion during loading. This allows midtarsal movement, arch collapse, and a mobile forefoot. The same pattern of rapid inversion seen in the single segment foot model takes place just prior to 50% of the gait cycle; this locks the midtarsal joints creating a rigid lever and a restored arch (Fig. 4.8B). Although the range of motion into further eversion is less than

that seen in the single segment model, this is due to metatarsal-to-calcaneus movement.

Transverse Plane (Internal–External Rotation or Pronation–Supination). The transverse plane movement of the calcaneal-to-tibia segment gives much the same pattern of movement as the foot-to-tibial movement, although the range of motion is less. This difference in the range is due to the small amount of movement in the transverse plane between the metatarsal and calcaneal segments (Fig. 4.8C). Therefore, the treatment of the foot as a signal segment (calcaneal and metatarsal segments) appears to give a good

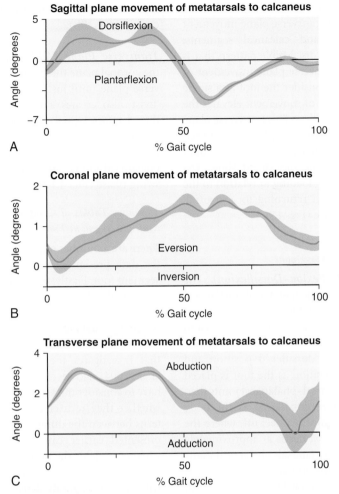

FIGURE 4.9 ■ (A) Metatarsal-to-calcaneus movement in the sagittal plane, (B) metatarsal-to-calcaneus movement in the coronal plane and (C) metatarsal-to-calcaneus movement in the transverse plane

indication of the movement pattern of the calcaneus in relation to the tibia.

Metatarsal-to-Calcaneus Movement

Sagittal Plane (Plantarflexion–Dorsiflexion). The sagittal plane motion shows metatarsal–calcaneal dorsiflexion during early stance phase, with plantarflexion during late stance phase. This corresponds to the dorsiflexion of the ankle joint as the body moves over the stance limb, and plantarflexion during propulsion. This would imply the metatarsal calcaneal movement has a role in both progression of the

body over the foot and in the push-off phases (Fig. 4.9A). This involvement causes the discrepancy in the plantarflexion–dorsiflexion pattern between the tibia-to-foot movement and the calcaneal-to-tibia movement.

Coronal Plane (Inversion–Eversion). At heel strike, the angle between the metatarsal and calcaneal segments is approximately in neutral and moves progressively into a slightly more everted position. The range of movement during stance phase is in the order of 1.5°, although it is possible that skin and soft-tissue movement could be responsible (Fig. 4.9B).

Transverse Plane (Adduction–Abduction). There is a relatively small amount of transverse plane movement between the metatarsal and calcaneal segments; however, this does follow a discernible pattern that is synchronized with the transverse plane movement of the ankle joint. When we consider the foot as a single segment, this small amount of movement elevates the value of transverse plane motion. This transverse plane movement equates to adduction–abduction between the metatarsal and calcaneal segments, which would be expected to be small. The movement shows the metatarsal abducting during loading in relation to the calcaneal segment and then returning to a 'neutral' position during swing phase (Fig. 4.9C).

Metatarsal-to-Phalangeal Movement

Sagittal Plane (Plantarflexion–Dorsiflexion). The sagittal plane movement of the metatarsal–phalangeal joints is perhaps the most notable omission from foot and ankle mechanics in the majority of textbooks and the research literature. At heel strike the metatarsal–phalangeal joints are in a dorsiflexed position and move towards a neutral position as the foot is placed flat on the floor. The metatarsal–phalangeal joints then become progressively more dorsiflexed as the body moves forwards over the foot. At heel off, where the foot pivots on the metatarsal heads at approximately 50% of the gait cycle, the metatarsal–phalangeal joints are forced into rapid dorsiflexion, or extension, just as the ankle starts its rapid plantarflexion during push off. This can be referred to as metatarsophalangeal dorsiflexion, which will tense the plantar fascia. After toe off, the phalanges return to a slightly less dorsiflexed position to aid foot clearance during swing phase (Fig. 4.10A). Although this movement may not have any function in power generation during push off, it may still be clinically important when considering the stiffness of the forefoot and the effect of foot orthoses designed to control or assist forefoot movement.

Coronal Plane (Inversion–Eversion). There is little movement in the coronal plane until just before heel off. At heel off, an inversion 'twist' occurs at the metatarsal–phalangeal joints, which corresponds with the calcaneal-to-tibial movement, and a pattern of rapid inversion is seen just prior to 50% of the gait cycle (Fig. 4.10B).

Transverse Plane (Adduction–Abduction). As with the coronal plane there is little movement in the transverse plane until just before heel off. At heel off, the 'twist' also creates a rapid abduction movement at the metatarsal–phalangeal joints, which corresponds to the calcaneal-to-tibial movement into a supinated position and restored arch. After toe off, the metatarsal–phalangeal joints return to their neutral position for swing phase (Fig. 4.10C).

Summary Tables of Comparison between Single and Multiple Segment Foot

There are a number of points we need to consider with the data presented here with the single and multi-segment foot. Firstly, these data were collected with the individual walking barefoot. This enabled the identification of the anatomical structures of the foot. Normative data do now exist (Carson et al., 2001; MacWilliams et al., 2003), and work is now appearing that investigates the pathological foot (Woodburn et al., 2004). However, this does not help us when we have to consider the action of orthoses on the foot, and whether they restrict or modify the movement patterns between the different foot segments. The model presented earlier can be applied to footwear, but whether it yields the same data and even the question *should it* yield the same data, as the foot function itself will almost certainly have changed when wearing shoes, has yet to be answered by the research literature. (See Tables 4.2A, B, C for relative position of segments in the different planes.)

4.3.3 Movement of the Tibial Segment

The movement of the tibial segment can give important information on pathological gait patterns. It can be the case, with various pathologies, that the ankle joint motion needs to be restricted with orthotics, due to a lack of muscle power or spasticity. However, if the ankle motion is restricted this will restrict the motion of the tibia over the stance limb, making it very hard for the subject to move over the stance limb. In these cases, rocker soles are sometimes fitted to the shoes to enable a movement over the stance limb without movement at the ankle joint. If this is the case, then

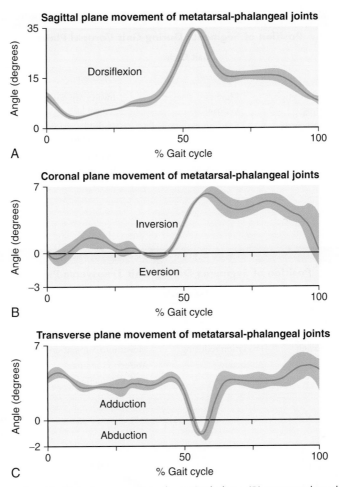

FIGURE 4.10 ■ (A) Metatarsal-to-phalangeal movement in the sagittal plane, (B) metatarsal-to-phalangeal movement in the coronal plane and (C) metatarsal-to-phalangeal movement in the transverse plane

TABLE 4.2A			
Position of Segments During Gait Sagittal Plane			
Gait Cycle			
Segment	*Contact/Loading*	*Midstance*	*Propulsion*	*Swing*
---	---	---	---	---
Foot	Neutral moving into plantarflexion	Dorsiflexing as leg moves over foot	Plantarflexing as foot pushes off	Dorsiflexed to aid foot clearance
Calcaneal	Neutral moving into plantarflexion	Similar amount of dorsiflexion to single segment	Reduced amount of plantarflexion	Dorsiflexed to aid foot clearance
Metatarsal	Dorsiflexion due to eccentric contraction	Maintains dorsiflexion as forefoot accepts weight	Rapid plantarflexion as heel lifts	Dorsiflexing
Phalangeal	Held in dorsiflexion/ extension	Moves into neutral	Dorsiflexes as heel lifts	Less dorsiflexed position as long extensors aid tibialis anterior

TABLE 4.2B
Position of Segments During Gait Coronal Plane

	Gait Cycle			
Segment	Contact/Loading	Midstance	Propulsion	Swing
Foot	Hits in inverted position rapidly everting	Maintains everted position	Inverts rapidly as heel lifts	Initially everts then moves towards inversion
Calcaneal	Hits everted and everts further	Eversion maintained	Rapid inversion as heel lifts	Everts before moving towards neutral
Metatarsal	Neutral	Everts a little	Everted	Everted
Phalangeal	Neutral	Neutral	Rapid inversion twist	Initially inverted moves towards neutral

TABLE 4.2C
Position of Segments During Gait Transverse Plane

	Gait Cycle			
Segment	Contact/Loading	Midstance	Propulsion	Swing
Foot	Abduction increasing	Abduction reducing	Rapid motion into adduction	Abducting
Calcaneal	Pronation increasing	Pronation reducing	Moves into Supination	Supinated through swing
Metatarsal	Abducting	Abducting	Abduction reduces	To neutral
Phalangeal	Adducted	Adducted	Rapid abduction	Rapid adduction

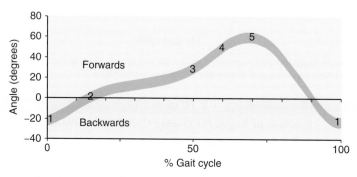

FIGURE 4.11 ■ Tibial movement in relation to the vertical axis

we need a measure which will indicate how well the tibia is moving, not in relation to the foot, as this motion has been restricted, but in relation to the vertical.

Fig. 4.11 shows the tibial movement in relation to the vertical axis. At heel strike (1), the tibia is at its most inclined backwards (the distal joint in front of the proximal joint) at 20°. The tibia then advances quickly past the vertical position (2). The motion then slows as the movement over the stance limb is controlled by eccentric activity of the calf muscles. At heel off (3), the rate of movement increases again, which is in preparation for swing phase. Toe off occurs at (4), where the tibia continues to incline forwards up to 60°

to ensure foot clearance (5), which is followed by a quick movement of the tibia forwards for the next heel strike. The point when the tibia is vertical (2) can be used as a reference for the ankle plantar–dorsiflexion pattern. As the tibia passes the vertical position, the foot is in the foot flat position; therefore, the ankle will be in an approximately neutral position, although care must be taken using this for pathological gait patterns.

4.3.4 Motion of the Knee Joint

During gait the knee joint moves in the sagittal, transverse and coronal planes. However, the majority of the motion of the knee joint is in the sagittal plane, which involves the flexion and extension of the knee joint. However, important functional and pathological movement patterns may be observed when considering both the coronal and transverse planes.

Motion of the Knee Joint in the Sagittal Plane

The flexion and extension of the knee joint is cyclic, and varies between 0° and 70°, although there is some variation in the exact amount of peak flexion occurring. These differences may be related to differences in walking speed, subject individuality, and the landmarks selected to designate limb segment alignments. The knee flexion extension pattern may be divided up into five phases (Fig. 4.12). At heel strike, or initial contact, the knee should be flexed. However, people's knee posture can vary between slight hyperextension −2° to 10° of flexion, with a mean value of 5°.

Phase 1. After the initial contact, there is a flexion of the knee joint to about 20° when the knee is flexed under maximum weight-bearing load. The knee joint flexes to absorb the loading at a rate of 150–200°/s. This occurs at the same time as the ankle joint plantarflexes, with a net effect of acting as a shock absorber during the loading of the lower limb. During this time the knee extensors are acting eccentrically.

Phase 2. After this first peak of knee flexion the knee joint extends at a rate of 80–100°/s to almost full extension. This relates to a smooth eccentrically controlled movement of the body over the stance limb.

Phase 3. The knee then begins its second period of flexion, which coincides with heel lift. During this second flexion, the lower limb is in the propulsive phase of the gait cycle. The knee undergoes a rapid flexion in preparation for swing phase, sometimes referred to as pre-swing.

Phase 4. Toe off occurs when the knee flexion is approximately 40°, at which time the knee is flexing at a rate of 300–350°/s. This flexion, coupled with the ankle dorsiflexion, allows the toe to clear the ground. During initial to mid swing the knee continues to flex to a maximum of 65–70°.

Phase 5. During late swing, the knee undergoes a rapid extension (400–450°/s) to prepare for the second heel strike.

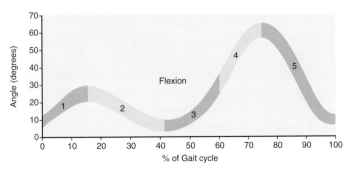

FIGURE 4.12 ■ Knee joint flexion pattern in the sagittal plane

Motion of the Knee Joint in the Coronal Plane

The knee angle in the coronal plane has long been an area of debate centred on our ability to truly isolate abduction and adduction. This debate has been very much fuelled by the variance that is evident between individuals and the susceptibility of this plane to the effects of cross talk from other planes. Various authors have attempted to compensate for this; however, in doing so they have used algorithms to artificially minimize the effect. This results in data which are essentially false. It is very rare that you will ever get the same profile for different individuals; however, the standard deviation between individuals will generally be very low.

The most reliable measurements of the coronal plane knee angle have been taken during stance phase. These imply that in normal individuals there will be very little movement besides a slight deformation during loading and the opposite deformation during terminal stance. The direction of this deformation is based on the anatomical alignment of the knee, i.e. adducted or abduction (varus or valgus). Typically, we would not expect to see more than 4° of movement during stance (Fig. 4.13). Swing phase is typically where the majority of movement is recorded; however, it is still unclear if this movement is real and due to the laxity of the joint or if it is an artefact due to the change in the orientation of the segment coordinate systems. We will typically see up to 10° of movement during swing phase. If movements that are anatomically impossible appear to be recorded, there may be the effects of planar cross talk present or there may be

pathology such as osteoarthritis, which can cause a greater deformation during loading.

Motion of the Knee Joint in the Transverse Plane

The motion of the knee in the transverse plane is dominated by the motion of the tibia rotating about the femur during both stance and swing phase (Fig. 4.14). This effect has been termed the screw home effect. The screw home movement of the tibia has been documented in a number of publications. One of the first documentations of the movement was identified in *Gray's Anatomy*, where the motion was attributed to length discrepancy between the medial and lateral condyles of the femur. Hallen and Lindahl (1966) conducted a study to investigate the degree of axial rotation and screw home movement in normal knees. This study took 'autopsy specimens' and evaluated the axial rotation of the tibia about the knee at full extension and 160° of extension, 20° of flexion. In addition, measurements were taken that accounted for the natural movement of the tibia during extension (passive extension). Hallen and Lindahl (1966) found 12° of rotation during full extension, 23° of rotation at 160° of extension and 7° of external rotation during passive extension. Andriacchi and colleagues (2005) provide a good clinical explanation of this movement as 'this motion externally rotates the tibia through swing into stance in order for the tibia and foot to be in the correct alignment at initial contact'. In addition to the joint geometry contribution to screw home, Andriacchi and colleagues (2005) noted that in the anterior cruciate ligament (ACL)-deficient knees, screw

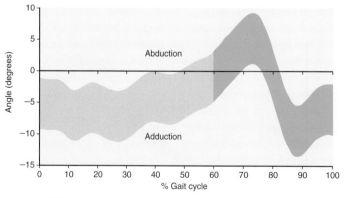

FIGURE 4.13 ■ The coronal plane motion of the knee

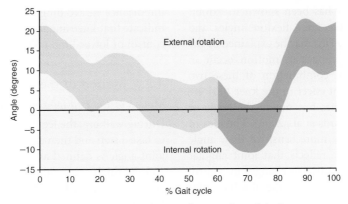

FIGURE 4.14 ■ Transverse plane motion of the knee

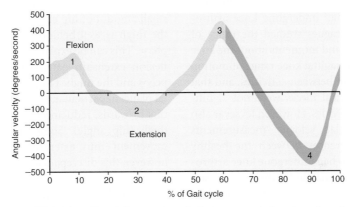

FIGURE 4.15 ■ Knee joint angular velocity patterns during the gait cycle

home is reduced, which was attributed to the passive tension developed during terminal stance within the structures of the ACL. Once this tension is released, the ACL passively causes the tibia to externally rotate with respect to the knee. It is important to note that it is often assumed that the ACL of a patient suffering from osteoarthritis (OA) is deficient, so the screw home should also be reduced, which supports the findings of Andriacchi and colleagues (2005). This movement essentially involves the passive generation of tension in the ACL during stance due to internal rotation of the foot and tibia. This tension is then released in swing phase, coupled with the discrepancy in condyle length, which results in the external rotation of the tibia.

Knee Joint Angular Velocity in the Sagittal Plane

Fig. 4.15 shows the knee joint angular velocity in the sagittal plane. At heel strike, the knee is already flexing. During the initial loading of the foot, the knee flexion peaks at 150–200°/s (1). The knee flexion velocity slows as the knee reaches its initial peak flexion. The thigh then moves over the tibia, which causes an extension velocity of between 80°/s and 100°/s (2). At 50% of the gait cycle, the heel begins to lift and the knee starts flexing again. At toe off, the knee is flexing at a rate of 300–350°/s (3); the knee velocity then slows as the knee reaches its maximum flexion. After maximum flexion, the knee then starts to extend rapidly at a rate of 400–450°/s (4).

Knee angular velocity has been shown to exhibit more sensitivity than the knee flexion angles and timing parameters alone. This may be explained by the fact that slightly larger ranges in motion occur in slightly lower times. The combination of these two factors causes a significant effect in the knee flexion–extension angular velocities but not in the angle or timing parameters (Richards et al., 2003). The idea of velocity being potentially more sensitive than angle and time measures alone reflects that joint angular velocity may be more sensitive to changes in control of joint repositioning rather than the absolute joint position itself (Richards et al., 2003).

The use of knee angular velocity has been reflected in other work. Jevsevar and colleagues (1993) studied knee kinematics during walking, stair ascent and descent, and arising from a chair in both healthy subjects and subjects who had undergone knee arthroplasty. Jevsevar and colleagues studied the range of motion, angular velocity and moments about the knee joint. They reported that sagittal knee range of motion was significantly different between activities, and that knee moments and vertical forces were not significantly different for all activities. However, Jevsevar also found that knee angular velocity measurements showed significant differences between the healthy subjects and subjects who had undergone knee arthroplasty. Jevsevar reported that both the flexion and extension velocities during swing phase (unloaded) and stance phase (loaded) in gait were significantly lower in the knee arthroplasty subjects. All the loaded angular velocities were also significantly lower for all activities in the knee arthroplasty subjects, with the exception of stair descent. Messier and colleagues (1992) investigated the kinematics of subjects with osteoarthritis of the knee during walking. Messier found that a number of characteristics showed a significant difference between the subjects with OA and normal subjects. These differences included a decrease in the knee range of motion, mean knee angular velocity, and maximum knee extension angular velocity in the subjects with OA. Chou and colleagues (1998) investigated the lower-limb kinematics of moving the foot over obstacles of different heights. Chou and colleagues found that knee angular velocity increased as toe–obstacle distance increased, and stated that angular velocity of knee flexion appears to be of primary importance in avoiding obstacle contact. This would indicate that knee angular velocity is important in the control of lower-limb tasks.

4.3.5 Motion of the Hip Joint

Motion of the Hip Joint in the Sagittal Plane

During walking, the leg flexes forward at the hip joint to take a step and then extends until push off. The hip joint angle is defined as the angle between the pelvis and the thigh segment. The motion of the hip forms an arc starting at heel strike and finishing at toe off (Fig. 4.16A).

Phase 1. After heel strike, the hip extends as the body moves over the limb at a rate of 150°/s. Maximum hip extension occurs just after opposite foot strike. The small amount of hip extension can cause confusion as the thigh is well behind the body during late stance phase. This can be explained quite simply: the hip joint flexion–extension pattern is the angle between the pelvis and the thigh. The pelvis also moves in the sagittal plane; in particular, at opposite foot strike the pelvis tilts forwards, reducing the angle between pelvis and thigh (hip angle). Some authors have reported no movement into extension at opposite heel strike; however, this discrepancy can, in part, be explained by how the pelvic tilt angle is defined (see Chapter 9: Anatomical Models and Marker Sets).

Phase 2. After maximum hip extension, weight is transferred to the forward limb and the trailing limb begins to flex at the hip. This is the pre-swing period. The toe leaves the ground at 60% of the gait cycle and the hip flexes rapidly at a rate of 200°/s. This can be seen from the increased slope of the angle against time plot below; this rapid flexion progresses the limb forward to take a step. The hip reaches maximum flexion just before the heel strike.

Phase 3. After the maximum hip flexion has been reached, there is often a small movement towards extension, i.e. the hip becomes slightly less flexed. This is concerned with the placement of the foot or pre-positioning just prior to heel strike. This is particularly prevalent during marching, where the hip flexion is exaggerated.

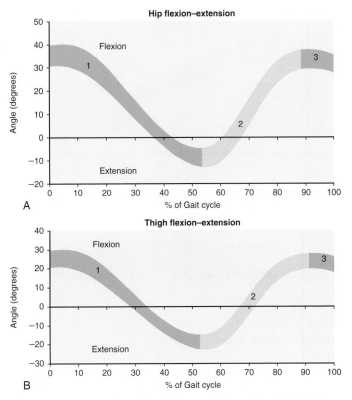

FIGURE 4.16 ■ (A) Motion of the hip joint in the sagittal plane and (B) motion of the thigh segment in the sagittal plane

Thigh Angle. Thigh angle has also previously been used in gait assessment. Thigh angle is simply the movement of the thigh segment in relation to the vertical (Fig. 4.16B). This may all seem confusing initially; however, the patterns of thigh and hip flexion extension remain almost the same during normal walking, although the maximum values may move up or down. This is because the range of the motion of the pelvis in the sagittal plane varies approximately 10° during gait, which accounts for the apparent 10° increase in extension of the thigh angle in comparison to the hip angle.

One parameter always worthy of examination for both thigh and hip angle is the active range of motion, i.e. the difference between the maximum and minimum values; this should be nearly identical whichever system is used and should be in the order of 45° for normal. However, in the study of pathological conditions that have pelvic involvement, the thigh angle

should not be considered alone, as this could mask clinically important movement strategies.

Motion of the Hip Joint in the Coronal Plane

The movement of the hip in the coronal plane may be described as hip abduction and adduction—abduction referring to a movement away from the centre line of the body, adduction referring to a movement towards the centre line of the body (Fig. 4.17).

Phase 1. At heel strike the hip is in a slightly abducted position. It then moves quickly into adduction as the limb is loaded and the body is supported. This adduction is also due to the dropping down of the pelvis on the contralateral or unsupported limb.

Phase 2. After the initial adduction, the hip moves into abduction as the pelvis levels out and the body

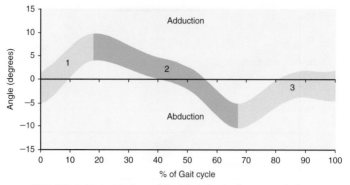

FIGURE 4.17 ■ Motion of the hip joint in the coronal plane

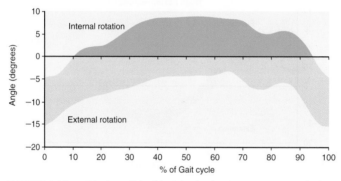

FIGURE 4.18 ■ Motion of the hip joint in the transverse coronal plane

progresses over the stance limb. The hip reaches maximum abduction shortly after toe off, as the pelvis drops down because the limb is now in swing phase.

Phase 3. The pelvis then levels off again as the limb swings through during mid-to-late swing phase.

The total range of motion is in the order of 15°, with equal amounts of abduction and adduction. This, not surprisingly, produces a very similar movement pattern to the pelvic movement in the coronal plane (see Section 4.3.6: Motion of the pelvis in the coronal plane [pelvic obliquity]).

Motion of the Hip Joint in the Transverse Plane

The movement of the hip in the transverse plane, the movement between the femur and the pelvis, may be described as internal and external rotation (Fig. 4.18). At heel strike the hip is in an externally rotated position of approximately 10°, but as the knee flexes and the body starts to move over the stance limb, the hip rotates internally to approximately 5° as the pelvis

rotates forwards on the swing side. The peak internal rotation occurs at approximately opposite heel strike. During late swing phase, the hip shows a quick movement back into external rotation. There is a considerable amount of variation in the pattern, which is as much to do with the pelvis positioning as with the femoral rotation itself.

4.3.6 Motion of the Pelvis

Motion of the Pelvis in the Coronal Plane (Pelvic Obliquity)

During early stance phase the contralateral side of the pelvis drops downward in the coronal plane. In normal gait the peak pelvic obliquity occurs just after opposite toe off, which corresponds to early stance phase on the weight-bearing limb (Fig. 4.19). Pelvic obliquity serves two purposes: to allow shock absorption and to allow limb length adjustments. To illustrate the second point when studying above-knee amputee gait, the pelvic obliquity does not always follow the normal pattern. As normal control of the knee joint had been lost, foot

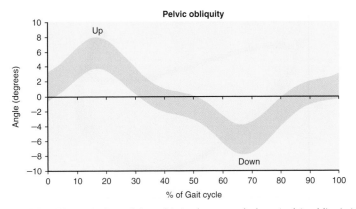

FIGURE 4.19 ■ Motion of the pelvis in the coronal plane (pelvic obliquity)

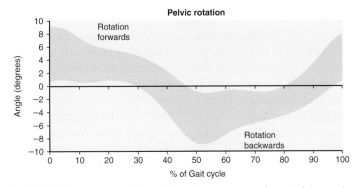

FIGURE 4.20 ■ Motion of the pelvis in the transverse plane (pelvic rotation)

clearance is ensured by hitching up the contralateral pelvis; in this way, pelvic obliquity can be used to shorten the effective limb length when required. However, this may have energy costs as it increases the excursion of the centre of mass of the entire body, therefore, increasing the work by producing this change in potential energy.

Motion of the Pelvis in the Transverse Plane (Pelvic Rotation)

During normal level walking the pelvis rotates about a vertical axis alternately to the left and to the right. This rotation is usually approximately 4° on either side of this central axis, the peak internal rotation occurring at foot strike and the maximal external rotation at opposite foot strike (Fig. 4.20). This rotation effectively lengthens the limb by increasing the step length and prevents excessive drop of the centre of mass of the whole body, making the walking pattern more efficient. Pelvic rotation also has the effect of smoothing

the vertical excursion of the centre of mass and reducing the impact at foot strike.

4.3.7 Angle–Angle Diagrams

Angle–angle diagrams are angle diagrams of two joints plotted on the same graph. Grieve (1968) proposed the use of angle–angle diagrams as a simple method of presenting joint angle data in relation to the cyclic nature of each stride or gait cycle. The most popular angle–angle diagram is that of knee flexion–extension versus hip flexion–extension. Knee flexion–extension is plotted on the ordinate (y-axis) and hip flexion–extension on the abscissa (x-axis). This gives a graphical representation of the relationship between the knee flexion–extension and the hip flexion–extension (Fig. 4.21). It should be noted that knee flexion is often considered as negative when drawing knee flexion–extension versus hip flexion–extension angle–angle diagrams.

Distinct parts of the angle–angle plot in Fig. 4.21 show the coordination between the knee and hip

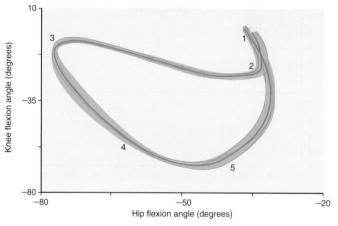

FIGURE 4.21 ■ Knee versus hip angle–angle diagram

during phases of the gait cycle. The loading response from point 1 to point 2 shows a smooth increase in the knee flexion, with a small amount of movement of the hip in the extension direction; at point 2 the knee is at its maximum flexion during loading response. Point 2 to point 3 shows an equal rate of knee extension to hip extension, indicating a smooth advancement of the body of the stance limb. Point 3 corresponds to heel lift. From point 3 to point 4, the knee starts its second wave of flexion as the hip starts to flex. Knee flexion then slows and the hip flexion becomes more rapid until point 5, mid swing. After mid swing from point 5 to point 1, the hip flexion slows and the knee flexion becomes gradually more rapid until heel strike at point 1. In late swing, the hip reaches its maximum flexion and begins to move in an extension direction.

This interaction between the knee and the hip joints gives valuable information not only on the coordination of the two joints but also on the smoothness of the transition between the phases of the gait cycle. Hershler and Milner (1980) discussed the use of visual inspection and quantification of single and multiple loops. Single loops represent a single gait cycle, whereas multiple loops represent many gait cycles. Charteris and Taves (1978) used multiple loop angle–angle diagrams as a measure of the repeatability of knee and hip motion during habituation during treadmill walking, which proved to be a very useful measure for whether a subject had reached a steady-state gait pattern. Hershler and Milner (1980) stated that the visual inspection of single loops gives a clinically

acceptable representation of the simultaneous movement of two joints. By using this technique various gait patterns may be characterized by the shape of the loop and described with respect to function. Although visual inspection is useful, single angle–angle loops may also be quantified. Parameters that may be obtained from these diagrams include: the ranges of motion of the two joints, the area contained within the loop, the length of the perimeter of the loop, and the position of the centroid of the loop.

Single loop studies are useful; however, they only represent one gait cycle and do not give any information about the repeatability of the movement pattern. To study the repeatability of a gait pattern, multiple loops may be studied. These allow a visual impression of the repeatability of the gait pattern. However, multiple loop diagrams can become confusing, with multiple traces running over one another. Sidway and colleagues (1995) demonstrated that it was possible to quantify the variability contained within angle–angle diagrams using correlation analysis. The mean and standard deviations of multiple loops are presented in Fig. 4.21.

Angle–angle diagrams have been used in the assessment of various gait patterns, including cerebral palsy (Hershler & Milner, 1980; De Bruin et al., 1982; Rine et al., 1992; Drezner et al., 1994) and above-knee amputees (Hershler & Milner, 1980). These investigations used electrogoniometers to study the angle–angle patterns for the knee and hip joints, and demonstrated the value of angle–angle diagrams in the description and quantification of gait patterns within clinical assessment. Hurley and colleagues (1990) used a similar technique to hip versus knee angle–angle plots to study the contralateral limb in below-knee amputee gait. They used angle–angle plots of thigh segment angle with leg inclination, which they felt would give them a better depiction of the action of the lower limbs during walking.

4.3.8 Vector Coding of Angle–Angle Diagrams

A technique that has been used to quantify measures from angle–angle diagrams is vector coding. This calculates the vector orientation between adjacent data points relative to the right horizontal, which has been used to give a quantitative measure of coordination and coordination variability over time. The measures taken are referred to as the 'coupling angle' (CA), and can range from 0° to 360° (Sparrow et al., 1987; Hamill et al., 2000) (Fig. 4.22). As the CA is a directional variable, circular statistics are required for averaging and angular deviation calculations (Batschelet, 1981). A step-by-step approach on the necessary procedures and calculations on vector coding are reported elsewhere (Needham et al., 2014). The CA can be classified to one of four coordination patterns based on the polar position (Fig. 4.23). These are in-phase with proximal dominancy (white), in-phase with distal dominancy (light grey), anti-phase with proximal dominancy (dark grey) and anti-phase with distal dominancy (black) (Needham et al., 2015).

An in-phase pattern suggests both the proximal and distal segments are rotating in the same direction (Fig. 4.23a, e), whereas an anti-phase pattern would imply that the two segments are rotating in an opposite direction (Fig. 4.23c, g). The polar position of the CA within a coordination pattern highlights the dominancy of either the proximal or distal segment towards relative movement (grey numbers around the circumference of Fig. 4.23 represent a percentage contribution). For instance, at 45°, both segments are rotating in the same direction at the same rate; thus the contribution to relative movement of the proximal and distal segment is 50% (In-phase, D50–P50). If the CA is 81°, both segments are still rotating in the same direction (In-phase) but the distal segment is contributing 90% towards relative movement (D90–P10).

4.3.9 Angle versus Angular Velocity Diagrams (Phase Plane Portraits)

Craik and Oatis (1995) described a technique to show the behaviour of the shank during walking. They isolated two variables that capture the state of the system at any one time and then studied how these variables change over time. The two variables they selected were shank angular displacement and shank angular velocity (Fig. 4.24A). The choice of variables is considerable within movement analysis and the selection of variables depends, in part, on the questions being asked of the data.

Hurmuzlu and colleagues (1994) stated that standard plots of joint angular displacements against time,

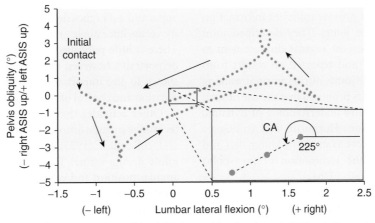

FIGURE 4.22 ■ Angle–angle plot representing pelvis obliquity and lumbar lateral flexion during gait. The inset provides an expanded view of one CA.

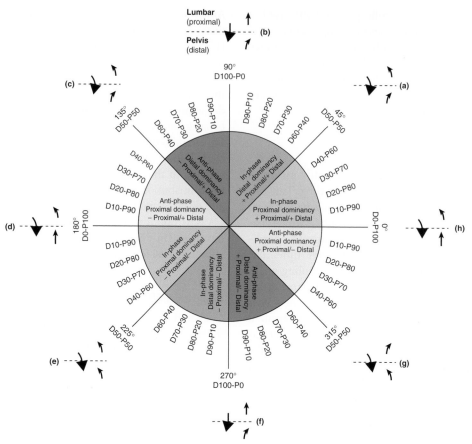

FIGURE 4.23 ■ Fig. 4.23 shows the coordination pattern classification. Segmental dominancy is shown around the circumference of the polar plot (grey text) with the inclusion of visual illustrations to show the coordination pattern between the lumbar region (proximal) and the pelvis (distal) at specific CA's (a–h). It is important to note that the proximal and distal segment angle data is displayed on the horizontal and vertical axis, respectively *(from Needham et al., 2015 – permission required)*

although useful, do not provide sufficient information about the dynamics of joints. They described joint angular velocity versus joint angular displacement as 'phase plane portraits' and reported on results from the hip, knee and angle joints during normal walking (Fig. 4.24B). The authors proposed that the use of such plots would give a greater understanding of dynamic changes of joint motion. Hurmuzlu and colleagues went on to use these plots to study the kinematics and dynamic stability of the locomotion of post-polio patients (Hurmuzlu et al., 1996).

Sojka and colleagues (1995) used hip versus knee angle–angle graphs and knee angular velocity versus knee angle graphs to assess the knee function in children with cerebral palsy who exhibit genu recurvatum.

Sojka and colleagues described the knee angular velocity versus knee angle graphs as knee phase plane plots. There is little published work using phase plane plots or portraits; however, the use of such plots has been shown to give important information about the coordination and control of movement of a particular joint.

Other activities that have been studied using this technique include lifting techniques. Burgess-Limerick and colleagues (1993) studied the coordination of joints during lifting. The authors stated 'The use of angular position and velocity information to describe joint movement on a phase plane is advantageous on theoretical grounds because the afferent information available from muscle receptors is effectively in terms of joint position and velocity.'

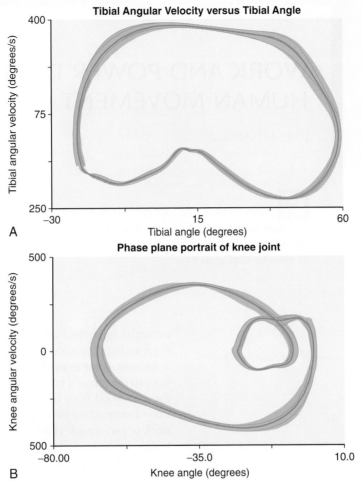

FIGURE 4.24 ■ (A) Shank angular velocity versus shank angle and (B) phase plane portrait of the knee joint

SUMMARY: MOTION AND JOINT MOTION

- Temporal and spatial parameters of gait relate to the foot contact times and distances, and are two of the simplest measures that can be taken. Despite this, they can yield very useful data in clinical assessment and may be used to monitor change during treatment and rehabilitation programmes.

- The movement patterns of all the joints of the lower limb occur in all three planes during walking. Although this makes the clinical assessment of an individual patient's gait difficult, it is possible to isolate the different movement patterns of the different joints and describe their movement in relation to function.

- Using new techniques, it is now possible to break the foot down into three separate segments. This allows us to consider the movement patterns of each foot segment separately and describe them in relation to function.

- There are several different methods of graphing the movement patterns of the lower limb and pelvis. Angle-versus-time graphs are the most common; however, some, such as angle–angle diagrams, allow the interaction between joints to be described, whereas angle–angular velocity graphs can be used to consider the relationship between joint position and control.

5 WORK AND POWER DURING HUMAN MOVEMENT

JIM RICHARDS

C hapter 5 covers the concepts of linear and angular work, energy and power, and how these can be determined from force and movement data. It also demonstrates the concept that angular work and power can be used to analyse the action of muscles during gait.

AIM

To consider the concepts of work, energy and power, and how these relate to muscle function.

OBJECTIVES

- To explain the differences between angular and linear work and power
- To calculate linear and angular work and power
- To describe how linear power may be found during the vertical jump test
- To explain which factors are important in the assessment of muscle and joint power
- To interpret and explain the interaction between moments, angular velocity and power during gait
- To explain what is meant by body segment energies and how they may be used

5.1 LINEAR WORK, ENERGY AND POWER

5.1.1 Linear Work

Work is a product of a force applied to a body and the displacement of the body in the direction of the applied force (Fig. 5.1). This does not refer to the muscular or mental effort. Work is basically a force overcoming a resistance and moving an object through a distance. If for example, an object is lifted from the floor to the top of a table, work is done in overcoming the downward force of gravity. On the other hand, if a constantly acting force does not produce motion, no work is performed. Holding a book steadily at arm's length, for example, does not involve any work, irrespective of the apparent effort required.

$$\text{Work} = \text{Force (F)} \times \text{Displacement (s)}$$
$$W = Fs$$

Units of Work

The units of work may be described in terms of the force multiplied by the displacement (newton × metre or Nm). However, work is usually considered in joules (J). There is an easily defined link between these two units with one joule of work being equal to a force of 1 N pushing an object 1 m. Therefore, in this case Nm and joule may be used interchangeably.

Differences between Work and Torque

Work is not the same as turning moments, even though they both have the same units of Nm. In the case of turning moments, the force acts perpendicular to the

104

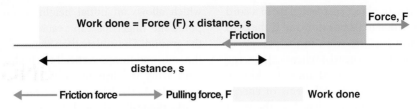

FIGURE 5.1 ■ Linear work done

distance. In the case of work, the force acts in the same direction as the displacement.

Work Is Also Done as an Object Accelerates

From Newton's second law the relationship between an object's mass, its acceleration and the applied force F is:

$$F = ma$$

However, Work = Force × Displacement.
   Therefore:

$$Work = Mass \times Acceleration \times Displacement$$

$$Work = m \times a \times s$$

Positive and Negative Work

Positive work is said to be done when a force acts parallel to the movement in the direction of the movement. Negative work is said to be done when a force acts parallel to the movement in the opposite direction to the movement.

Linear Power

Power is the rate of performing work or transferring energy. Work is equal to the force applied to move an object multiplied by the distance the object travels. Power measures how quickly the work is done:

$$Power = \frac{Work\ done}{Time\ taken}$$

For example, suppose a person wants to push a heavy box across a room. To overcome the friction between the bottom of the box and the floor, the person must apply force to the box to keep it moving. Now suppose the person pushes the box from one end of the room to the other, let's say 5 metres, in 10 seconds and the frictional force was 100 N, then they push the box back to its original position in 5 seconds. In each trip across the room, the same force was applied over the same distance, so the work done in each case is the same. However, the second time the box is pushed across the room the person has to apply more power than in the first trip because the same amount of work is done in 5 seconds rather than 10 seconds.

$$Work\ done = Fs$$
$$Work\ done = 10 \times 5 = 50\ J$$

However, the power would be:

$$Power = \frac{50}{10} = 5\ W$$

or

$$Power = \frac{50}{5} = 10\ W$$

The units of power are joules per second (J/s) or watts (W).

5.1.2 Linear Energy

When work is done on a body, there is a transfer of energy to the body and so work can be said to be energy in transit. Energy has the same units as work, as the work done produces a change in energy in joules (J). Energy is the capacity of matter to perform work as the result of its motion or its position in relation to forces acting on it. Energy related to position is potential energy and energy associated with motion is known as kinetic energy. If we consider a swinging pendulum, this has a maximum potential energy at the terminal points; at all intermediate positions, it has both kinetic and potential energy in varying proportions.

   Energy can be transformed, but it cannot be created or destroyed. In the process of transformation either kinetic or potential energy may be lost or gained, but the sum total of the two remains always the same.

Examples of this are: an object suspended from a cord has potential energy due to its position and if the cord is cut the object will perform work in the process of falling; if a gun is fired the potential (chemical) energy of the gunpowder is transformed into the kinetic energy of the moving projectile. All forms of energy tend to be transformed into heat, which is the most transient form of energy.

5.1.3 Potential Energy

This is stored energy possessed by a system as a result of the relative positions of the components of that system. For example, if a ball is held above the ground, the system comprising the ball and the earth has a certain amount of potential energy; lifting the ball higher increases the amount of potential energy the system possesses.

$$\text{Potential Energy} = \text{Mass} \times \text{Gravity} \times \text{Height}$$

$$\text{Potential Energy} = mgh$$

Work is needed to give a system potential energy. It takes effort to lift a ball off the ground. The amount of potential energy a system possesses is equal to the work done on the system. Potential energy also can be transformed into other forms of energy. For example, when a ball is held above the ground and released, the potential energy is transformed into kinetic energy.

5.1.4 Kinetic Energy

This is energy possessed by an object, resulting from the motion of that object. The magnitude of the kinetic energy depends on both the mass and the speed of the object according to the equation:

$$\text{Kinetic Energy} = \tfrac{1}{2} mv^2$$

The relationships between kinetic and potential energy can be illustrated by the lifting and dropping of an object.

5.2 THE RELATIONSHIP BETWEEN FORCE, IMPULSE AND POWER

5.2.1 The Vertical Jump Test

The vertical jump test, or Sargent jump test, involves a subject performing a jump from an initially static position. This is usually performed next to a wall, which allows an initial height measurement of the point the fingertips can reach with the feet flat on the ground. The athlete then stands slightly away from the wall, and jumps vertically as high as possible using both arms and legs to assist in projecting the body upwards. At the highest point the subject touches or marks the wall and the difference in distance between the static reach height and the jump height is the score. This score gives a useful indication of the levels of explosive muscular power production of the subject. This is used in activities often described as stretch shortening cycle movements, where muscles are first loading eccentrically to produce a larger concentric power production.

Variations of this test have been used in various sporting activities including basketball, volleyball, netball and rugby. The vertical jump test has also been used in the understanding of the motor development process and the functional performance in the injured athlete. However, the simple measure of the height jumped does not necessarily tell us all that we need to know about power production. Section 5.6: Joint power during the vertical jump test, shows how the power production from the individual joints may be found during the vertical jump test. But first we will consider what information may be drawn from force platforms alone when performing a vertical jump test to enable objective measurements of performance.

5.2.2 Maximum Force at Take Off and Landing

The values for maximum take-off force and landing force may be measured straight from the force–time graph (Fig. 5.2).

5.2.3 Velocity During the Jump

The instantaneous velocity of the body during the vertical jump may be identified by finding the area under the force–time graph. The area under the force–time graph gives us the impulse:

$$\text{Impulse} = \text{Mass} \times \text{Velocity (v)}$$

The area under the force–time graph is found after each time division, allowing the instantaneous velocity to be found (see Section 3.2: Impulse and Momentum). To find the velocity we must first subtract body

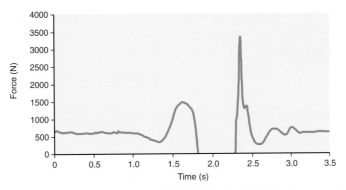

FIGURE 5.2 ■ Vertical GRF during a vertical jump test

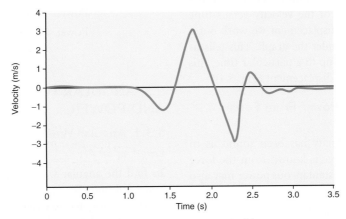

FIGURE 5.3 ■ Velocity during a vertical jump test

weight from the original force data. This gives us a force trace that now dips below zero. From:

$$F = ma$$

we can say:

$$F = m \times \frac{\text{change in velocity}}{\text{time}}$$

$$F = m \times \frac{(v - u)}{t}$$

So:

$$Ft = m(v - u)$$

where u is the initial velocity of the jump (which must be zero as the person has yet to start moving). Thus,

$$Ft = mv$$

We want to find the velocity v, and we know the mass because the mass is the body weight divided by 9.81 m/s². Thus,

$$Ft = \frac{v}{m}$$

So how do we find velocity from these data? We first work out the area of each trapezium for each time slice of the force–time curve. Then to find the velocity we work out a rolling sum of the area under the graph (see Section 3.5: Integration and the area beneath data curves). This tells us the area under the graph up to a particular time; this will be the instantaneous velocity (Fig. 5.3)!

5.2.4 Calculation of Height Jumped From Force Plate Data

This may be found by measuring the total time of flight. The time may be measured off the graph (during the time when the person is off the ground, the force will be zero). The height jumped may also be calculated from the take-off velocity (see earlier), again using the equations of motion. The instantaneous displacement of the centre of mass of the subject may be found by integrating the velocity:

$$\text{Velocity} = \frac{\text{Displacement}}{\text{time}}$$

So:

$$\text{Displacement (s)} = \text{Velocity} \times \text{Time}$$

(or the area under the velocity–time graph, the integral of velocity)

So How Can We Find Displacement From These Data? Again, we first work out the area of each trapezium for each time slice of the velocity-versus-time curve. Then to find the displacement we work out a rolling sum of the area under the graph. This tells us the area under the graph up to a particular time; this will be the instantaneous displacement (Fig. 5.4)!

5.2.5 Calculation of Power From Force Plate Data

If the instantaneous velocity has been found as in Section 5.2.3, and the force is known from the force platform readings, then instantaneous power may also be found:

$$\text{Power} = \frac{\text{Work done}}{\text{time}}$$

However, the work done may be expressed as:

$$\text{Work} = \text{Force} \times \text{Displacement}$$

$$\text{Work} = F \times s$$

Therefore, power may be express as:

$$\text{Power} = \frac{F \times s}{t}$$

And velocity may be expressed as:

$$\text{Velocity} = \frac{\text{Displacement}}{\text{time}}$$

$$v = \frac{s}{t}$$

Therefore, power may be found by multiplying force and velocity (Fig. 5.5):

$$\text{Power} = \text{Force} \times \text{velocity}$$

$$\text{Power} = F \times v$$

5.3 ANGULAR WORK, ENERGY AND POWER

5.3.1 Angular Work

Length of an Arc

To find the angular work, the length of an arc must first be found. To do this we will first consider the circumference of a circle. The circumference of a circle may be found from the relationship between pi (π) and the circumference and diameter of a circle.

$$\pi = \frac{\text{Circumference of a circle}}{\text{Diameter}}$$

$$\pi \times \text{Diameter} = \text{Circumference of a circle}$$

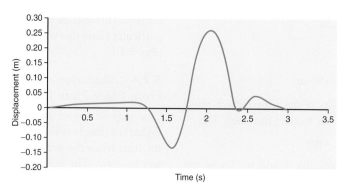

FIGURE 5.4 ■ Vertical displacement during a vertical jump test

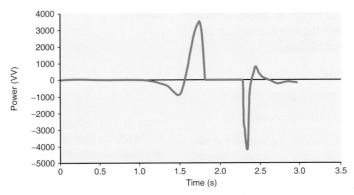

FIGURE 5.5 ■ Power during a vertical jump test

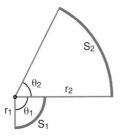

FIGURE 5.6 ■ Lengths of two arcs with different radii and different angles

The diameter of the circle will be twice the radius and so this may be rewritten:

$$2 \times \pi \times \text{Radius} = \text{Circumference}$$

However, the circumference may also be considered as the length of a very large arc. If we consider two arcs with different radii and different angles, these will produce different lengths of arc (Fig. 5.6).

As the angle increases, so will the length of the arc, and as the radius increases, the length of the arc will also increase. Therefore, we can say that the length of the arc is dependent on both the radius and the angle. This may be described by the following equation:

$$\text{Length of an arc} = \text{Angular displacement} \times \text{Radius}$$

$$\text{Length of an arc} = \theta \times r$$

We previously described the circumference of a circle as the length of a very large arc. Where:

$$\text{Circumference} = 2 \times \pi \times \text{Radius}$$

This implies that the angular displacement, or angle moved through may be described as $2 \times \pi$ one complete revolution. π is a constant (3.1415926……); therefore, $2 \times \pi = 6.2831852$ approximately. But what are the units?

The units for the angular displacement (angle travelled through) are radians (rad) and θ needs to be specified in radians when being used in calculations of angular work done or power. So, if 6.2831852 radians is one revolution (360°), then:

$$1\,\text{radian} = \frac{360}{6.2831852}$$

$$1\,\text{radian} = 57.295°$$

This is often rounded to 57.3°.

The need to use radians can be demonstrated by considering a leg raise activity. If a subject has a leg length of 1 m, and they move their leg through 90°, what is the length of the arc their foot has moved through?

If we use degrees in this calculation:

$$\text{Length of an arc} = \theta \times r$$

$$\text{Length of an arc} = 90 \times 1$$

$$\text{Length of an arc} = 90\,\text{m}$$

This is clearly **not** the case; however, if we consider the angle 90° in radians, we get 1.57 radians.

$$\text{Length of an arc} = \theta \times r$$

$$\text{Length of an arc} = 1.57 \times 1$$

$$\text{Length of an arc} = 1.57\,\text{m}$$

This gives us the correct length of arc, or angular displacement through which the foot has moved. This can **only** be found if we use radians.

Calculation of Angular Work

The definition for 'work done' as given earlier is:

$$\text{Work} = \text{Force} \times \text{Distance moved}$$

However, when we are considering angular work, the distance moved is the distance the force is moved through an arc, and the length of the arc is **r × θ**, therefore:

$$\text{Work} = \text{Force} \times r \times \theta$$

However:

$$\text{Force} \times r = \text{Moment (M)}$$

Therefore:

$$\text{Work} = \text{Moment (M)} \times \theta$$
$$\text{Work} = M \times \theta$$

Where θ is in radians (rad), as in Fig. 5.7.

5.3.2 Angular Power

If we consider the definition for power from earlier:

$$\text{Power} = \frac{\text{Work done}}{\text{time taken}}$$
$$\text{Power} = \frac{F \times r \times \theta}{t}$$

or

$$\text{Power} = \frac{\text{Moment (M)} \times \theta}{\text{time taken}}$$
$$\text{Power} = \frac{M \times \theta}{t}$$

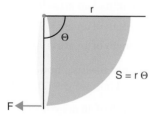

Angular work = F r Θ

    FIGURE 5.7 ■ Angular work done

however:

$$\text{Angular velocity } \omega = \frac{\theta}{t}$$

so:

$$\text{Power} = M\omega$$

where ω is in radian/s (rad/s).

Therefore, to study power we need to consider the net moments about a joint and the angular velocity of the joint. Power may be positive and negative. In the following sections, we will consider a positive power as power generation and a negative power as power absorption.

In the following three sections, we will explore the interaction between the joint moments, angular velocity and power during common movement tasks, including normal gait, running and jumping.

5.4 THE RELATIONSHIP BETWEEN MOMENTS, ANGULAR VELOCITY AND JOINT POWER DURING NORMAL GAIT

5.4.1 Joint Moments, Velocity and Power During Normal Gait

Inverse dynamics combines the study of kinetics and kinematics. This may be used to calculate moments about joints and the angular power produced. Moment calculations are covered in Chapter 2: Forces, moments and muscles; however, these calculations assume that the body segments are not accelerating. If the body segments are accelerating, then the inertial effects have to be taken into consideration in the calculation of turning moments and power about joints (see Chapter 6: Inverse dynamics theory).

We will now consider the relationship between joint moments, angular velocity, and the power absorption and generation about the ankle, knee and hip joints during normal walking. The joint power is found by combining the joint moments (Nm or Nm/kg) and joint angular velocities (rad/s). Positive values for joint power show power generation and negative values show power absorption. It is also beneficial to consider whether the muscle action is either concentric or eccentric. Concentric activity is associated with power development, whereas eccentric activity is associated with power absorption.

These graphs can be hard to interpret; however, it will help if you consider whether the moments are trying to flex or extend, and what movements are occurring at these time points. The following examples consider the power absorption and generation of the ankle, knee and hip joints in the sagittal plane during walking.

Careful Note

The angular velocities are presented in degree/s to aid understanding; however, these must first be converted into rad/s, to calculate power, i.e. divide by 57.3 (the number of degrees in 1 radian).

5.4.2 Ankle Moments, Velocity and Power During Normal Gait

Ankle Moments

At heel strike, the ground reaction force (GRF) passes very close to the ankle joint centre, producing a very small moment. In some cases, this will be behind the ankle joint, giving rise to a plantarflexion moment. After heel strike, the GRF moves in front of the ankle joint, producing a dorsiflexion moment. This increases as the force moves under the metatarsal heads and the force increases during push off (Fig. 5.8A).

Ankle Angular Velocity

At heel strike, the ankle plantarflexes to foot flat, which is controlled by an eccentric action of the muscles in the anterior compartment of the ankle joint. After this point the tibia begins to move over the ankle joint, which is also controlled by an eccentric action by the muscles in the posterior compartment. At 50% of the gait cycle, the heel begins to lift and the ankle rapidly plantarflexes at up to 250°/s (4.4 rad/s). This is produced by a concentric contraction of the posterior group and provides propulsion for the body (Fig. 5.8B).

Ankle Power

If the ankle has a dorsiflexion velocity and the moment is dorsiflexing about the ankle joint, then the posterior muscles must be working eccentrically and absorbing power. If the ankle has a plantarflexion velocity and a dorsiflexion moment, then the posterior muscles must be working concentrically and therefore generating power.

At heel strike, there is a plantarflexion moment and angular velocity; therefore, at heel strike there is eccentric power absorption from the dorsiflexors. This is followed by an eccentric power absorption by the plantarflexors as the body moves forwards over the foot. During push off there is a dorsiflexion moment and a plantarflexion velocity; therefore, power is generated by concentric activity of the plantarflexors (Fig. 5.8C).

5.4.3 Knee Moments, Velocity and Power During Normal Gait

Knee Moments

At heel strike, the GRF initially passes anterior to the knee joint, giving rise to an extension moment. The GRF then quickly passes behind the knee joint causing a flexion moment. After midstance, the force passes in front of the knee again until toe off. During swing phase, the knee also has significant moments due to the acceleration and deceleration of the foot and tibia (Fig. 5.9A).

Knee Angular Velocity

At heel strike the knee is already flexing. During the initial loading of the foot the knee flexion peaks at 150–200°/s. The knee flexion velocity slows as the knee reaches its initial peak flexion. The thigh now moves over the tibia (at the same time the tibia moves over the ankle joint), which causes an extension velocity of between 80 and 100°/s. At 50% of the gait cycle, the heel begins to lift and the knee starts flexing again. At toe off the knee is flexing at a rate of 300–350°/s; after this point the knee velocity slows as the knee reaches its maximum flexion. After the point of maximum knee flexion, the knee then starts to extend rapidly at a rate of 400–450°/s (Fig. 5.9B).

Knee Power

At heel strike the knee shows power generation, which is due to the GRF passing anterior to the knee joint, producing an extension moment while the knee is flexing. This initial power generation or concentric

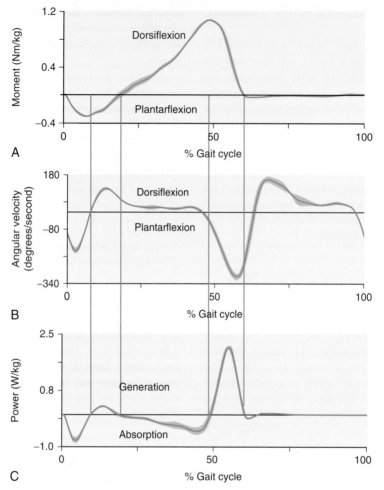

FIGURE 5.8 ■ Ankle. (A) Moments, (B) angular velocities and (C) power during normal gait

contraction of the hamstrings ensures that the knee does, indeed, flex at heel strike, rather than moving into a hyper-extended position. After this point the GRF falls behind the knee, creating a flexion moment whilst the knee is flexing; therefore, the quadriceps will be working eccentrically to act as a shock absorber. The knee then shows power generation at approximately 20% of the gait cycle. At this point the GRF is behind the knee and this, therefore, relates to the quadriceps acting concentrically pulling the femur over the tibia. As the knee extends, the GRF passes through the knee joint, producing no moment and, therefore, little or no power is generated or absorbed. It is interesting to note the involvement of the knee during push off at

50–60% of the gait cycle. During this time the GRF falls behind the knee, creating a flexion moment, but at this time the knee is flexing and, therefore, power is being absorbed and not generated. So, the knee has little or no involvement in the power production during push off (Fig. 5.9C).

5.4.4 Hip Moments, Velocity and Power During Normal Gait

Hip Moments

At heel strike the GRF passes quite far anterior to the hip joint, producing a peak flexion moment. After heel strike the GRF still passes anterior to the hip; however,

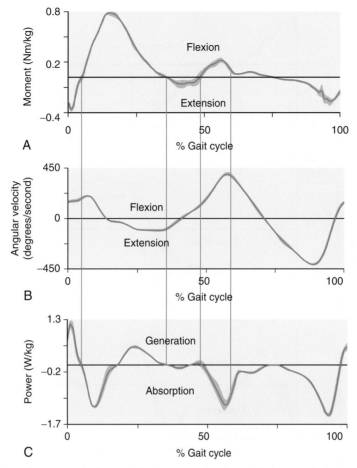

FIGURE 5.9 ■ Knee. (A) Moments, (B) angular velocities and (C) power during normal gait

the distance from the force to the hip is much reduced. After midstance, the force passes posterior to the hip, giving rise to an extension moment. As with the knee, there are significant moments during swing phase due to the acceleration and deceleration of the lower limb (Fig. 5.10A).

Hip Angular Velocity

The hip flexion velocity graph is far simpler. At heel strike the hip is flexed and stationary; it then moves into extension as the body moves over the limb. At 50% of the gait cycle, the heel lifts, pushing the thigh forwards, which starts the hip flexion velocity.

The hip extension velocity is slower than its flexion velocity, which indicates a controlled movement over the stance limb, followed by a more rapid flexion velocity to move the leg forwards to take a step (Fig. 5.10B).

Hip Power

At heel strike the hip shows power absorption, which is due to the hip having a small period of flexion velocity coupled with a flexion moment. The hip then extends to start to move the body over the stance limb whilst the moment is still trying to flex the hip; this is achieved by power generation by the

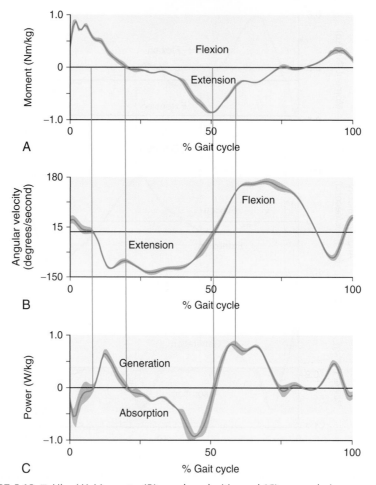

FIGURE 5.10 ■ Hip. (A) Moments, (B) angular velocities and (C) power during normal gait

hip extensors. At approximately 25% of the gait cycle, the moment passes behind the hip, changing from a flexion moment to an extension moment; however, the hip is still extending and this relates to power absorption or eccentric control of the hip flexors as the body moves over the stance limb. After 50% of the gait cycle, the hip reaches its maximum extended position. After this point, there is a rapid power generation during push off. This power generation is due to the GRF creating an extension moment as the hip changes from an extending angular velocity to a flexing angular velocity and, therefore, contributes to power production during push off (Fig. 5.10C).

5.5 THE RELATIONSHIP BETWEEN MOMENTS, ANGULAR VELOCITY AND JOINT POWER DURING RUNNING

The data presented here show the interaction of joint angles, moments, angular velocities and powers of the ankle, knee and hip joints for a runner with a forefoot strike pattern (please note the patterns for rearfoot strikers will be entirely different).

5.5.1 Ankle Moments, Velocity and Power During Running

At foot strike the ankle is in a near neutral position. The ankle then becomes progressively more dorsiflexed

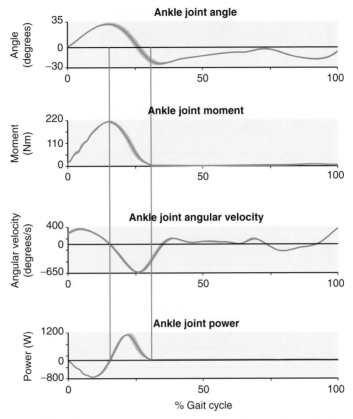

FIGURE 5.11 ■ Ankle angles, moments, angular velocities and power during forefoot running gait

and reaches a peak dorsiflexion velocity of 400°/s. During this time the calf muscles are acting eccentrically, which can be seen from the 800 watts of negative power as the muscles act as a shock absorber. This eccentric action also has a functional benefit of providing a stretch in the muscle which allows for a better push off by initiating a 'stretch shortening cycle'. After the eccentric phase, the tibia moves forwards over the foot and at the same time the ankle undergoes rapid plantarflexion at up to 650°/s. During this time the GRF moves forwards under the foot, which increases the moments at the ankle, resulting in a peak moment of 220 Nm and power of 1150 W. This power is provided by a concentric contraction of the calf muscles, and is used to propel the person forwards and upwards and to maintain forward momentum. During swing phase the subject moves the ankle joint towards dorsiflexion to ensure foot clearance and to prepare for the next foot strike (Fig. 5.11).

5.5.2 Knee Moments, Velocity and Power During Running

The knee is flexed at foot strike and continues to flex up to 50° to absorb the shock of the impact. During this time the knee has a peak knee flexion moment of 200 Nm and a peak power absorption of 500 W. The knee then starts to extend rapidly at up to 550°/s to advance the body over the stance limb and to produce power of up to 700 W. The peak power production of the knee occurs fractionally earlier than that of the ankle and is less than half that produced by the ankle. The knee reaches maximum extension at toe off at 30% of the gait cycle; at this point the knee angular velocity is zero. The knee then flexes rapidly up to 110°

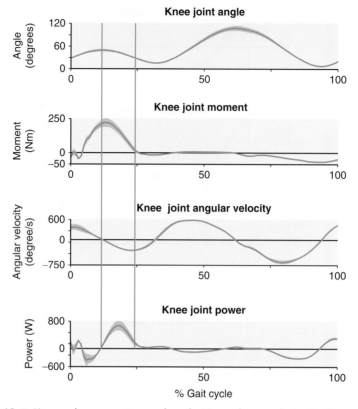

FIGURE 5.12 ■ Knee angles, moments, angular velocities and power during forefoot running gait

to aid foot clearance, which also has the effect of reducing the inertial moment required to accelerate the leg forwards. The knee then extends rapidly to prepare for the next foot strike at 100% of the gait cycle (Fig. 5.12).

5.5.3 Hip Moments, Velocity and Power During Running

The hip is flexed at foot strike and moves into extension at 350°/s as the body moves over the stance limb, which also has a significant role in the power production. The peak power production of the hip reaches 450 W, and this occurs earlier than the knee power production, which in turn is fractionally earlier than the ankle power production. This indicates a power delivery mechanism working proximal to distal. After toe off the hip flexes at approximately 400°/s, with a total range of motion of the hip of 45° during swing phase. During swing phase, there is still significant

power production to accelerate the limb forwards to the next foot strike (Fig. 5.13).

5.6 JOINT POWER DURING THE VERTICAL JUMP TEST

We have looked at the interaction of moments, angular velocity and power. We will now consider the individual contributions of the ankle, knee and hip joints during a vertical jump test (Fig. 5.14A,B).

5.6.1 Preparation and Propulsion

The ankle and knee motion show the knee joint flexing and the ankle dorsiflexing as the subject moves into the crouched position. The knee and ankle then rapidly extend and plantarflex together, generating the power to take off. During this phase the knee angular velocity is approximately 600°/s, slightly less than that of the ankle joint, which peaks at approximately 700°/s

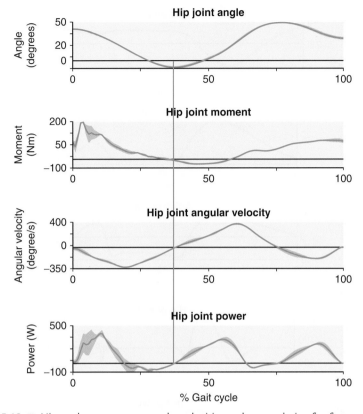

FIGURE 5.13 ■ Hip angles, moments, angular velocities and power during forefoot running gait

plantarflexion. The knee produces 550 W of power during propulsion, similar to that of the ankle at 600 W, with the hip producing the least power with 390 W. The peak power production of the hip occurs fractionally earlier than that of the knee, which in turn is fractionally earlier than the peak ankle power production. This indicates a power delivery mechanism working proximal to distal, which is also seen during running.

5.6.2 Flight

During flight the knee remains slightly flexed and the ankle remains plantarflexed in preparation for landing.

5.6.3 Landing

During landing the knee and ankle undergo rapid flexion and dorsiflexion to absorb the shock of the impact. Again, the ankle angular velocity is in the order of 1000°/s, and is greater than the knee angular

velocity, in the order of 500°/s. The ankle has the largest power absorption, in the order of 1500 W, more than twice that of the knee at approximately 700 W, with the hip having the smallest contribution to power absorption at 200 W.

5.7 BODY SEGMENT ENERGY

5.7.1 What Are Body Segments Energies?

Body segment energies tell us about the mechanical energy required to move each body segment. Body segment energies consists of three components: potential energy, translational kinetic energy and rotational kinetic energy. Once the anthropometric data have been estimated, and angular and linear displacements and velocities found, the segment energies can be calculated.

We will now consider how the different segment energies may be found. This includes:

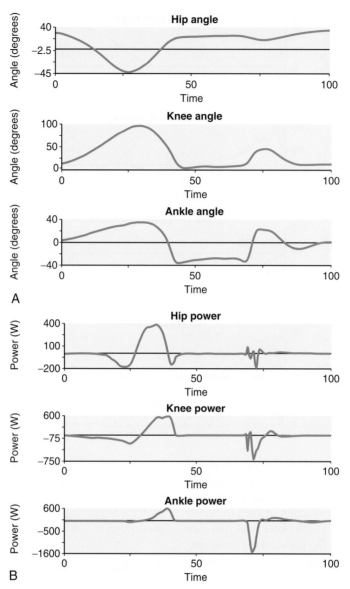

FIGURE 5.14 ■ (A) Joint angles and (B) joint power during a vertical jump test

1. Translational kinetic energy of body segment
2. Rotational kinetic energy of body segment
3. Potential energy of body segment
4. Total energy of body segment.

5.7.2 Calculation of Translational Kinetic Energy

Initially, linear displacements and velocities can be found for each joint in the three directions. From this

the velocity of the centre of mass of a body segment in a given direction may be found (Fig. 5.15).

5.7.3 Calculation of Rotational Kinetic Energy

If we consider an object rotating with a constant angular velocity (ω), any position away from the point of rotation will have both a linear and angular velocity. One such position is called the radius of gyration, k

To find the rotational kinetic energies the angular velocities are calculated for each segment in the xy, xz, and yz planes. Fig. 5.16 shows the angular velocities in the xy and xz planes. The moment of inertia of the body segment about the proximal end can be found using anthropometry. From this information, the rotational kinetic energy can, therefore, be calculated using the following equation (Fig. 5.16A):

$$\text{Rotational kinetic energy} = I(\omega_{xy}^2 + \omega_{xz}^2 + \omega_{yz}^2)$$

5.7.4 Calculation of Potential Energy

The potential energy may be found by finding the location of the centre of mass for a particular segment in the vertical direction in relation to a datum, e.g. the ground (Fig. 5.16B).

5.7.5 Calculation of Total Segment Energy

The total energy for a given body segment may be calculated from the sum of the translational kinetic energy, rotational kinetic energy, and the segment potential energy.

$$\text{Total energy} = PE + KE_{\text{rotational}} + KE_{\text{translational}}$$
$$E_{\text{total}} = mgh_{\text{com}} + I(\omega_{xy}^2 + \omega_{xz}^2 + \omega_{yz}^2) + m(v_x^2 + v_y^2 + v_z^2)$$

5.7.6 Calculation of Total Body Energy and Power

Total body energy may simply be found by summing all the total segment energies. The instantaneous body power may be found by finding the rate of change of total body energy.

5.7.7 Body Segment Energy Patterns During Normal Walking

Segment energies are the mechanical energies involved in the movement of body segments. Cavagna and colleagues (1963) referred to the external work done as that associated with the displacement of the centre of mass of the body. Inman (1966) studied the energy required for human locomotion by considering the GRFs and the kinetic and potential energy of the centre of mass of the body. Inman stated that the centre of

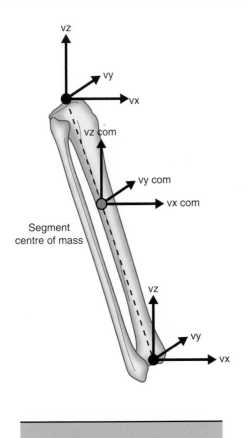

FIGURE 5.15 ■ Calculation of translational kinetic energy: translational kinetic energy $= \tfrac{1}{2} m (v_{x\,\text{com}}^2 + v_{y\,\text{com}}^2 + v_{z\,\text{com}}^2)$

(see Section 6.2.2). As a result, the linear velocity at the radius of gyration may be found by:

$$\text{Linear velocity (v)} = \text{Angular velocity } (\omega) \times \text{radius of gyration (k)}$$

$$\text{Velocity} = \omega k$$

Linear kinetic energy is given by:

$$\text{Kinetic energy} = \tfrac{1}{2} m v^2$$

Therefore:

$$\text{Rotational kinetic energy} = m(\omega k)^2$$
$$\text{Rotational kinetic energy} = m\omega^2 k^2$$
$$I_{\text{com}} = mk^2 \text{ (see Chapter 6)}$$
$$\text{Rotational kinetic energy} = I\omega^2$$

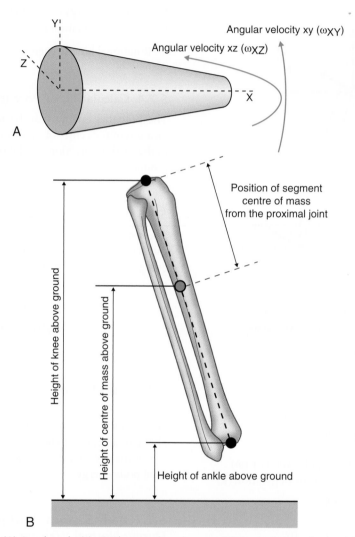

FIGURE 5.16 ■ (A) Angular velocities in the xy and xz planes and (B) calculation of potential energy = mgh_{com}

gravity of the body rises and falls, converting energy from potential to kinetic and back to potential.

Ralston and Lukin (1969) defined 'total external positive work,' as work measured by increases in the sum total of the energy levels of the body segments. They carried out a study of energy levels of human body segments during level treadmill walking, and concluded that the simultaneous measurement of the mechanical energy levels of the principal body segments and the metabolic expenditure during walking provided a powerful tool for the analysis of human locomotion.

Winter and colleagues (1976) reported further results for the instantaneous energy of body segments during normal walking. This included the total body energy, torso energy and energy of both legs. They tested normal individuals whilst free walking under laboratory conditions. The data were gathered from three strides in five subjects in the sagittal plane, ignoring any movement in the coronal and transverse planes. However, Winter and colleagues did calculate the rotational as well as the translational kinetic energy of each body segment. Pierrynowski and colleagues (1980) also investigated the transfer of mechanical energy within the total body and mechanical efficiency during treadmill walking. Pierrynowski and colleagues (1981) used the same technique to study load carriage using different devices.

The calculation of potential energy varies in different studies. To calculate potential energy a reference height or datum is required, which may either be relative to the ground (Quanbury et al., 1975; Winter et al., 1976; Pierrynowski et al., 1980) or some other reference height, such as the centre of mass of the body while standing (Ralston & Lukin, 1969). Whichever system is used, the same pattern of changing energies can be demonstrated. However, the magnitudes of the values will differ.

Winter (1978) reported on the use of energy in the assessment of pathological gait. Winter stated that: 'This technique permits identification of precise sources of high energy cost, locates the period when they occur in the gait cycle, indicates the segment(s) that are responsible and the type of energy involved.' Mansour and colleagues (1982) studied segmental mechanical energy changes of normal and pathological human gait. Subjects included healthy individuals and a group of subjects with pathologically impaired gait. In the healthy subjects, there was a greater exchange between potential and kinetic energy than in those with a pathological gait. The authors also stated that the patterns of energy changed in subjects with pathologically impaired gait, but this varied with the type of pathological disorder. This indicated that body segment energy patterns may give further information as to changes in movement and function of movement in different pathologies (Fig. 5.17A–D).

Olney and colleagues (1986) studied the mechanical energy of walking in 10 subjects who had suffered a cerebral vascular accident. The authors commented on the statement by Winter (1978): 'This precision is important: It allows identification of changes in the movement that are required to obtain energy savings. It is therefore a sophisticated method of analysis but has the advantage over many such methods by not requiring information from force plates.' Olney and colleagues identified that the body segment energy approach does have limitations in predicting energy costs. The technique does not consider the energy involved in maintaining a static position by isometric action of postural muscles present in slower pathological walking. One muscle group may generate energy at one joint at the same time as it is absorbed at another, resulting in metabolic costs that do not appear in a mechanical energy analysis. Despite these limitations, the authors concluded that the use of mechanical energy analysis could be valuable in pinpointing specific causes of high energy costs of walking in different pathological gait patterns and be used to assist in determining approaches in treatment. Olney and colleagues (1987) studied the gait of cerebral palsied children with hemiplegia, and also applied the same technique to the mechanical energy patterns for slow speed walking in normal older adults to determine any differences from slow walking to normal walking speeds (Olney et al., 1989).

Other uses of the body segment energy approach include work carried out by Miller and Verstraete (1996), who used the body segment energy approach to calculate the total body energy during gait initiation, and determined the number of steps necessary to attain a net mechanical work over one stride of zero to define a steady-state gait pattern.

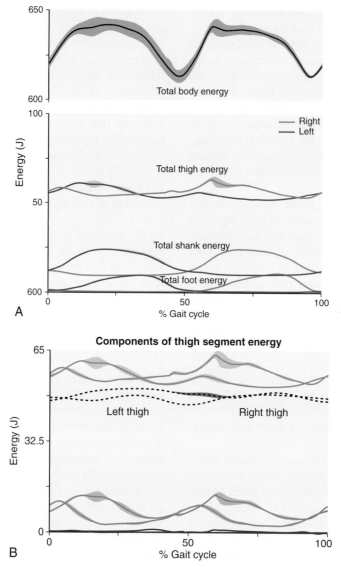

FIGURE 5.17 ■ (A) Total body segment energy, (B) thigh segment energy,

SUMMARY: WORK, ENERGY AND POWER

- Work and power depends on the applied force and the distance moved over time. With linear work and power, the force and the distance act in a straight line. With angular work and power, the force is moved through the arc of movement.

- As linear power involves movement in a straight line it is possible to calculate a useful estimate of the whole body power production during a vertical jump using force data alone.

- Angular power about specific joints may be found by knowing the moments acting and the angular velocity of the joint. This allows the calculation of positive and negative power, which can be

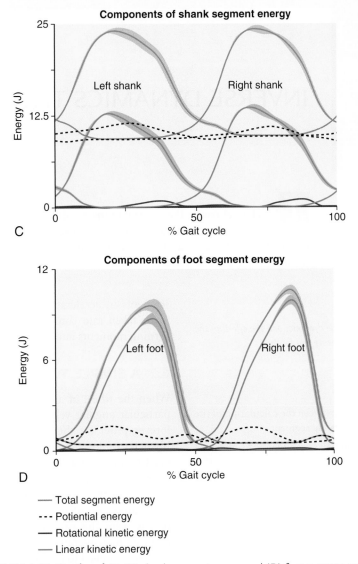

FIGURE 5.17, Continued ■ (C) shank segment energy and (D) foot segment energy

related to the concentric and eccentric muscle action about the joints during different tasks.

■ Body segment energies may be used to assess the amount of work associated with a particular movement. This allows the amount of energy used by each body segment to be assessed separately, which may be compared to the total body energy. This can give a useful estimate of the energy expenditure and physiological cost of different movement patterns.

6

INVERSE DYNAMICS THEORY

JIM RICHARDS

This chapter covers the concept of inverse dynamics. This includes the nature of radius of gyration and inertial moments. Examples of how the dynamic joint moments and forces may be found and the consequences of not considering dynamic force are also covered.

AIM

To consider the concepts of inverse dynamics and the effect on the calculation of dynamic joint moments and forces.

OBJECTIVES

- To explain the differences between the calculation of the I-value for a wheel and a body segment
- To explain why the inertial components need to be found
- To explain what forces are present and how they can be calculated
- To explain what moments are present and how they can be calculated
- To compare the simple and advanced techniques of finding joint moments and forces.

6.1 INTRODUCTION TO INVERSE DYNAMICS

Inverse dynamics combines the study of kinetics and kinematics. This may be used to calculate turning moments about joints and the power absorption–generation about those joints. Moments calculations are covered in Chapter 2: Forces, moments and muscles; however, these calculations assume that the body segments are not accelerating. If the body segments are accelerating, then the inertial effects have to be taken into consideration in the calculation of turning moments and power about joints.

6.2 A SIMPLE WHEEL

When the wheel of a car or bike is spun round at a particular angular velocity, it will continue spinning forever, providing there is no friction at the bearings. If you try to slow down or speed up the wheel, you will need to produce a torque or turning moment with your hands. So far, we have considered an object's mass in terms of the centre of mass, a point where all the mass can be considered to be concentrated (Fig. 6.1A). If this were the case, no turning effect would be required to accelerate or decelerate the wheel, i.e. no force would need to be applied.

This case can clearly not occur as the wheel's mass is not all concentrated at the centre of mass; therefore, there will be a turning moment twisting your arms as you try to accelerate or decelerate the wheel. The wheel's unwillingness to change angular velocity, or deceleration, depends on the distribution of mass around the centre of mass and the mass, m, of the wheel itself. The distribution of mass around the centre of mass of the wheel may be given a value called 'the radius of gyration, k'.

This is a fictitious distribution of the mass around the centre of mass. In the same way, the centre of mass

Acceleration of wheel with all mass
concentrated at the centre of mass

Acceleration of wheel with its mass distributed
and acting at the radius of gyration

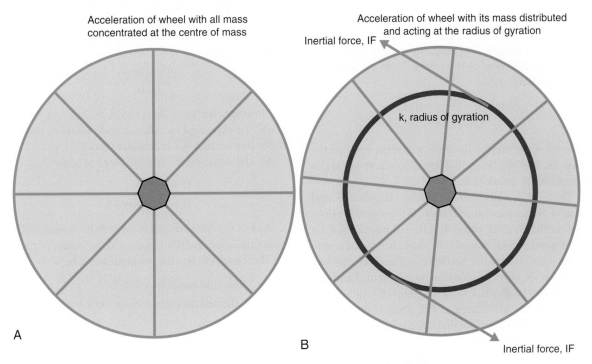

FIGURE 6.1 ■ (A and B) Inertial properties of a wheel

assumes that all the object's mass acts at one place for linear motion. The radius of gyration, k, assumes a distribution of the object's mass is a known distance away from the centre of mass for rotational motion. How the radius of gyration is calculated need not concern us here, as tables of values for the position of the radius of gyration exist (see Chapter 2: Anthropometry) (Fig. 6.1B).

If the mass is now considered to be concentrated at the radius of gyration, then any accelerations of this mass will need to overcome an inertial force (Newton's second law) as the radius of gyration will have a tangential linear acceleration–deceleration. Fig. 6.1 shows the inertial force acting at the radius of gyration perpendicular to the radius of the wheel.

6.2.1 Moment of Inertia

The moment of inertia, I, is a physical property which is used to describe an object's unwillingness to change its angular velocity. The moment of inertia of an object depends on the mass and the distribution of mass (radius of gyration) around a point of rotation.

$$I_{com} = mk^2$$

where: I_{com} = moment of inertia, m = mass of the object, k = radius of gyration.

Example (Part 1). If a wheel had a mass, m, of 5 kg and a radius of gyration, k, of 0.2 m, then we would have a moment of inertia about its centre of gravity of:

$$I_{com} = mk^2$$
$$I_{com} = 5 \times 0.2^2$$
$$I_{com} = 0.2 \text{ kg/m}^2$$

As with mass and length, the moment of inertia of an object is fixed and cannot change unless the mass or the dimensions of the object are changed.

6.2.2 Inertial Torque or Moment

Inertial torque or moment is not the same as the moment of inertia. The inertial moment is the moment (force × distance) required to create a rotational acceleration of an object. Fig. 6.1B shows the inertial force that has to be overcome to accelerate the wheel. This force acts at the radius of gyration and will, therefore, produce an inertial torque or moment. This is purely as a result of the change in angular velocity, or angular acceleration of the wheel. From earlier work, we know that:

$$\text{Moment} = \text{Force} \times \text{distance}$$
$$M = F \times d$$

where: F = the inertial force, IF, which depends on the mass, m, and the tangential linear acceleration, a; d = the radius of gyration, k.

There is a link between linear (tangential) and angular velocity and acceleration. If we consider velocities on the rotating wheel at different positions, for a given constant angular velocity the closer to the centre of the wheel we are the smaller our linear velocity will be, but as we move towards the outer rim of the wheel, or increase the distance from the point of rotation, our linear velocity will increase. The same is true of the relationship between angular acceleration and linear acceleration (Fig. 6.2).

This relationship between the tangential linear velocity and acceleration depends on the magnitude of the angular velocity and acceleration, and the perpendicular distance of the point being considered away from the point of rotation, **r**. In the previous example, this is the centre of the wheel; however, this may not always be the case. So, the linear velocity and acceleration of a point may be found using the following equations:

$$v = \omega r$$
$$a = \alpha r$$

However, we must be careful as both ω and α MUST be expressed in radian/s and radian/s^2, respectively (see Section 5.3.1: Angular work).

So, the equation for inertial force becomes:

$$IF = \text{mass} \times a$$
$$IF = \text{mass} \times \alpha \times k$$

As **k** is the distance, the mass may be considered to be acting as it is distributed around the centre of mass. Therefore, the inertial moment must be:

$$\text{Inertial moment} = IF \times k$$
$$\text{Inertial moment} = \text{mass} \times \alpha \times k \times k$$

or

$$\text{Inertial moment} = \text{mass} \times \alpha \times k^2$$

or

$$\text{Inertial moment} = \text{mass} \times k^2 \times \alpha$$

However:

$$I_{com} = mk^2$$

So, a simpler way of finding the inertial moment is:

$$\text{Inertial moment} = I_{com}\,\alpha$$

Therefore, if the wheel is turning at a constant angular velocity and is then slowed down, or decelerated, it produces a moment that depends on:

1. The distribution of the mass around the centre of mass, k, and the mass, m, of the wheel, which we have called the moment of inertia.
2. The angular acceleration or deceleration of the wheel, α.

The calculation to find the amount of turning moment (inertial torque) derived earlier is:

$$M = I_{com}\,\alpha$$

where: M = inertial moment as a result of the change in angular velocity, I_{com} = moment of inertia, α = the angular acceleration.

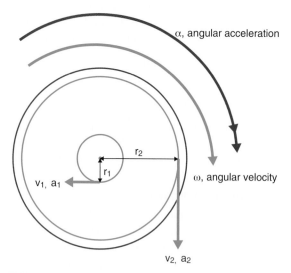

α, angular acceleration

r_2

r_1

v_1, a_1

ω, angular velocity

v_2, a_2

FIGURE 6.2 ■ Linear and angular velocity and acceleration

Example (Part 2). So, if a wheel had a mass, m, of 5 kg and a radius of gyration, k, of 0.2 m as before, then it would have a moment of inertia about its centre of gravity of:

$$I_{com} = 0.2 \text{ kg/m}^2$$

If the wheel was initially turning at an angular velocity of 2 rads/s, and was slowed down to 0 rads/s in 1 s, the angular acceleration–deceleration would be:

$$\alpha = 2 \text{ rad/s}^2$$
$$M = I_{com} \, \alpha$$
$$M = 0.2 \times 2$$
$$M = 0.4 \text{ Nm}$$

6.3 BODY SEGMENTS

6.3.1 Rotation About the Centre of Mass

A wheel turns about its centre of mass, a leg does not. When the point of rotation of an object changes so does its inertial properties (Fig. 6.3).

If we accelerate a bar about its centre of mass to 1 rads/s over a 1 second period, it will produce (or we will have to overcome) an inertial moment and/or unwillingness to change its angular velocity. This will act in the opposite direction to the angular acceleration. We have seen this depends on:

1. the distribution of mass around the centre of mass, k;
2. the mass, m, itself; and
3. the angular acceleration/deceleration.

$$I_{com} = mk^2$$

where: I_{com} = moment of inertia, m = mass of the object, k = radius of gyration.

$$M = I_{com} \, \alpha$$

where: M = turning moment as a result of the change in angular velocity, I_{com} = moment of inertia, α = the angular acceleration.

6.3.2 Rotation About One End

If we now rotate the bar about one end, point P, the inertial properties will change. If we consider holding a pool cue at one end and try to rotate it, we find it is harder than rotating it about its centre. Therefore, as we accelerate the bar in the two positions, to the same angular velocity over the same period of time, a larger resistance would be felt if the bar were rotated about one end. But the only thing that has changed is the distance of the centre of rotation (the point about which you are swinging the bar) to the centre of mass.

In the first case, we are rotating the bar about the centre of mass, so the distance to the centre of mass is obviously zero; therefore, there will be no tangential linear acceleration of the centre of mass.

However, now we are considering rotating the bar about a point away from the centre of mass, so you now not only have the inertia caused by the distribution of the mass about the centre of mass I_{com} but also the moment caused by an inertial force at the centre of mass. This inertial force is due to a tangential linear acceleration at the centre of mass, caused by a change in linear velocity of the centre of mass, i.e. an inertial force acting perpendicularly to the bar at the centre of mass as the mass of the bar tries to resist the change in velocity (Fig. 6.4).

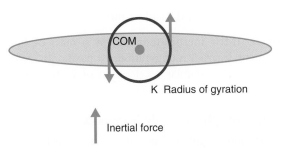

FIGURE 6.3 ■ Rotation about the centre of mass

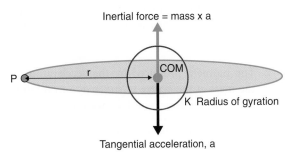

FIGURE 6.4 ■ Inertial properties when rotation about one end

If the body segment shown in Fig. 6.4 accelerates in a clockwise direction, then there must be an inertial force acting in an opposite anticlockwise direction. As the body segment decelerates the 'acceleration' acts in the opposite (clockwise) direction, causing the inertial force to also act in the opposite (anticlockwise) direction. In both cases the inertial force will try to resist any change in angular velocity. Fig. 6.4 also shows the tangential linear acceleration at the centre of mass, which will always act in the opposite direction to the inertial force. See the following:

The inertial force IF
$$= \text{mass}\,(m) \times \text{tangential linear acceleration}\,(a)$$

However, there is a link between linear (tangential) and angular velocity and acceleration:

$$v = \omega r$$
$$a = \alpha r$$

So, the equation for inertial force becomes:

$$IF = \text{mass} \times \alpha \times r$$

where: r = distance from the proximal end to the centre of mass.

This internal force at the centre of mass will produce a moment at the proximal end of the body segment, which can simply be found by force × distance.

$$M = IF \times r$$

or

$$M = \text{mass} \times \alpha \times r^2$$

6.3.3 Total Inertial Torque

So, the total inertial moment can be found by adding:

1. The moment produced by the acceleration of the mass at the radius of gyration, due to rotation about the centre of mass (see Section 6.3.1)

and

2. The moment produced by the acceleration of the mass at the centre of mass, due to rotation about the proximal end (see Section 6.3.2).

$$M = mk^2\alpha + mr^2\alpha$$

or

$$M = (mk^2 + mr^2)\,\alpha$$

which can be written:

$$M = I_{total}\,\alpha$$

Therefore, the total moment of inertia or I value when an object is rotated about one end may be found by the equation:

$$I_{total} = mk^2 + mr^2$$

To find the I value of a body segment we rely on previous work on anthropometry, where the radius of gyration (k), the distance of the centre of mass from the proximal joint (r) and the mass of the body segment are found by studying the proportions of mass and height of an individual, or by measurements of segment lengths.

Therefore, to find the internal moment all we need to do is find I_{total} from anthropometry and multiply the answer by the angular acceleration of the segment, making sure we have converted to rads/s^2 first!

6.3.4 Inertial Forces and Inertial Moment

Whenever we move there will be both linear and angular accelerations of our body segments. To be able to accelerate the body segments we must overcome inertial forces, i.e. forces associated with the segment's unwillingness (inertia) to change linear and angular velocity. These inertial forces act in the opposite direction to the accelerations as they are resisting the object's motion. The inertial forces will affect joint forces and produce inertial moment about proximal joints. Inertial forces should be considered in all directions (x, y, z), and inertial moments should be considered in all planes (xy, xz, yz).

$$IF = ma$$

where: IF = inertial force, m = mass of body segment, a = linear acceleration of centre of mass (x, y, z).

$$IM = (mk^2 + mr^2)\,\alpha$$

where: IM = inertial moment, m = mass of body segment, k = radius of gyration, r = position of centre of mass, a = angular acceleration of body segment (xy, xz, yz).

6.3.5 Weight of Body Segments

Every body segment has a mass (m), which has an acceleration due to gravity. Therefore, every body segment has a weight that will affect the joint forces and the moment about joints. The weight of the body segments will always be acting straight down; therefore, the position of the body segment will have an effect on the magnitude of the moment about the joint and the joint forces. The rotary component of the weight will always be acting perpendicular to the body segment. This force will cause a moment about the proximal joint (see Fig. 2.19):

$$\text{Weight acting vertically} = mg$$

$$\text{Rotary component of weight} = mg\cos\theta$$

where: m = mass of body segment, g = acceleration due to gravity, θ = angle of inclination of body segment from horizontal.

6.3.6 Centripetal Force

When an object is moving in a circular path there has to be a force acting on it, otherwise it will move in a purely linear path. This force must be acting towards the centre of rotation. This force is known as the 'centripetal force'. If there is a force and the object has a mass, then from Newton's second law the object must also have an acceleration; again, this must be acting towards the centre of rotation and this is known as the 'centripetal acceleration'.

$$\text{Centripetal acceleration} = \omega^2 r$$

where: ω = instantaneous angular velocity (radian/s), r = radius arm (distance from the centre of rotation to the centre of mass of the segment).

If we know the centripetal acceleration and the mass of the object, we can find the centripetal force.

$$F_{cen} = \text{mass} \times \text{centripetal acceleration}$$

$$F = m \times \omega^2 r$$

During normal walking the angular velocity of the shank during swing phase can get up to 400°/s or 7 rad/s. If the centre of mass is 0.2 m from the knee joint and the mass of the shank is 4 kg, then the centripetal force will be 39.2 N. This force will be acting on the centre of mass and along the body segment, so it will produce no moment directly.

STUDENTS' NOTE

Should these be included in the calculation of joint forces in inverse dynamics? If measurements are

taken from the inertial frame (e.g. using a motion analysis system), the linear accelerations and the inertial forces acting on the centre of mass are calculated, which will automatically include centripetal acceleration and centripetal force, so they should not be included twice. However, if joint rotations are being studied exclusively in terms of rotational kinematic variables (e.g. calculating the joint forces during an exercise on an isokinetic dynamometer where the angular velocity and acceleration are known), then centripetal force should be included separately.

6.4 JOINT FORCES

6.4.1 Terminology

Joint forces may be found by considering all the forces previously mentioned. Each force needs to be considered acting along a sensible frame of reference. In this case, we consider the forces acting vertically and horizontally:

6.4.2 Forces on the Foot and Ankle (Fig. 6.5)

$$Fy_{ankle} = V_{GRF} - mg_{foot} + IF_y$$

$$Fx_{ankle} = H_{GRF} + IF_x$$

$$V_{GRF} = \text{Vertical Ground Reaction Force}$$

$$H_{GRF} = \text{Horizontal Ground Reaction Force}$$

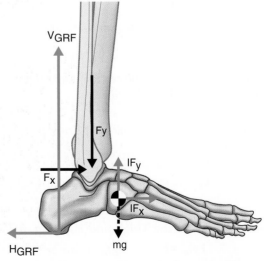

FIGURE 6.5 ■ Forces on the foot and ankle

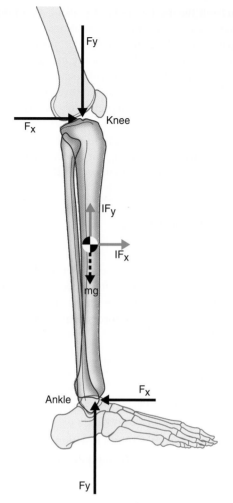

FIGURE 6.6 ■ Forces on the shank and knee

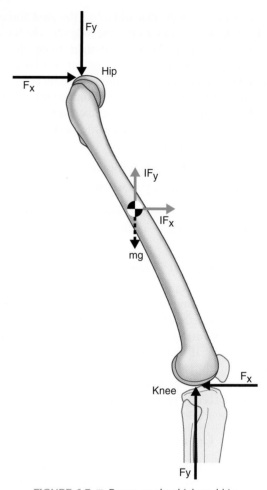

FIGURE 6.7 ■ Forces on the thigh and hip

where V_{GRF} = vertical ground reaction force and H_{GRF} = horizontal ground reaction force

6.4.3 Forces on the Shank and Knee (Fig. 6.6)

$$Fy_{knee} = Fy_{ankle} - mg_{shank} + IF_y$$
$$Fx_{knee} = Fx_{ankle} + IF_x$$

6.4.4 Forces on the Thigh and Hip (Fig. 6.7)

$$Fy_{hip} = Fy_{knee} - mg_{thigh} + IF_y$$
$$Fx_{hip} = Fx_{knee} + IF_x$$

6.5 JOINT MOMENTS

This requires the consideration of all the parameters mentioned in Section 6.4: Joint forces. This leads us to the following general formula. Each component of the total moment can either be clockwise or anticlockwise. Therefore, the symbol (±) has been used:

M = moment
V_{GRF} = vertical ground reaction force
H_{GRF} = horizontal ground reaction force
X_{ankle} = horizontal distance from ground to ankle joint centre
Y_{ankle} = vertical distance from ground to ankle joint centre

weight = component of weight acting perpendicular to body segment

X_{com} = distance to the centre of mass in X direction from the proximal joint

com = distance to the centre of mass from the proximal joint

I_{com} = moment of inertia about the centre of mass

α = angular acceleration of body segment

m = mass of body segment

IF = inertial force acting perpendicular to the body segment

segment length = length of body segment

joint force = rotational component of horizontal and vertical joint force.

6.5.1 Ankle Joint Moment (Fig. 6.5)

$$M_{xy,ankle} = \pm(V_{GRF} \times X_{ankle}) \pm (H_{GRF} \times Y_{ankle})$$
$$\pm (\text{weight of foot} \times X_{com}) \pm I_{com}\, \alpha_{foot}$$
$$\pm (IF \times com)_{foot}$$

6.5.2 Knee Joint Moment (Fig. 6.6)

$$M_{xy,knee} = \pm(\text{joint force at ankle} \times \text{segment length})$$
$$\pm M_{ankle} \pm (\text{weight of shank} \times X_{com})$$
$$\pm I_{com}\, \alpha_{shank} \pm (IF \times com)_{shank}$$

6.5.3 Hip Joint Moment (Fig. 6.7)

$$M_{xy,hip} = \pm(\text{joint force at knee} \times \text{segment length})$$
$$\pm M_{knee} \pm (\text{weight of thigh} \times X_{com})$$
$$\pm I_{com}\, \alpha_{thigh} \pm (IF \times com)_{thigh}$$

6.6 SO WHY DOES IT HAVE TO BE SO COMPLEX? A COMPARISON OF THE SIMPLE AND ADVANCED MODELS

6.6.1 Simplified Model

If we know the magnitude, the direction and the point of application of the ground reaction force (GRF) on the force platforms, the position of the force platforms relative to the position of the joints in the sagittal, coronal and transverse plane, then we can calculate the moment about each joint due to this force (see Chapter 2: Forces, moments and muscles) (Fig. 6.8).

If a GRF of 600 N acts at 80° to the horizontal, and the point of application of the force and the distances to the joint centres are known, the moments about the joints can be estimated using the simple techniques covered in previous chapters.

$$M_{ankle} = (600 \sin 80 \times 0.02) + (600 \cos 80 \times 0.05)$$

$$\mathbf{M_{ankle} = 17\,Nm}$$

$$M_{knee} = -(600 \sin 80 \times 0.08) + (600 \cos 80 \times 0.38)$$

$$\mathbf{M_{knee} = -7.7\,Nm}$$

$$M_{hip} = -(600 \sin 80 \times 0.3) + (600 \cos 80 \times 0.82)$$

$$\mathbf{M_{hip} = -91.6\,Nm}$$

However, this assumes that the body segments are not accelerating and decelerating during the activity.

6.6.2 Advanced Model

Previously we have simplified the method of calculating moment assuming the GRF was the only force producing a moment. Although this is a useful technique to get a quick answer, it will only give an approximate answer, and can underestimate the true value; this is unacceptable for many applications.

Foot Segment (Fig. 6.9A)

Acceleration of the com in x direction, $a_x = 0$ m/s^2
Acceleration of the com in y direction, $a_y = 0$ m/s^2
Angular acceleration, $\alpha = 2$ rad/s^2
Mass, m = 1 kg
Distance to centre of mass from joint, $X_{com} = 0.06$ m
Distance to centre of mass from joint, $Y_{com} = 0.03$ m
Moment of inertia, $I_{com} = 0.007$ kg/m^2
Ground reaction force = 600 N at 80°
Distances to ankle joint centre = as earlier.

Forces from foot to shank (joint forces at ankle)

$$F_{ay} = V_{GRF} - mg + ma_y$$

$$F_{ay} = 600 \sin 80 - 9.8 + 0$$

$$\mathbf{F_{ay} = 581\,N}$$

$$F_{ax} = -H_{GRF} + ma_x$$

$$F_{ax} = -104 + 0$$

$$\mathbf{F_{ax} = -104\,N}$$

Moments from foot to shank (moment at ankle)

$$Ma = -I\,\alpha + 600 \sin 80 \times 0.02 + 600 \cos 80 \times 0.05$$
$$+ mg\, X_{com} + ma\, X_{com} + ma\, Y_{com}$$

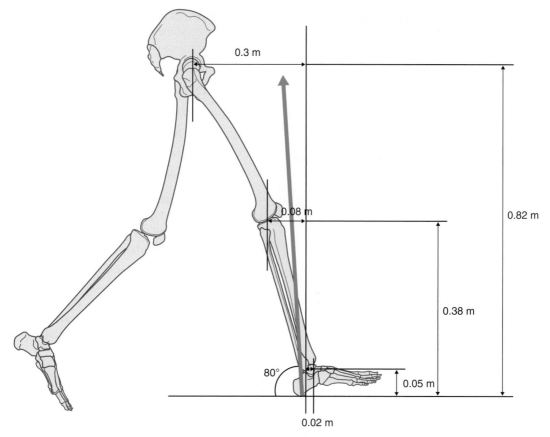

FIGURE 6.8 ◼ Simplified model

$$Ma = -0.007 \times 2 + 600 \sin 80 \times 0.02 + 600 \cos 80$$
$$\times 0.05 + 10 \times 0.06 + 0 + 0$$
$$Ma = -0.014 + 11.8 + 5.2 + 0.6$$
$$\textbf{Ma} = \textbf{17.6 Nm}$$

Shank Segment (Fig. 6.9B). The shank is beginning to move over the ankle joint. Therefore, its centre of mass will have a linear acceleration up and to the right. The angular acceleration will be in the clockwise direction:

Acceleration of the com in x direction, $a_x = 5$ m/s^2
Acceleration of the com in y direction, $a_y = 2$ m/s^2
Angular acceleration, $\alpha = 4$ rad/s^2
Mass, m = 4 kg
Distance to the centre of mass from proximal joint, com = 0.13 m

Moment of inertia, $I_{com} = 0.07$ kg/m^2
Length of body segment, l = 0.345 m
Angle of inclination from vertical, $\theta = 16.8°$.

$$F_{ky} = -mg - ma_y + F_{ay}$$
$$F_{ky} = -39.2 - 8 + 581$$
$$F_{ky} = \textbf{533.8 N}$$
$$F_{kx} = -ma_x - F_{ax}$$
$$F_{kx} = -20 - 104$$
$$F_{kx} = \textbf{−124 N}$$

Moments from shank to thigh (moment at knee)

$$M_k = M_a - F_{ay} \, l \sin\theta + F_{ax} \, l \cos\theta - I\,\alpha + mg \text{ com} \sin\theta$$
$$+ ma_y \text{ com} \sin\theta + ma_x \text{ com} \cos\theta$$
$$M_k = 17.6 - 581 \times 0.345 \sin 16.8 + 104 \times 0.345 \cos 16.8$$
$$- 0.07 \times 4 + 4 \times 9.81 \times 0.13 \sin 16.8 + 4 \times 2$$
$$\times 0.13 \sin 16.8 + 4 \times 5 \times 0.13 \cos 16.8$$

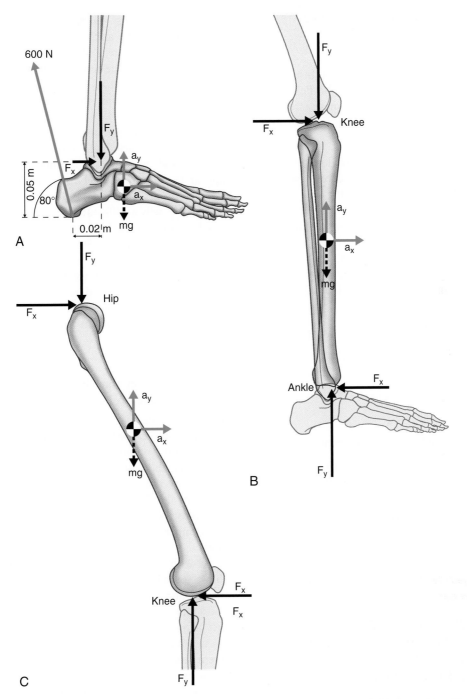

FIGURE 6.9 ■ (A) Foot segment, (B) forces from shank to thigh (joint forces at knee) and (C) forces from thigh to pelvis or joint forces at hip

$$M_k = 17.6 - 57.9 + 34.3 - 0.28 + 1.5 + 0.3 + 2.5$$

$$M_k = -1.98 \, \text{Nm}$$

Thigh Segment (Fig. 6.9C). The thigh is beginning to move forward; however, the knee is also flexing. This causes the centre of mass to have a linear acceleration down and to the right. The angular acceleration will be in the clockwise direction:

Acceleration of the com in x direction, $a_x = 3 \, \text{m/s}^2$
Acceleration of the com in y direction, $a_y = -2 \, \text{m/s}^2$
Angular acceleration, $\alpha = 2.5 \, \text{rad/s}^2$
Mass, $m = 6 \, \text{kg}$
Distance to the centre of mass from proximal joint, com = 0.22 m
Length of body segment, $l = 0.46 \, \text{m}$
Moment of inertia, $I_{com} = 0.18 \, \text{kg/m}^2$
Angle of inclination from vertical, $\theta = 26.6°$.

$$F_{hy} = -mg + ma_y + F_{ky}$$

$$F_{hy} = -6 \times 9.81 + 6 \times 2 + 533.8$$

$$F_{hy} = 486.9 \, \text{N}$$

$$F_{hx} = -ma_x - F_{kx}$$

$$F_{hx} = -6 \times 3 - 124$$

$$F_{hx} = -142 \, \text{N}$$

Moments from thigh to pelvis (moment at hip)

$$M_h = M_k - F_{ky} \, l \sin\theta + F_{kx} \, l \cos\theta - I\alpha + mg \, com \sin\theta$$
$$+ ma_y \, com \sin\theta + ma_x \, com \cos\theta$$

$$M_h = -1.98 - 533.8 \times 0.46 \sin 26.6 + 124 \times 0.46 \cos 26.6$$
$$- 0.18 \times 2.5 + 6 \times 9.81 \times 0.22 \sin 26.6 - 6 \times 2$$
$$\times 0.22 \sin 26.6 + 6 \times 3 \times 0.22 \cos 26.6 = 0$$

$$M_h = -1.98 - 109.9 + 51 - 0.45 + 5.8 - 1.18 + 3.54$$

$$M_h = -53.1 \, \text{Nm}$$

6.7 SO WHAT EFFECTS DO THE SIMPLE AND ADVANCED METHODS HAVE ON MOMENTS AND POWER CALCULATED DURING GAIT?

6.7.1 The Effect the Simple and Advanced Methods Have on Moments

The graphs in Fig. 6.10 show the relative effect of the moments caused by the GRF and the inertial and weight components. These demonstrate that the moments about the ankle joint are not significantly affected by the moment of inertia and the weight of the distal segment, the foot. This is not surprising as the mass, centre of mass and radius of gyration are comparably small and, therefore, have a minimal effect.

With the moments about the knee we begin to see some differences with the introduction of the inertial and weight components, although these appear to be only at heel strike, early loading and just before toe off.

The hip tells a different story, as the inertial and weight components have a significant effect, and can account for up to 25% of the moment during heel off to toe off. This increase is not surprising, as the further proximal we travel, the greater the mass, centre of mass and radius of gyration distal to the joint of interest, and, therefore, the greater the inertial and weight moment components (Fig. 6.10A, B, C).

6.7.2 The Effect the Simple and Advanced Methods Have on Power

The graphs in Fig. 6.11 show the relative effect of the power caused by the GRF and the inertial and weight components. These demonstrate that the power about the ankle joint is not significantly affected by the inertia and the weight of the distal segment, the foot. Again, this is not surprising as the mass, centre of mass and radius of gyration are comparably small and therefore have minimal effect.

As with the moments, the power about the knee shows some differences with the introduction of the inertial and weight components; again, these differences only appear at heel strike and early loading and just before toe off.

As with the moments, the hip power tells a different story with a much greater difference or error when only considering the effect of the GRF. Again, this increase is not surprising as the further proximal we travel the greater the mass, centre of mass and radius of gyration distal to the joint of interest, and therefore the greater the inertial and weight moment components.

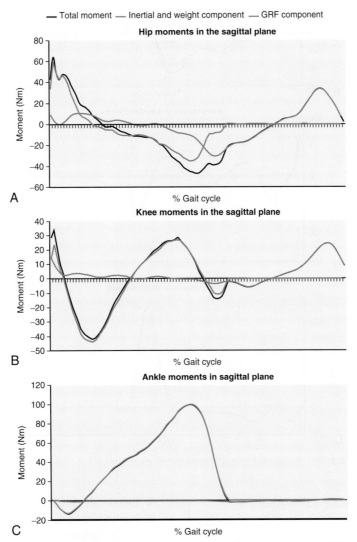

FIGURE 6.10 ■ (A) Hip moments in sagittal plane, (B) knee moments in sagittal plane and (C) ankle moments in sagittal plane

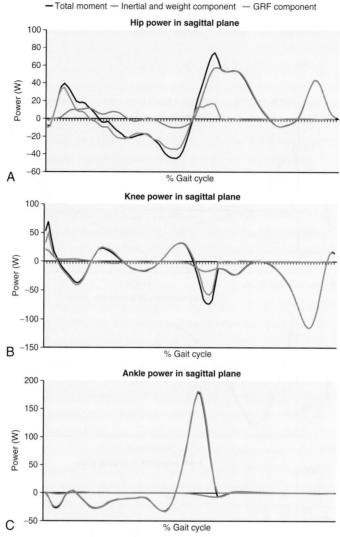

FIGURE 6.11 ■ (A) Hip power in sagittal plane, (B) knee power in sagittal plane and (C) ankle power in sagittal plane

SUMMARY: INVERSE DYNAMICS THEORY

- Inertial forces are present whenever the body segments accelerate or decelerate during a movement. For slow movements, the inertial forces will be minimal; however, for walking and running inertial forces can be considerable.
- Inertial forces have a direct effect on the joint forces, the moments acting around joints and the muscle power production.

- Inverse dynamics is one of the most complex methods used in movement analysis. However, in not including inertial forces, joint moments and power may be significantly underestimated.
- When conducting an analysis of joint moments and power it is important to consider inertial forces and moments wherever possible.

MEASUREMENT OF FORCE

Section 2 **METHODS OF MEASUREMENT AND MODELLING**

7 MEASUREMENT OF FORCE AND PRESSURE

JIM RICHARDS ■ AOIFE HEALY ■ DOMINIC THEWLIS ■ NACHIAPPAN CHOCKALINGAM

7.1 METHODS OF FORCE MEASUREMENT

Force platforms are considered as a basic but fundamentally important tool for gait analysis. The first force measurements date back as far as the late 19th century, when Marey used a wooden frame on rubber supports. Elftman (1939a) used a similar method with a platform on springs. However, it was not until the advancement of computers and electronic technology that the readings could be accurately measured. In 1965, Peterson and colleagues developed one of the first strain-gauge force platforms. A plethora of publications now exists on the applications of such devices in both clinical research and sports.

Since 1965, forces platforms have undergone considerable development by three internationally accepted manufacturers: Kistler Instruments, AMTI and the Bertec Corporation. Advances have been in the form of making the platforms more accurate (reducing crosstalk), increasing sensitivity (increasing the natural frequency) and making the platforms portable.

Force platforms measure and record the ground reaction forces (GRFs) and their point of application (centre of pressure, COP). A GRF is made up of three components acting at the centre of pressure. The three components can be categorized in an anatomical sense as vertical forces (the weight of the body and how it progresses over the supporting limb), anterior–posterior forces (the accelerating and breaking forces) and medial–lateral forces (the force acting from side to side).

7.1.1 Force Platform Types

Generally, force platforms come under two categories: strain gauge (AMTI, Bertec) or piezoelectric (Kistler) (Fig. 7.1A, B). It is accepted that piezoelectric force platforms are more sensitive and allow a larger range of force measurements. They are also able to measure higher frequency content of up to 1000 Hz, their natural frequency, in all three directions, whereas strain gauge platforms generally have a natural frequency of 400–500 Hz, although strain gauge platforms are available up to 1000 Hz in the vertical and 500 Hz in the horizontal directions. However, piezoelectric force platforms are generally more expensive and offer no real advantage in clinical research. Therefore, for general use strain gauge platforms are more than adequate; however, for activities with a higher frequency content, piezoelectric platforms are recommended.

7.1.2 How Force Platforms Work

Strain gauge force platforms are based on the principle that when a force is applied to a structure, the structure changes in length. Strain is the ratio of changes between the original dimensions and the deformed dimensions. Strain gauges contain material which, when distorted, produces a resistance. So, by measuring the resistance we can measure the strain. In order for the strain gauge to work correctly it must be wired up in what is known as a Wheatstone bridge arrangement, which simply arranges the gauges or resistors in an arrangement with four arms. For strain gauge platforms to work a power supply is required, as the strain

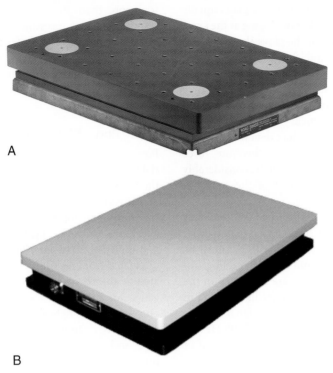

A

B

FIGURE 7.1 ■ (A) Kistler and (B) AMTI Force Platforms

is based on the resistance provided to the electrical current. The strain gauges are grouped in triplets in the pylons of the platform, situated in the corners. The resistance is normally relatively small, so signals produced by strain gauge force platforms require amplification; this can either take place in the platform or in a separate amplifier.

Piezoelectric platforms are based on the use of piezoelectric crystals such as quartz. The basis for the generation of the signal is the same as strain gauges, in terms of the deformation of the crystal. However, when piezoelectric crystals are deformed they generate what is known as an electric dipole moment, which in turn generates an electric current. As piezoelectric crystals generate their own current, there is no need for a power supply. The coordinate system of the platform is determined by the alignment of the piezoelectric crystals on each pylon, which are aligned with the x, y and z axis of the platform (Fig. 7.2).

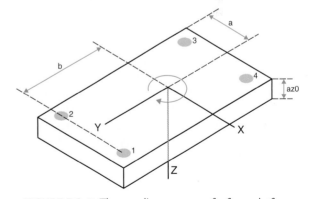

FIGURE 7.2 ■ The coordinate system of a force platform

7.1.3 Frequency Content and Force Platforms

Every mechanical system has a particular frequency at which it will vibrate when excited in the right way. A wine glass, for instance, may be made to make a tone by running a finger around the rim of the glass, equally an opera singer can break a glass by singing at the exact pitch or natural frequency of the glass; the glass then vibrates at this frequency and, according to the urban myth, breaks! The exact natural frequency will depend on the mass and the dimensions of the object. Force platforms are no exception and also have a natural frequency. This natural frequency should be avoided during testing as this would produce data which would not represent the task being studied; therefore, force platform's natural frequencies need to be considerably higher than the frequency content of the signal of interest.

Antonsson and Mann (1985) studied the frequency content of GRFs during gait; it was found that in order to preserve 99% of the signal fidelity, 15 Hz must be maintained. This requires a minimum sampling frequency of 30 Hz, although if we are interested in the transient effects at heel strike these are likely to fall in the 1% rather than the 99% of the signal fidelity. Therefore, higher sampling frequencies of 100 Hz and above are generally chosen. The frequency content of sporting tasks has, not surprisingly, been found to have much higher frequency during applications such as analysis of take off and landing in jumping and running.

Since the early work on frequency content of GRFs, this is now recognized as providing useful insights into changes due to age (Stergiou et al., 2002) and offers information as a differential diagnosis for different conditions such as multiple sclerosis (Wurdeman et al., 2011) and peripheral arterial disease (McGrath et al., 2012). This is an interesting area of work which offers additional outcome measures over and above looking at the magnitude and timing of GRFs, which in turn might help develop new assessment tools to determine different levels of impairment and monitoring of treatment.

7.1.4 Signal Drift

Although piezoelectric platforms have the advantages of sensitivity, range and natural frequency, they do have the disadvantage of signal drift. Before each trial, it is recommended that the platform is reset, which is a simple process of flicking a switch which drops all the signals to zero. However, if work is being carried out when the test subject is standing on the platform

for more than 30 s, for example, when looking at standing balance assessment or static measurements, piezoelectric platforms suffer from the signal drifting, which can introduce errors. Strain gauge platforms, however, are more stable due to the technology they use and are, therefore, better suited to these types of study.

7.1.5 Force Plate Scaling

Piezoelectric and strain gauge force platforms essentially measure the same forces and moments; however, the ways in which they accomplish this are different. Strain gauge force platforms produce an output of 6 channels of analogue data. These are forces in the x, y and z directions (Fx, Fy, Fz) and the moments in the x, y and z (Mx, My, Mz). The moments and forces are calculated from readings taken at the four pylons situated at the corners of the force platform. So, strain gauge force platforms are relatively simple in their output, with a single analogue channel representing a single force or moment that is then simply multiplied by a scaling factor to convert the raw voltage into forces and moments in newtons (N) and newton metres (Nm), respectively. Fig. 7.3 A and B show the raw voltage data and final output data from a typical strain gauge force platform.

Piezoelectric force platforms produce essentially the same data as strain gauge platforms, but the way in which they generate the data is different. Piezoelectric platforms generate 8 channels of analogue output, none of which contain any information about the moments acting on the platform (Mx, My or Mz). The channels are simply used to calculate the force data. Eight channels produce the following voltage data, which can then be used to calculate the force data in newtons:

Fx
  Ch1: fx12 – force in the x direction measured by pylons 1 and 2
  Ch2: fx34 – force in the x direction measured by pylons 3 and 4
Fy
  Ch3: fy12 – force in the y direction measured by pylons 1 and 2
  Ch4: fy34 – force in the y direction measured by pylons 3 and 4

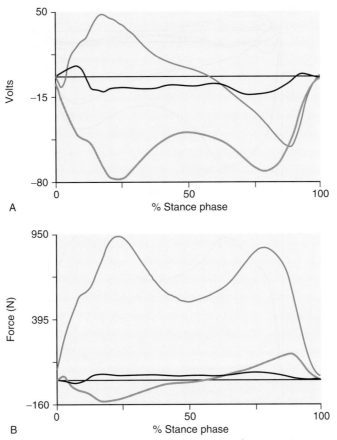

FIGURE 7.3 ■ (A) Raw voltage and (B) final output data from a strain gauge platform

Fz

    Ch5–8: fz1, fz2, fz3, fz4 – force in the z direction measured by pylons 1 to 4

The final force data are calculated by simply adding the x channels for the x data, the y channels for the y data and the z channels for the z data. Fig. 7.4 shows the raw voltage data and the subsequent force data.

7.1.6 Calculating Moments on a Force Plate

Piezoelectric platforms do not provide a direct measurement of the moments about the platform, but these can be calculated if we know the location of the centre of the platform in the medial–lateral and anterior–posterior directions (Fig. 7.2).

$$Mx = b\,(fz1 + fz2 - fz3 - fz4)$$
$$My = a\,(-fz1 + fz2 + fz3 - fz4)$$
$$Mz = b\,(-fx12 + fx34) + a\,(fy14 - fy23)$$

We now know how to find the moments and forces produced from both piezoelectric and strain gauge force platforms; the next stage in the processing of force data is to calculate the centre of pressure. The following set of equations can be used with either type of force platform. The first stage we must address is the calculation of the moments about the platform (Mx′ & My′): for this we need to know the moments about the platform axis, the forces and the offset between the vertical position of the centre of the platform and the top of the platform.

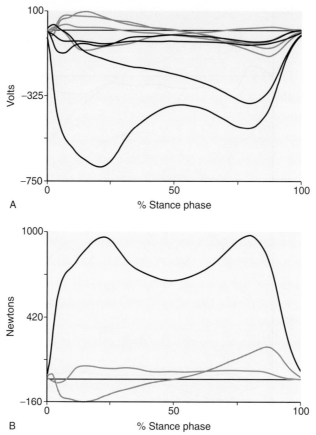

FIGURE 7.4 ■ (A) Raw voltage and (B) final output data from a piezoelectric platform

Moment in the sagittal plane $(Mx') = Mx + Fy \times az0$

Moment in the coronal plane $(My') = My + Fx \times az0$

We now know the force, the direction and the turning effect of the force and from these we can calculate the centre of pressure by simply dividing the moment about the plate by the vertical force.

$$COPy = \frac{Mx'}{Fz}$$

$$COPx = \frac{My'}{Fz}$$

Finally, we can calculate the free moment about the platforms, or the vertical moment. These data along with the centre of pressure data are important to consider during the calculation of joint kinetics using inverse dynamics.

$$\text{Moment in the transverse plane}\,(Mz')$$
$$= Mz - Fy \times COPx + Fx \times COPy$$

Example of Moment Calculations Using a Piezoelectric Platform

We will now consider an example of an individual applying a known force of 800 N vertical force, 200 N posterior force and 50 N medial force. There are moments of 150 Nm in the sagittal plane (X), 110 Nm in the coronal plane (Y) and 20 Nm in the transverse plane (Z). In this example a clockwise moment is considered as positive and anticlockwise negative.

Moments and COP in the Sagittal Plane Mx′ (Fig. 7.5A)

Moment in the sagittal plane (Mx′)

$$Mx + Fy \times az0 + Fz \times COPy = 0$$
$$150 - 200 \times 0.04 - 800 \times COPy = 0$$
$$142 - 800 \times COPy = 0$$
$$142 = 800 \times COPy$$
$$COPy = \frac{142}{800}$$
$$COPy = 0.1775 \text{ m}$$

Moments and COP in the Coronal Plane My′ (Fig. 7.5B)

Moment in the coronal plane (My′)

$$My + Fx \times az0 + Fz \times COPx = 0$$
$$110 - 50 \times 0.04 - 800 \times COPx = 0$$
$$108 - 800 \times COPx = 0$$
$$108 = 800 \times COPx$$
$$COPx = \frac{108}{800}$$
$$COPx = 0.135 \text{ m}$$

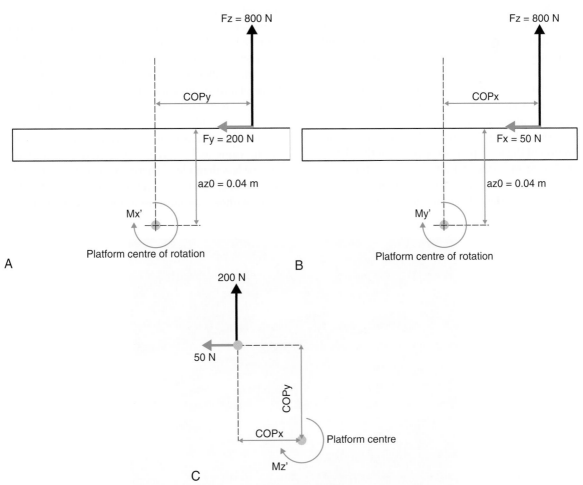

FIGURE 7.5 ■ (A) Moment calculations in the sagittal plane, (B) moment calculations in the coronal plane and (C) moment calculations in the transverse plane

So, the centre of pressure coordinates are:

$$COPy = 0.1775 \, m$$

and

$$COPx = 0.135 \, m$$

Moments in the Transverse Plane Mz′ (Fig. 7.5C). Finally, we can now calculate the moment in the transverse plane, which is sometimes referred to as the free moment (Mz′).

$$Mz' = 20 + (200 \times 0.1775) + (-50 \times 0.135)$$
$$Mz' = 48.75 \, Nm$$

7.1.7 Considerations for Force Platform Fitting and Positioning

When a movement analysis laboratory is designed, it is essential that consideration is given to future funding becoming available to expand equipment. So, designs of a new movement analysis laboratory where two force platforms are available should leave sufficient space for an additional two platforms; otherwise fitting additional platforms in the future may become a very expensive job. Fig. 7.6 shows the holes for the usual position of the force platforms, but with the platforms lifted to form a step. An alternative solution is to fit a false floor, which is a somewhat more expensive solution; however, it allows the platforms to be moved anywhere in the laboratory, provided the fixings are in place.

It is very important to note that all force platforms (besides some of the new USB platforms) have quite substantial cabling associated with them. This cable must run from the platform to the amplifier (normally situated by the control PC), so it is essential to either have the ability to pass these cables under a false floor or through a conduit with a large enough inner diameter to allow the cables and end connectors to pass.

7.1.8 Force Platform Location and Configurations

In order to best use the space in your laboratory it is essential that the force platform pit(s) be located in a position that will allow for cameras to be easily positioned around the platforms and to ensure that steady-state gait (walking or running) can be achieved. It has been suggested in the past that laboratories that are near the smaller end of the spectrum should set up the force platforms in a way that runs diagonally across the laboratory. This works by creating a larger distance to achieve steady-state gait. However, when using a multicamera system you may reduce the effectiveness of the cameras by forcing them into positions that may be less than ideal. Therefore, if your laboratory is of a smaller nature, concentrate on getting the best setup without compromising the movement analysis

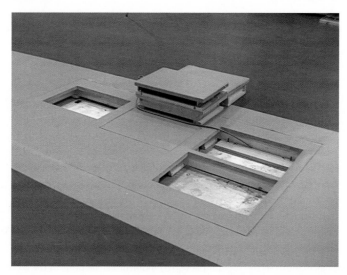

FIGURE 7.6 ■ AMTI force platforms set for a step-down task showing the 4-platform configuration

measurements. The best configuration often is to position the force platforms in the centre of the laboratory, allowing ample room either side to position cameras. If you plan on having doors at either end of the laboratory, try to line these up with the force platform pit. This will be useful if you ever want to investigate running gait, as it will allow the participant to start and finish outside of the laboratory, without having to stop by running in to a wall.

Once the position of the platform pit has been decided upon, it is essential that the possible configurations of platforms will maximize the number of foot contacts collected, or will allow the laboratory to be as adaptable as possible by considering all possible movement tasks that may be of interest. There have been many different configurations that have been suggested over the years; in the next section, we will consider some of these. We will consider an example of four platforms, but highlight how these configurations can be used with fewer platforms as well.

Force Platform Configuration 1

The configuration shown in Fig. 7.7 (*bottom*) is particularly suited to laboratories that are interested in gait, gait initiation, postural stability, child gait and running gait. The first two platforms are positioned next to each other to allow the participant to have one foot on each platform during postural stability and gait initiation assessment, where not only the force beneath each foot can be obtained but also the subsequent foot contacts.

In addition, if the two platforms are used along the x axis, they will allow for child gait or individuals with a shorter step length. The subsequent two platforms are ideally positioned for normal gait, collecting a total of three heel strikes. This is the current configuration we use at the University of Central Lancashire. A similar configuration, which includes the inset second platform, can be used if only two platforms are available; this again is particularly suited to gait, Fig. 7.7 (*top*). We would recommend using this configuration as it is adaptable and covers a variety of activities.

Force Platform Configuration 2

This configuration position two platforms sideways at either end of two platforms that are orientated

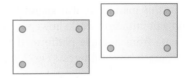

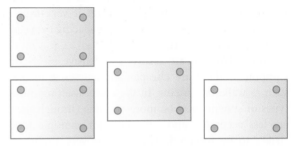

FIGURE 7.7 ■ Force platform configuration 1

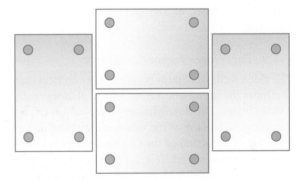

FIGURE 7.8 ■ Force platform configuration 2

lengthways. This is particularly suited to gait analysis in children, as the relative distance between the platforms is small, which will increase the chance of getting a clean strike. Again, two platforms can be removed from this configuration to provide one which still allows for child gait (Fig. 7.8).

Force Platform Configuration 3

This configuration is particularly useful when working with larger systems, so typically four platforms would be the minimum. Normally this configuration would be used when dealing with children or pathologies where step length may be limited, as it increases the chance of a clean foot contact. With the platforms orientated in this manner it also allows for postural stability and gait initiation work; however, this will be limited to data from a single platform if the subsequent foot contacts are of interest (Fig. 7.9).

Force Platform Configuration 4

This configuration is very simple and is generally used for collecting walking and running gait data. We would not recommend this configuration as it very much limits the type of work by the relative spacing between the platforms (Fig. 7.10).

Force platforms are relatively expensive pieces of equipment; there are cheaper alternatives available, such as walkmat systems. In the correct setting these types of systems may be more versatile (though they are less accurate), and will allow for multiple foot strikes to be recorded. They can be very useful in studies that concentrate on simpler gait variables, such as step time and gait velocity, as the heel-strike and toe-off phases of the gait cycle can be easily recorded.

7.1.9 The Video Vector Generator

The video vector generator is a piece of equipment that combines the information from a force platform with a video image. The force platform information can be superimposed on top of the video information, giving a picture of the action of the GRFs with respect to the joints of the lower limb (Fig. 7.11). This information may be used to identify biomechanical pathologies and monitor changes due to treatment.

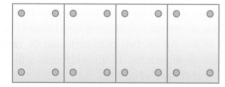

Before pathological gait can be studied using this technique, the patterns of 'normal' gait should be studied and related to function. Using this technique, any deviation away from an expected gait pattern may be identified and related to a change in function. This information may be used to identify biomechanical pathologies and monitor changes due to treatment. One particular technique that has found its way into clinical practice is using the video vector generator. This has been used in the fine tuning of prosthetic legs, where adjustments can be made to the sagittal plane and coronal plane alignments of the socket and prosthetic joints, and in the fine tuning of orthotic management in cerebral palsy.

Commercial systems that allow the visualization of the video vector, whilst also allowing the recording of the force data to allow more objective measurements to be taken, have been available for approximately 20 years. One of the first systems able to do this was the Orthotic Research and Locomotor Assessment Unit (ORLAU) video vector generator, which used electronic hardware to project a vector line onto a video screen. Since then, software versions, such as ProVec (MIE, Leeds, UK), which can work with any commercial platform (e.g. Kistler, AMTI or Bertec), have been developed, and are capable of collecting data from multiple platforms. More recently the TEMPLO Video Vector Lab has become available and is beginning to appear in clinical orthotics and prosthetics hospitals. All these systems allow video data to be superimposed, so that real-time video vectors can be displayed and recorded. Although these can be a very useful assessment tool on their own, they do not allow any accurate measurements of joint moments to be taken.

7.2 METHODS OF MEASURING PRESSURE

7.2.1 What Is the Difference between a Force Platform and a Pressure Platform?

Whereas force platforms provide GRF and the path of the centre of gravity data, they do not provide information on the stress acting on individual areas of the foot. Pressure measurement systems provide detailed information on the pressures between the contact points of the foot and the ground during static and dynamic activities.

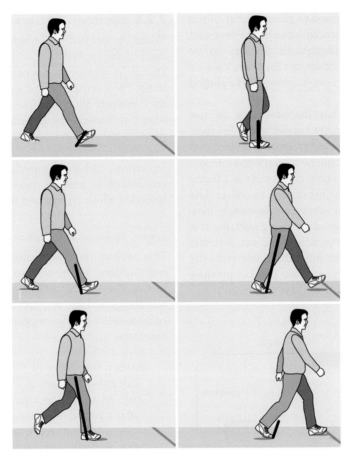

FIGURE 7.11 ▪ Force Vector Generator

7.2.2 Pressure Measurement Technologies

The application of various techniques to measure pressures under the foot can be traced back to the late 1800s. The following section will provide information on some of the earlier pressure measurement systems and a detailed description of the different types of electronic sensors which are used within the current commercially available pressure measurement systems.

7.2.3 Pressure-Sensitive Mats and Film

These systems consist of a film or mat on which a patient walks across, after which an image of the foot remains. The image of the foot will have varying levels of colour or amounts of ink depending on the amount of pressure exerted as they walked. The Fuji Prescale Film, Shutrak, and Harris-Beath mat are examples of pressure-sensitive mats and film. All these systems give an indication of the pressure exerted; however, they cannot put a precise value to it. The resulting image is useful to get an impression of whether there are any areas of high pressures and abnormalities. However, they are limited as they cannot show how the pressure dynamically changes beneath the foot at different stages during the stance phase.

7.2.4 Pedobarograph (Optical)

A pedobarograph comprises an elastic mat laid on top of an edge lit glass plate (Fig. 7.12). When a subject walks on the mat it compresses and loses reflectivity, becoming progressively darker with increasing pressure. The amount of darkening gives a quantitative measurement. The underside of the plate is viewed by a camera, usually via a mirror set to 45°, the image from which may be processed to give a display of the different pressures experienced beneath the foot and to allow the dynamic changes in pressure to be assessed. Different areas under the foot can then be identified and the pressure-versus-time graphs plotted (Fig. 7.13).

The pedobarograph has the advantage of not relying on individual pressure sensors under the foot; therefore, the resolution of the pressure is far better as it is dependent on the camera resolution. However, most pedobarographs can only collect data at a frame rate of 25 Hz. This is only just enough to collect data from walking and it is not suitable for faster activities. It also has the disadvantage of not being portable as it requires being set into the floor. Whereas dynamic pedobarograph systems are still utilized, currently the majority of commercially available plantar pressure measurement devices utilize electronic sensors.

7.2.5 Electronic Sensors

When compressed, these sensors calculate the variations of the applied load, measuring the proportional change in voltage, conductance, or resistance, respectively. These sensors do not directly measure pressure; they measure force and use the known area of the sensor to convert the recorded force to pressure. At present, resistive and capacitive technologies are the most widely used sensor technologies. Both have their limitations and it is important to consider what is required of the pressure measurement system in order to decide which system best fits your requirements.

Sensor Technical Specifications

This section provides definitions for the technical specifications of sensors used for pressure measurements. There are limitations to each type of sensor technology and it is important to understand these in order to select the right system for your measurement requirements.

> *Accuracy:* the degree of closeness of a measured or calculated quantity to its actual (true) value.
> *Creep:* related to the variation of pressure reading after a period of constant loading at a fixed pressure; ideally, there would be no variations in pressure over time, i.e. no creep.
> *Hysteresis:* relates to the maximum difference in pressure readings when a comparison is made between loading to unloading of the sensors; ideally, a sensor should exhibit low hysteresis.

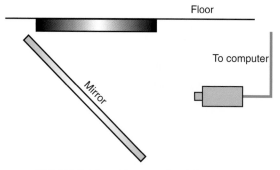

FIGURE 7.12 ■ Diagram of pedobarograph

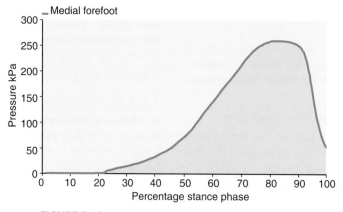

FIGURE 7.13 ■ Pressure-time graph for medial forefoot

Linearity: relates to the relationship between the force applied to the sensor and the output signal; if the relationship is proportional (i.e. a straight line), the sensor is considered linear, which is desirable.

Pressure range: relates to the range of pressures the system is capable of measuring; the system should be capable of measuring the expected peak pressures of the activities it will be recording.

Reliability: is the consistency of the measuring system.

Sampling rate/frequency (temporal resolution): is the number of samples the system is capable of measuring per second. The unit for sampling rate is the hertz (Hz). Systems with a 50–100 Hz sampling frequency are considered acceptable for recording walking; higher speed activities (e.g. running) require a minimum sampling frequency of 200 Hz.

Sensitivity: describes the smallest absolute amount of change that can be detected by a measurement system.

Sensor crosstalk: is undesired activation of unloaded sensors when pressure is applied to neighbouring sensors; sensors with low crosstalk are desirable.

Spatial resolution: refers to the size of the sensors; they need to be small enough to be capable of measuring pressures over small anatomical structures such as the metatarsal heads. This is especially important when measuring small/children's feet. It is essential to consider the size of the anatomy being measured (i.e. metatarsal heads and toes).

Calibration

Calibration of pressure sensors across systems varies considerably, and different manufacturers have a variety of ways of doing this. These range from standing on the sensors and inputting the person's weight (usually in the wrong units of kg or worse, lbs), to calibration under a pressurized bladder. This calibration is then related to subsequent measured readings.

The first method of standing on the sensors assumes that all the sensors have exactly the same properties,

which is not an adequate basis upon which to undertake meaningful research. However, in its defence, this technique is fast, and if the purpose is to look for approximate values of disproportionate pressures, say to predict possible site of ulceration, rather than accurate measurements, then it could be considered as fast and useful, and it is clearly better than walking over carbon paper. The counter-argument, however, is 'when does a pressure become too high', in which case we need to know and be confident in the measurements taken.

The second method of calibration uses a pressurized bladder to provide an even distribution of pressure over all sensors. This is usually repeated at a number of pressures, so a calibration curve is found for each sensor. Therefore, any variation in the sensors is taken into account and only then can the system be considered to be properly calibrated. Some manufacturers offer a service of recalibration where not only the magnitudes of the pressures but also the speed of application of pressure is recorded. This is sometimes referred to as ramp loading. Ramp loading offers the advantage of also examining for any time dependence, or time lag, in the sensors.

Capacitive

This technology operates on the change of capacitance when two electrical plates with a dielectric material between them are squeezed together. When a force is applied, the distance between the plates changes, causing a change in the properties of the dielectric material, and the capacitance varies. Commercial manufacturers, such as Novel and AM Cube, utilize this technology within their pressure measurement systems.

Resistive/Force-Sensing Resistors (FSRs)

These systems typically consist of a pressure-sensitive ink contained within two layers of film or similar resistive technology. When the force sensor is unloaded, its resistance is very high. When a force is applied to the sensor, this resistance decreases. This technology is utilized within the pressure measurement systems supplied by commercial manufacturers such as Medilogic, RSScan and Tekscan. An advantage of the technology is that, when used within in-shoe pressure measurement systems, the insoles can be thin (2 mm); however, these in-shoe systems suffer from fast ageing, as their sensitivity changes after repeated use, and need to be replaced frequently. As these sensors are thin, they can easily deform to accommodate foot structure, so care is needed to ensure the sensors do not crease within the shoe, as this may cause an overestimation of peak pressures.

Sensors utilizing conductive rubber technology (e.g. Inventables and Zoflex), which are predominantly used in industrial applications, can also be adapted for use in clinical settings.

Hydrocell

The hydrocell is a fluid-filled cell with a sensor contained within the fluid. It is proposed that this design allows for pressure measurement that incorporates the normal and tangential components of the applied force. This technology is used by one commercial company (Paromed, Germany). Limitations of this system are that the insoles are thicker than other manufacturers and the sensors do not cover the entire surface of the insole.

7.2.6 Sensor Arrangements

This section provides information on the different sensor devices available (individual sensor, mat, in-shoe, treadmill) which utilize the different electronic sensor

principles detailed previously. Whereas this section focuses on sensor arrangements for the assessment of foot pressures, there are a wide range of other arrangements available, which allow for the assessment of pressures while gripping with the hand, sitting/laying in a chair or bed, and to assess the fit of prosthetics (Cochrane et al., 2008, Dumbleton et al., 2009).

Individual Sensors

Small individual sensors may be placed under specific areas of interest to give an indication of the pressure under that area (Branthwaite et al., 2013). Examples of such sensors include the FlexiForce sensor (Tekscan, USA) (Fig. 7.14A) and the WalkinSense (Kinematix, Portugal) (Fig. 7.14B) which can record pressures for eight individual sensors. The WalkinSense systems currently include a small data logger worn on at the ankle; this allows for continuous monitoring of pressures and spatio-temporal gait parameters during the day to get an overview of what the foot experiences in a typical day.

Benefits of these sensors over matrix systems are that they are cheaper and require less computer processing power; however, they cannot provide as much detail as matrix systems. Individual sensors are also useful for measuring pressures on the margins or dorsal surface of the foot. If, due to foot deformity,

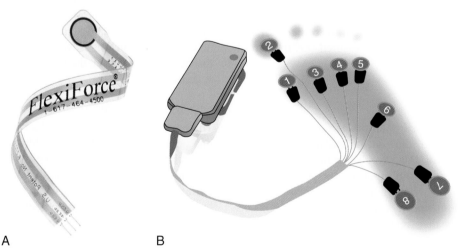

A   B

FIGURE 7.14 ■ Individual pressure sensor (A) FlexiForce (Tekscan Inc., Boston, MA, USA); (B) WalkinSense (Kinematix, Porto, Portugal)

the area of interest cannot be measured with matrix sensors, then individual sensors are a suitable option. These systems require the researcher/clinician to define the area of interest and secure the sensor to this area; adhesive tape is commonly used to secure the sensors. Careful application of the sensors is needed to ensure they do not move off the area of interest during dynamic activity (i.e. walking, running, etc.). It is important to consider the size and thickness of these sensors to ensure comfort within the shoe; it may be advisable to embed sensors in insoles of similar thickness.

Matrix Sensors

These systems consist of an arrangement of single sensors to allow measurement of plantar pressures across the entire contact area of the foot/shoe under investigation. Most of these systems allow synchronization with video, whilst some offer synchronization options for other systems, such as electromyography (EMG) and motion capture. Matrix arrangements are utilised in pressure mats/platforms, in-shoe pressure systems and pressure treadmills, which are discussed in the following section.

7.2.7 Pressure Mats/Platforms

These are generally used for barefoot measurements of plantar pressure distribution. Initially these systems were available as a mat or platform capable of recording one foot step at a time (Fig. 7.15A and B); however, they are now available as walkways which allow for the capturing of multiple steps (Fig. 7.15C). Table 7.1 provides information on the technical specifications of some of the commercially available pressure mats/platforms. These systems use an array of load sensors, which usually range from 1 to 4 sensors/cm^2, and are available in a range of sizes. They can record at a variety of different frequencies that vary from 20 Hz up to 500 Hz. There is much debate on which is the best system; however, this generally comes down to sensor resolution (sensors/cm^2), sampling frequency, stability of calibration and, perhaps the decisive factor for clinical use, the cost of the different systems. All systems offer valuable data for research and clinical assessment, although the majority of scientific publications to date have used the Novel and Tekscan systems.

7.2.8 In-Shoe Pressure Systems

So far, we have looked at floor-mounted devices, but these have the major drawback in that they do not measure the pressure acting on the foot, unless the subject is walking barefooted. In-shoe pressure systems are able to measure the interface pressures between the foot and the shoe, or the foot and a foot orthosis. This gives important clinical data on the nature of pressure offloading and force distributions. The main difficulties which in-shoe devices face are the curvature on the foot and the shape of the foot orthoses, and the lack of space for the transducers. The transducers are therefore generally thin enough to fit in the shoe without affecting the very thing they are trying to measure (see Fig. 7.16A–D for examples of these systems).

The first in-shoe systems were wired but recent systems use wireless technology, utilizing SD cards, Bluetooth technology or other wireless means to store/transfer the data to a computer. In wireless systems, a portable data logger is usually worn around the waist, which can allow for measurements outside the lab/clinic environment. In-shoe systems can record a large number of continuous steps which is only limited by the size of the area available for measurement and, if the system is wired, the length of the cables.

Information is provided in Table 7.2 for some of the different currently available commercial in-shoe pressure systems. The standard insole sizes available from manufacturers are listed, although some manufacturers can supply extra-large/wide insoles and supply custom-designed insoles to suit customer needs. Whereas the insoles for some systems consist of rows and columns of sensors which cover the entire sensor surface (e.g. F-Scan and pedar-x), others arrange the sensors in positions according to anatomical structure, with spaces between sensors, and therefore some areas of the insole do not measure pressure (e.g. Parotec). The majority of the systems use insoles which come in standard shoe sizes; however, the Tekscan F-Scan insoles are trimmed by the user to the required size. Shortly, a new version of the WalkinSense system will be available where the data logger is attached to the shoe, and with the option of using a pressure insole or the individual sensors (Fig. 7.13B).

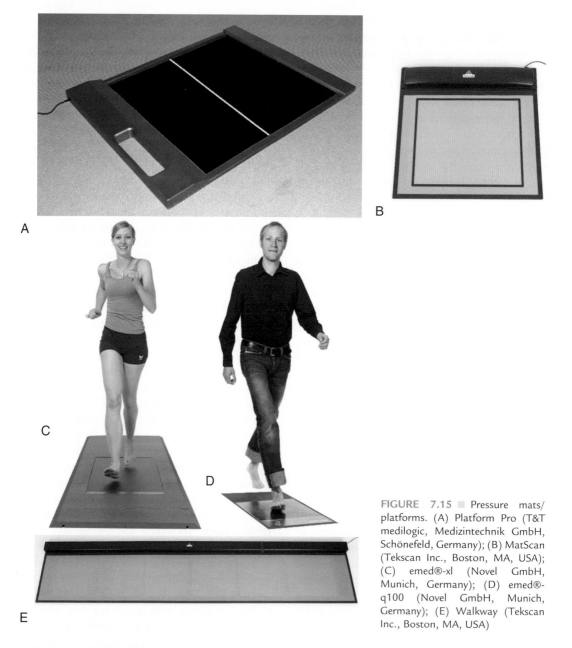

FIGURE 7.15 ■ Pressure mats/platforms. (A) Platform Pro (T&T medilogic, Medizintechnik GmbH, Schönefeld, Germany); (B) MatScan (Tekscan Inc., Boston, MA, USA); (C) emed®-xl (Novel GmbH, Munich, Germany); (D) emed®-q100 (Novel GmbH, Munich, Germany); (E) Walkway (Tekscan Inc., Boston, MA, USA)

7.2.9 Pressure Treadmills

There is now an increased availability of instrumented treadmills which allow for plantar pressure measurement. These systems have similar local resolution (sensors per cm^2) as mat/platform systems and are directed at clinicians with limited space for patient assessment. With these treadmills, it is possible to assess the effect of incline/decline on plantar pressures. In addition to pressure assessment some of these systems can be synchronized with EMG, inertial sensors and video. Some systems also include software which allows gait training to be carried out using a virtual environment projected onto a screen in front of the treadmill. These systems might also be

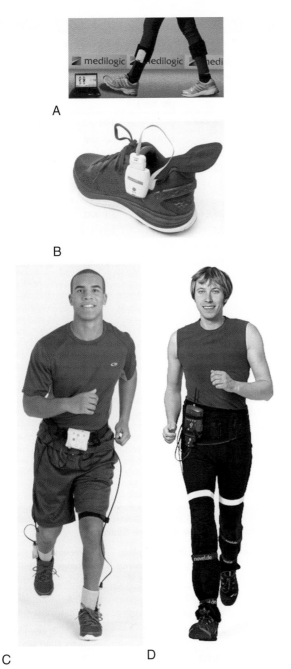

A

B

C   D

FIGURE 7.16 ■ In-shoe pressure measurement system (A) WLAN insole (T&T medilogic Medizintechnik GmbH, Schöne-feld, Germany); (B) new version of WalkinSense with insole (Kinematix, Porto, Portugal); (C) F-Scan® wireless (Tekscan Inc., Boston, MA, USA); (D) pedar®-x (Novel GmbH, Munich, Germany)

useful during rehabilitation which uses partial weight-bearing protocols.

7.2.10 Recommendations for Data Collection

When using pressure platforms, it is important to consider factors which may affect the measurements such as targeting, step protocol and walking speed. Targeting relates to the patient alternating their walking pattern (e.g. shortening or lengthening their stride) in order to make contact with the platform. The use of a walkway of mats (with the same thickness as the height of the platform) or embedding the platform in the floor are techniques used to reduce targeting and prevent tripping. The patients are instructed to walk across the walkway with their head up and looking ahead so they don't know when they are approaching the platform.

The step protocol (number of steps the patient takes before they contact the platform) is an important consideration, as research has found the number of steps taken before contacting the platform can affect the measurement values (Naemi et al., 2012). The midgait protocol, where the platform is positioned midway in either an 8 or 10 m walkway, was commonly used; however, it may not always be possible due to space limitations, and many people now use the two-step protocol. The two-step protocol (where the patient contacts the plate on their second step) has been recommended when recording barefoot plantar pressure data in patients with diabetes and neuropathy (Bus & de Lange, 2005). Arts and Bus (2011) recommended recording 12 midgait steps per foot to obtain valid and reliable in-shoe plantar pressure data for patients with diabetes and neuropathy. The type of socks and footwear worn during the data capture will also affect the results, so this should be noted, and the same footwear should be used for future testing sessions if comparison are to be made between sessions.

It is important to monitor walking speed when recording pressure measurement as, in general, an increase in walking speed causes an increase in pressures under the foot in adults (Taylor et al., 2004; Chung and Wang, 2012) and children (Rosenbaum et al., 2013). Rosenbaum and colleagues (1994) reported an increase in peak pressures under the heel and medial forefoot and a significant decrease under

TABLE 7.1
Technical Specifications of Commercially Available Pressure Mat/Platform Systems

Manufacturer	Sensor Type	Height from Floor (cm)	Measurement Area (cm)	Number of Sensors	Local Resolution (Sensors per cm^2)	Sampling Frequency (Hz)	Pressure Range (kPa)
AM Cube[a]	Capacitive	0.4–1.5	40 × 40–200 × 49	2704–16,384	2	100–200	10–1200
LorAn[b]	Resistive/ Capacitive	N/A	48 × 48–1600 × 48	N/A	N/A	50–100	30–500
Medicapteurs[c]	Resistive	0.4–1.0	40 × 40–150 × 50	1600–12,288	1–1.6	100–200	N/A–1000
Medilogic[d]	Resistive/ Capacitive	0.8	38.4 × 38.4–48 × 48	2048–4096	1.8–3.3	20–100	5–1280
Novel[e]	Capacitive	1.55–2.1	38.9 × 22.6–144 × 44	1760–25,344	1–4	50–400	10–1270
Paromed[f]	Hydrocells	1–4.5	40 × 40	1600–2304	1–1.4	100–150	7–343
RSscan[g]	Resistive	1.2–1.8	48.8 × 32.5–195 × 32.5	4096–16,384	2.6	200–500	1–1270
Tekscan[h]	Resistive	0.57–0.76	43.6 × 36.9–341.4 × 44.7	2288–59,136	1–4	25–440	N/A – 862
Zebris[i]	Capacitive	2.5	149 × 54.2–298.1 × 54.2	11,264–22,528	1.4	100–300	10–1200

N/A: information not available.
[a]AM Cube, France (Three models: Footwalk Pro, Footwork Pro and Footwork).
[b]LorAn Engineering Srl, Bologna, Italy (Three models: Modular pressure platform, EPS/R1 and EPS/C1).
[c]Medicapteurs France SAS, Balma, France (Three models: Win-Track, Win-Pod and S-Plate).
[d]T&T medilogic Medizintechnik GmbH, Schönefeld, Germany (Four models: Basic, Pro, wireless and NX).
[e]Novel GmbH, Munich, Germany (Six models: emed-xl, -x400, -q100, -n50, -c50 and -a50).
[f]Paromed Vertriebs GmbH & Co. KG, Neubeuern, Germany (Two models: Paragraph Pro and S).
[g]RSscan International NV, Paal, Belgium (Three models: footscan Hi-End, Advanced and Entry Level).
[h]Tekscan Inc., Boston, MA, USA (Six models: Walkway HRV, Walkway WE, Walkway WV, HR Mat, MatScan and MobileMat).
[i]Zebris Medical GmbH, Isny im Allgäu, Germany (Three models: FDM 3, 2 and 1.5).

TABLE 7.2
Technical Specifications of Commercially Available in-Shoe Pressure Systems

Manufacturer	Sensor Type	Wired/Wireless	Insole Thickness (mm)	Number of Sensors per Insole	Insole Sizes (European)	Sampling Frequency (Hz)	Pressure Range (kPa)
LorAn[a]	Resistive	N/A	N/A	600–1024	36/39/42/45	50	N/A
Medilogic[b]	Resistive	Wireless	1.7	35–240	19/20–49/50	100–400	6–640
Novel[c]	Capacitive	Wired or Wireless	1.9	84–99	22/23–48/49	100	15–600; extendible to 1000 kPa
Paromed[d]	Hydrocell	Wireless	3.5	24–36	23/24–47/48	300	N/A
Tekscan[e]	Resistive	Wired or wireless	0.15	Max 960	Up to 48.5	100–750	N/A–862

N/A: information not available.
[a]LorAn Engineering Srl, Bologna, Italy (Model FPS III).
[b]T&T medilogic Medizintechnik GmbH, Schönefeld, Germany (Three models: insole, WLAN insole and insole Sport).
[c]Novel GmbH, Munich, Germany (pedar-x).
[d]Paromed Vertriebs GmbH & Co. KG, Neubeuern, Germany (Parotec).
[e]Tekscan Inc., Boston, MA, USA (Three models: F-Scan tethered, wireless and datalogger).

the midfoot and lateral forefoot with increasing walking speed. As pressure is calculated from force, the increase in the acceleration of the body from standing to walking to running will result in increases in pressures.

An important consideration when using pressure measurement is that previous research has shown it is not recommended to compare pressure measurement values between different pressure measurement systems or between platform and in-shoe measurements (Larose Chevalier et al., 2010). Pressure systems measure the force that is perpendicular to the sensor surface; so for platforms this can be considered the vertical force. However, in in-shoe systems, the force can only be considered the vertical force for the portion of the stance phase when the foot is parallel to the ground. It should also be noted that it is not advisable to compare in-shoe pressure results recorded in different footwear.

As it is not recommended to compare across different measurement protocols, it is important to select the protocol, considering what suits the research/ clinical question best, and stick with it. This then allows for comparisons to be made over repeated testing sessions for the same patient and/or between patients.

7.2.11 Future Developments

Currently is it only possible to measure the vertical forces using pressure-measuring devices, although work is being carried out by a number of companies to find a way of reliably measuring the shear forces.

Developments are also being made to allow integration across different measurement systems: for example, the ability to overlay pressure data onto a 3D scan of the foot with the aim of improving orthotic prescription process (OrthoModel, Delcam, Birmingham, UK) and the ability to integrate kinematic and pressure measurement systems to allow anatomy-based identification of regions of interest and to assess correlations between the measurements (Giacomozzi et al., 2014).

With advances in sensor technology, in the not too distant future it will be possible to measure pressure-using sensors contained within socks or similar wearable accessories. Although there are technologies now available which can continually monitor these pressures, they still use traditional pressure-sensing technologies. The future wearable technology will be beneficial for high-risk patients (e.g. rheumatoid and diabetic patients), allowing continuous monitoring of pressures with the potential to identify the threat of tissue breakdown before it happens, giving the patient the opportunity to take measures to prevent tissue damage.

SUMMARY: MEASUREMENT OF FORCE AND PRESSURE

- Pressure plates are only capable of measuring the vertical force; however, they are able to measure the distribution of the force, or pressure, beneath the foot or contact area. High pressures can cause tissue breakdown and injuries, such as ulceration.
- Pressure plates/insoles can calculate the centre of pressure, giving information as to the location of the centre of pressure in relation to the foot shape.
- Force platforms are more useful in the assessment of the overall function during movement tasks, whereas pressure plates are better at determining variations in the loading or pressure patterns acting on specific areas of the foot.

METHODS OF ANALYSIS OF MOVEMENT

JIM RICHARDS ■ DOMINIC THEWLIS ■
JONNIE SINCLAIR ■ SARAH JANE HOBBS

This chapter covers the measurement of movement. This includes the different methods of assessing movement, the processes required to collect and analyse movement data, and the consideration of possible errors.

AIM

To consider the different methods of collecting and analysing movement data, and to consider the different measures that may be taken.

OBJECTIVES

■ To describe the different methods of assessing movement data

■ To summarize how movement analysis methods work

■ To distinguish what parameters can be measured with different systems

■ To describe the processing required when analysing movement data

■ To identify possible sources of error and how these may be allowed for or controlled.

8.1 EARLY PIONEERS OF MOVEMENT ANALYSIS EQUIPMENT

In the late 19th century the first motion picture cameras recorded patterns of locomotion for both humans and animals. In 1877 Muybridge demonstrated, using a series of photographs from 24 cameras in a line, that when a horse is moving at a fast trot there is a moment when all the animal's feet are off the ground. The work was commissioned by Leland Stanford, the founder of Stanford University, as he was intrigued to know whether a suspension phase existed at trot in his own horse, Occident. Muybridge later used the same cameras to study the movement patterns of a running man, and in 1901 published *The Human Figure in Motion*. At about the same time, Marey, a French physiologist, used a photographic rifle (la fusil photographique) to photograph the movement of animals, and in 1882 and 1885 to record displacements in human gait to produce a stick figure of a runner.

The 20th century saw the development of systems capable of automated and semi-automated computer-aided motion analysis. One of the first systems to become commercially available was the Ariel Performance Analysis System, which required the operator to manually identify the location of each marker used for each frame. Since then the problems of automatic marker identification have been at the forefront of computer-aided motion analysis development. In 1974 SELSPOT became commercially available, which allowed automatic tracking of active light-emitting diode (LED) markers. Later, Watsmart Optotrak and Codamotion used a similar technique. VICON, a camera-based system, became commercially available in 1982. Other systems based on camera technology have followed, including the Motion Analysis Corporation system, Elite, and ProReflex, then later Oqus by Qualisis.

In the latter half of the 20th century other methods of recording movement were developed. These include instrumented walkmats, accelerometers, electrogoniometers, and most recently inertial sensors using microelectromechanical systems (MEMS) technology, which have all contributed to our current knowledge of normal and pathological movement.

8.2 SIMPLE MEASUREMENT OF TEMPORAL AND SPATIAL PARAMETERS

Spatial parameters can be measured in a variety of simple ways, which include putting ink pads on the soles of the subject's shoes and walking on paper (Rafferty & Bell, 1995; Rennie et al., 1997), and using marker pens attached to shoes (Gerny, 1983). Although very cheap, these systems can involve awkward and time-consuming analysis. Temporal parameters can be measured by timing how long it takes an individual to walk a set distance, and counting the number of steps it took to cover that distance. At best this will only give average velocity and cadence, and will give no value to the symmetry of these parameters. This technique is extremely susceptible to human error.

In the last two decades of the 20th century advances in computer technology led to the development of a number of instrumented walkmat systems. These allow fast collection of temporal and spatial gait data. Using a computer also allows easier, less time-consuming analysis. These systems can be found in work by Al-Majali and colleagues (1993), Arenson and colleagues (1983), Crouse and colleagues (1987), Durie and Farley (1980) and Hirokawa and Matsumura (1987).

8.2.1 Temporal and Spatial Parameters in Clinical Assessment

The relationship between length of stride, step frequency, time of swing and speed of walking for children and adults has been studied by many authors, both in relation to normal (Grieve & Gear, 1966; Murray et al., 1966; Andriacchi et al., 1977; Hirokawa, 1989) and abnormal gait patterns (Gardner & Murray, 1975; Wall & Ashburn, 1979; Mizahi et al., 1982; Wall et al., 1987; Leiper & Craik, 1991; Lough, 1995; Isakov et al., 1996).

Once normal gait patterns became available in the literature they could be used for comparison to abnormal gait patterns. Gardner and Murray (1975) studied temporal parameters of gait from several pathologies, including patients with unilateral hip pain, Parkinson's disease and hemiparesis, and compared the results with those obtained from normal walking. Wall and Ashburn (1979), and Mizahi and colleagues (1982), used temporal and spatial parameters in the assessment of hemiplegic patients. Wall and colleagues (1987) examined a number of pathological gait patterns using a walkmat system developed at Dalhousie University. They highlighted some of the major difficulties encountered in applying normal gait nomenclature in a precise and unequivocal manner to the description of some pathological gait patterns. Despite this, the authors concluded that the study of temporal and spatial parameters of gait is a quick and easy way of assessing many pathological gait patterns.

Current uses of walkmat systems include quantification of the difference in temporal and spatial parameters of gait between frail and non-frail elderly subjects, changes over time in patients with peripheral vascular disease, the recovery of persons following hip fracture, and the quantification of motor performance in patients with peripheral neuropathy undergoing treatment.

8.2.2 Walkmat Systems

One of the first computer-controlled walkmat systems was developed by Crouse, Wall and Marble at the University of Dalhousie, Canada in 1984/5 and published in 1987. The system consisted of nine active mats and two dummy mats, each mat being 0.8 m long and 0.76 m wide. The two dummy mats were placed at each end of the walk mat to allow 0.8 m for the subject to start and finish walking before and after reaching the active mats. The active mats used metal rods running perpendicular to the direction of walking to record temporal and spatial parameters, but these sometimes interfered with the subject's walk.

This system was later modified by using printed circuit boards divided into a left and a right side with 87 copper tracks etched onto each side. Alternate tracks were connected by a 10 W resistor, and both ends of the mat were connected to a constant current source of 1 mA. To record a walk, self-adhesive

FIGURE 8.1 ■ Sensor Array of GAITRite system

aluminium tape was placed on the soles of the subject's shoes. When the subject walked along the walk mat, the metal tape created a potential divider and the constant current flowing through the resistors produced a voltage that was directly proportional to the position on the mat. The system was capable of measuring step and stride time, step and stride length, swing time, and double and single support time, but gave no indication of medial–lateral or transverse plane foot position.

Systems are now commercially available that do not require any footwear modifications and, as such, interfere much less with the gait cycle. One such system is the GAITRite system, which uses pressure sensor arrays to determine the foot positions (Fig. 8.1). These pressure sensors offer six levels of pressure assessment and give a rough guide to the pressures that determine the nature of the foot contact, i.e. heel striking or toe walking, as well as the temporal and spatial parameters. The resolution of the sensors also allows the measurement of medial–lateral and transverse plane foot position.

The design of this sensory array allows the system to be thin and light so it can be rolled up and transported easily (Fig. 8.2A, B). The validity and reliability of the GAITRite system's measurements were investigated and published by McDonough and colleagues (2001), who concluded that the GAITRite system is a valid and reliable tool for measuring selected spatial and temporal parameters of gait. Since then, nearly 300 papers have been published using the system for a wide variety of pathologies, including Parkinson's,

Alzheimer's, and Huntington's diseases; stroke; Charcot–Marie–Tooth, cerebellar- and basal-ganglia-related motor disorders; ataxia; and multiple sclerosis.

It is clear that the study of the temporal and spatial parameters of foot contact in gait gives extremely valuable information for both research and clinical assessment; however, the movement of the different body segments in time and space reveals more detailed information about the nature of movement patterns and movement disorders.

8.3 POTENTIOMETERS, ELECTROGONIOMETERS, ACCELEROMETERS, AND INERTIAL MEASUREMENT UNITS

8.3.1 Goniometers, Potentiometers, and Electrogoniometers

What is a Goniometer?

A goniometer is a simple hand-held device for measuring joint angles. A goniometer is, in essence, a protractor that features two arms: one that remains fixed and one which is positioned, accordingly allowing the desired angle to be measured. There are several types of goniometer, all giving a crude but useful measure of angles and range of motion. Clinically these allow a quick and useful assessment of static angles. However, these devices are of little use in measuring angles dynamically during different movement tasks. There are two main types of goniometer, the hinged 'ruler'

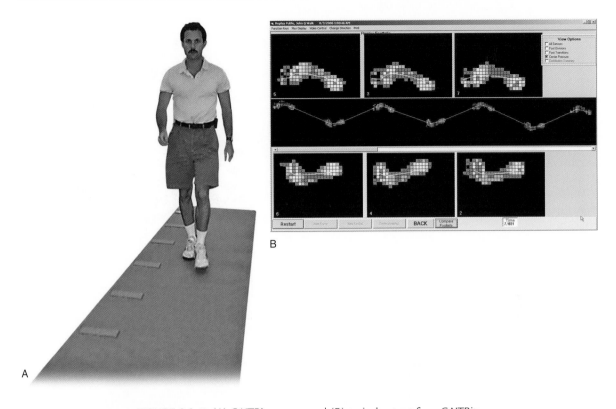

FIGURE 8.2 ■ (A) GAITRite system and (B) typical output from GAITRite

and the fluid filled. The fluid-filled types are generally more versatile in the angles they can measure, but they measure relative to gravity so, if not used correctly, can be prone to errors if the subject moves whilst taking readings (Fig. 8.3A, B).

What are Potentiometers and Electrogoniometers?

Potentiometers measure the change in linear or angular displacement by recording the change in voltage output. The potentiometer is positioned over a joint centre of rotation, with each arm orientated along the long axis of the proximal and distal segments. Changes in the angle of the joint will cause the voltage output of the device to change proportionally to the change in joint angle. Angular displacement potentiometers allow movement about one plane to be measured, and potentiometers are typically employed to measure sagittal plane movement only. Potentiometers are generally a robust 'student proof' and low-cost method of collecting kinematic data, which can allow real-time data to be observed. However, they are suitable only for hinge joins and, if more than one joint is being assessed at one time, then they may encumber the movement slightly, thus reducing the validity of the data being acquired. One example of potentiometer is produced by MIE (Leeds, UK) (Fig. 8.4A).

Electrogoniometers are very thin pieces of wire which are sensitive to bending. The amount of bending changes the output voltage. These can be sensitive to angular movement in up to three planes simultaneously, although biaxial and uniaxial electrogoniometers are more common (biaxial referring to the measure of flexion–extension and abduction–adduction of a joint simultaneously). These are relatively inexpensive, accurate and reasonably unintrusive and so minimize any gait modification (Fig. 8.4B).

8.3.2 Development of Electrogoniometers

Electrogoniometers have been in development since the early 1970s (Marciniak, 1973; Tata et al., 1978).

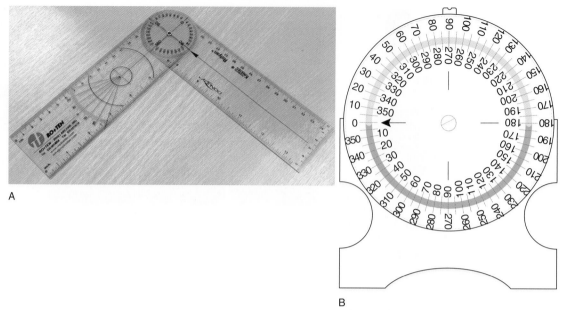

FIGURE 8.3 ■ (A) 'Ruler' goniometer and (B) fluid-filled goniometer (MIE, UK)

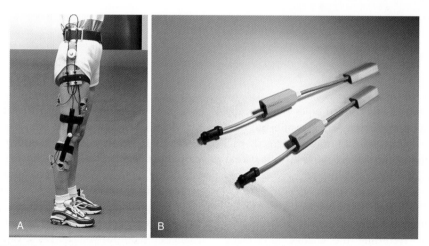

FIGURE 8.4 ■ (A) Potentiometers (MIE, Leeds, UK) and (B) Electrogoniometers (Biometrics Ltd)

Early devices were bulky potentiometers, sometimes referred to as 'potentio-metric goniometers', which were attached to the body segments with rigid bars. Various studies were carried out on normal subjects but they were cumbersome to wear, especially if multiple joints were to be studied. For these reasons the clinical use of such devices was limited.

Nicol (1987) described a new flexible electrogoniometer which used strain gauge wire. This eliminated the use of rigid bars and made it possible for the electrogoniometers to be very light. This breakthrough in relatively cheap and unintrusive joint motion analysis gave rise to a plethora of publications on movement analysis of both normal subjects and patients on a variety of joints. These included studies of the hip and knee motion in pre- and postoperative hip-joint-replacement patients (Rowe et al., 1989), dynamic analysis of wrist circumduction in normal subjects and subjects with wrist disorders (Ojima et al., 1991), the assessment of subtalar motion (Ball & Johnson, 1993) and active and passive ranges of movement of the ankle and knee motion in children with hemiplegic cerebral palsy (Hazlewood et al., 1994).

8.3.3 Accuracy of Electrogoniometers and Potentiometers

It is not uncommon for the accuracy of angular displacement of video-based motion analysis to be assessed in comparison with electrogoniometric techniques (Klein et al., 1992; Growney et al., 1994; Batavia & Garcia, 1996). However, for this to be useful the possible sources for error using electrogoniometers need to be considered. Possible sources of error when assessing human motion include:

- Movement of the end blocks of the electrogoniometer on soft tissue around a joint;
- Placement of the electrogoniometer in the correct plane of interest to eliminate crosstalk between two planes of motion, e.g. adduction and abduction, or rotation with sagittal plane movement; or
- The limiting mechanical properties of the strain gauge wire giving a finite accuracy of angular displacement measurements.

When used to study the accuracy of video-based motion analysis systems, if set up carefully, all but the last source of error should be eliminated. Nicol (1987, 1989) reported that the error in the measurement of angular displacements due to the mechanical properties of the strain gauge wire was within 1°. Therefore, accurate joint angular motion may be attained with careful placement of electrogoniometers.

8.3.4 Accelerometers

Accelerometers are electromechanical devices which measure acceleration from the movement of a small mass mounted to the case of the accelerometer itself. The measurements can be made using capacitive, mechanical, piezoelectric or piezoresistive methods. In capacitive accelerometers, the mass presses two capacitor plates together and acceleration is measured from the difference in capacitance. Mechanical accelerometers have the mass attached to a spring and the amount that the spring deflects measures the amount

of acceleration. In a piezoelectric accelerometer, the mass is attached to a piezoelectric crystal (typically quartz). When the body of the accelerometer experiences acceleration the piezoelectric crystals deform, resulting in the development of an electric charge which is proportional to the induced acceleration. Piezoresistive accelerometers use a piezoresistive substrate in place of the piezoelectric crystal and the force exerted by the mass changes the resistance across a whetstone bridge, which changes the voltage output. The voltage output is proportional to the acceleration. Acceleration measurement can be reported in either SI units of ms^{-2}, but often they are converted into a gravity ratio (g) by dividing the acceleration in ms^{-2} by $9.81\ ms^{-2}$.

Accelerometers are available commercially as either uniaxial, in which accelerations can be quantified in one direction, or triaxial, whereby accelerations can be measured in each of the orthogonal axes. Most accelerometers used in biomechanics are very light, weighing less than 10 g. Accelerometers typically serve two purposes in biomechanical analysis: either to measure movement of the body or body segments, or to measure impact transients. For movement of the body or body segments, the accelerometers only need to measure at a lower frequency, typically 60–100 Hz for walking, and with a smaller range, typically 6–9 g (nine times the acceleration due to gravity). For measuring impact transients, modern devices are capable of measuring accelerations at much higher frequencies, typically 1–5 kHz and up to 1000 g (one thousand times the acceleration due to gravity).

Accelerometers allow a direct and immediate signal output which can provide real-time visualization and biofeedback. They tend to be relatively low cost, although if data are required from multiple segments, the cost can become more significant. They can provide clinically important information on shock attenuation or the acceleration and deceleration of body segments, although they do not give any direct information on segment angles and joint positions. They have been used widely in clinical studies to track movement patterns during walking and are also in use in clinical practice. A review of studies that have used accelerometry to measure movement patterns was produced by Kavanagh and Menz (2008), which outlines the methods used in each study, their validity and reliability.

For measurement of acceleration and deceleration patterns of the body during walking, an accelerometer is usually fixed to the trunk with straps or a tunic, or the head with a cap or band. There are many different commercial products available and each product will provide outputs from the acceleration signal in different ways. Usually these include temporal variables (such as stance time, step time and stride time), peak amplitudes of acceleration, and/or frequency components contained within the signal. With the advances in smartphone sensor technology, applications are now available that utilize the signals from accelerometers in a mobile phone. These technologies are now used in research and clinical settings to monitor human activity to extrapolate periods of activity and rest, providing an estimation daily function and energy expenditure.

The most common application of accelerometers with respect to measuring shock attenuation is to attach the device to the distal end of the tibia and measure vertical accelerations of this segment during walking or running activities generated as a result of the foot striking the floor (Fig. 8.5). When the foot strikes the floor, its velocity is reduced to zero in a very short amount of time. The rate of change of the velocity of the foot allows the impact associated with footstrike during walking and running to be characterized: the rate of change in velocity is acceleration. When measuring accelerations of the tibia, the key difficulty that may be encountered is the attachment of the device sufficiently well to represent the acceleration of the underlying bone. Some researchers have attached the accelerometer to Steinmann pins, which are inserted directly into the tibia under local anaesthesia (such as Lafortune, 1991; Lafortune & Hennig, 1991). Many other studies using non-invasive methods have mounted the accelerometer on the anteriomedial aspect of the tibia with non-stretch tape. Often the skin is stretched prior to placing the accelerometer to reduce soft tissue movement and provide a more rigid attachment between the tibia and the device. Peak acceleration can be overestimated using non-invasive methods, particularly during running, although the frequency components of the signal may produce similar results to bone-mounted methods (Lafortune et al., 1995).

Impact shock has been measured using accelerometers to study changes following rehabilitation in knee osteoarthritic patients (Turcot et al., 2008), knee functionality in anterior cruciate ligament (ACL) deficient and ACL reconstruction patients (Bryant et al., 2003) and, most commonly, to investigate the risks of chronic injury development in runners.

8.3.5 Inertial Measurement Units

Sensors that use a combination of MEMS technology to determine the change in relative orientation of a body or body segment over time are called inertial

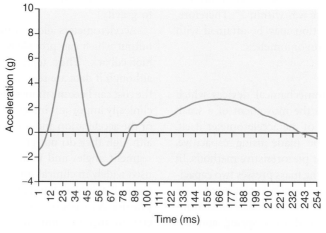

FIGURE 8.5 ■ Tibial acceleration measured during running

measurement units (IMUs). IMUs usually include a combination of gyroscopes and accelerometers, but may also include magnetometers or a global positioning system (GPS) to provide attitude, heading and global positioning. Triaxial sensors are used in IMUs, so that three-dimensional motion of segments relative to other segments can be calculated.

MEMS gyroscopes usually take the form of a tuning-fork, an oscillating wheel, a Foucault pendulum or a wine glass resonator, and take advantage of what is known as the Coriolis effect (Bernstein, 2003). In principle, each gyroscope has a vibrating mass which is suspended on a support and, as the vibrating mass rotates, a force is produced (Coriolis force), which is proportional to the rate of rotation of the mass. So, the amount of rotation is calculated by integrating the angular velocity of the plane of vibration. MEMS accelerometers, of a similar construction to those described in Section 8.3.4 are positioned in an orthogonal configuration to the gyroscope on a printed circuit board that makes up the IMU, together with any other sensors that are used by that system. Accelerometers are used in IMUs to calculate the change in position of segments relative to other segments, which is determined by double integration of the acceleration information.

The gyroscopes provide direct measures of the angular velocity in the local coordinate system of the sensor in the three planes. The accelerometers measure the acceleration, which can also be used to determine the inclination of the sensor with respect to gravity (Fig. 8.6A, B). The magnetometers measure the magnetic field, which can be used to determine the sensor inclination with respect to the earth's magnetic field.

Although IMUs are increasingly miniaturized and often combined with wireless data acquisition capabilities, which makes them extremely attractive for biomechanical analysis, they do suffer from inherent measurement issues. To measure the position and orientation of a segment, the IMU must integrate the raw data outputs from the accelerometer and the gyroscope. The position of the segment in space is calculated from the accelerometer, and then the orientation is calculated once the position is known, so the accuracy of the gyroscope depends on the accuracy of the accelerometer to determine the correct position. Accelerometers tend to be sensitive to pressure,

temperature and height changes and if these are not corrected, this can introduce a bias in the data which can cause drift. Drift is an accumulation of an error in the signal over time. Some IMUs compensate for temperature drift by using temperature sensors that correct the bias, and some eliminate pressure drift by creating a sealed environment at a fixed pressure (Shaeffer, 2013). Magnetometers can also be used to assist with the correction of drift bias, as they provide information on attitude and heading that can be included as part of a correction algorithm. Another issue with IMUs is that accelerometers suffer from noise. This results in overestimated, oscillatory acceleration measurements, which are usually corrected by filtering the raw signals before the data is integrated. Finally, the capture frequency of the IMU is important because the sensor can only estimate what happens in relation to position and orientation between each data point, which may mean that some of the real movement is lost if the sensor only captures at low frequency. For more on IMU models please see Chapter 9.

Inertial sensors can be used individually, but for biomechanical applications these have been incorporated into Mocap suits (e.g. the XSENS MVM suit) and systems that can use two or more sensors to measure specific joints (e.g. Delsys). These systems use an inertial sensor to represent each body segment. They are an exciting development for clinicians, as they are able to measure three-dimensional movement outside of a laboratory setting. Their accuracy is, however, dependent upon the quality of the combined sensor system and the algorithms used to calculate the three-dimensional position and orientation of each segment. Inertial sensors have been used in sports, such as downhill skiing, speed skating, horse riding and baseball pitching, and in studies involving clinical gait assessment and clinical tasks, such as sit-to-stand (Fong & Chan, 2010; Roetenberg et al., 2013).

8.4 CAMERA MOVEMENT ANALYSIS SYSTEMS

Movement analysis systems use either a single camera or multiple cameras to reconstruct two- or three-dimensional movement data, which allows quantification of the kinematics of different movement tasks. To

— Sagittal plane — Coronal plane — Transverse plane

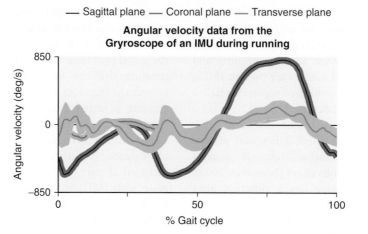

— Medial-lateral direction — Anterior-posterior direction — Vertical direction

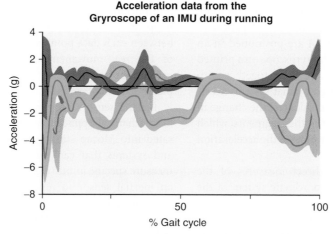

FIGURE 8.6 ■ Angular velocity and acceleration in three planes/directions during over 80 running cycles measured with a Delsys IMU placed on the tibia

do this accurately and effectively, it is important to consider the collection of the camera data, the processing methods to get two- or three-dimensional kinematic data, and, finally, the methods of analysis and modelling the data. These may be broken down into the following aspects that need to be considered when conducting any movement analysis:

- Camera positioning
- Camera speed, sampling frequency and shutter speed
- Synchronizing the cameras
- Calibrating image space
- Data capture

- Digitizing and transformation
- Data filtering
- Anatomical models and marker sets.

8.4.1 Camera Positioning

Data collection consists of filming an activity using cameras. The number and position of the cameras dictate whether the study is two- or three-dimensional. For a two-dimensional study, only one camera is needed, which has to be positioned in the plane of interest; for example, viewed from the side for sagittal plane analysis. The position of the camera relative to the movement of interest should be orthogonal (90° apart) to obtain the greatest accuracy. For a three-dimensional study, at least two cameras are needed and it is generally agreed that the cameras may be set between 60° and 120°, although the orthogonal positioning yields the best results (Woltring, 1980). When using two cameras to study human movement patterns, markers or points of interest on the body might not be in view from both cameras at all times. In this case these points cannot be tracked, as two cameras must see each marker at all times. For this reason, it is common to see four or more cameras used for three-dimensional movement analysis, as this increases the chance of tracking markers or points of interest through the entire movement.

The number and positioning of cameras do not just affect the identification and tracking of markers, but also the accuracy of the calculation of the final coordinates of the markers or points of interest. Work has been carried out using different numbers of cameras to study the effect of reducing or increasing the number of cameras on the accuracy of the data produced. Woltring (1980) studied multi-camera calibration and body marker trajectory reconstruction in three-dimensional gait studies. It was found that, as the number of cameras was increased, the errors in the calculation of three-dimensional coordinates decreased. The most noticeable decrease in error was found when moving from two to three cameras. Subsequent increases in the numbers of cameras used yields a smaller reduction in error. However, this finding should be treated with care as, if a camera system consists of only three cameras, then it is possible that not all the markers will be visible from all cameras. The number of cameras necessary, therefore, depends very much on the movement tasks being analysed, the anatomical models, and the marker set used (see Chapter 9). For instance, good gait data may be collected with a very simple marker set from four cameras; however, if you were to try the same with a more complex marker set, the data quality would be poor. It is for these reasons that camera systems containing 10 or more cameras are becoming more common in research laboratories (Fig. 8.7A, B). This raises rather important space considerations if more complex anatomical models and marker sets are required. So, to determine the most appropriate camera system one has to look at cost, space and complexity of models likely to be needed.

8.4.2 Camera Speed, Sampling Frequency and Shutter Speed

Standard video equipment, with electronically shuttered video cameras, has been used extensively in human movement analysis because of the price, immediacy and accessibility (Bartlett et al., 1992). Frame rates from standard video cameras are often a limiting factor, as they operate at 25 to 30 frames per second, providing a maximum sampling rate of 50 Hz for Phase Alternate Line (PAL) based systems and 60 Hz for National Television Standards Committee (NTSC) based systems. New developments in camera technology include simple-to-operate, lower-cost, high-speed cameras.

These faster camera systems allow faster movement patterns, such as sprinting, to be recorded; the faster the activity, the faster the sampling frequency or the camera speed must be. Cameras exist that can provide sampling frequencies up to 10 kHz, but it is well accepted that 50 Hz is adequate for studying many aspects of human walking. Nyquist's sampling criterion states that the sampling rate must be at least twice the maximum frequency of the signal, although this only gives the minimum usable sampling frequency (Antonsson & Mann, 1985).

Shutter speed or shutter factor is also extremely important if a clear image is to be achieved. The shutter speed is the amount of time the camera shutter is open. If the shutter is open too long, then the image will become blurred or smeared. For normal walking a shutter speed or shutter factor of 1/250 of a second or higher is required. If faster activities are being recorded, such as sprinting, a shutter speed of at least 1/1000 of a second is required.

8.4.3 Synchronizing the Cameras

When filming an activity with more than one camera, it is essential that all cameras record the event simultaneously as only then can the data from one view be combined with another to form a three-dimensional picture of the motion. The one requirement that must be met to combine simultaneous camera views is that all cameras must record a single distinct event, called the synchronizing event. This event varies from system to system; a flash of light, an electronic beep, and signals from a computer to start the cameras recording have all been used successfully. In addition to this,

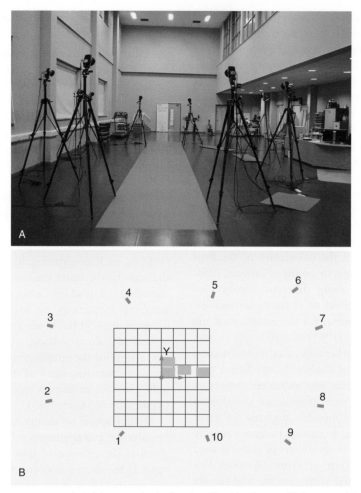

FIGURE 8.7 ■ UCLan's movement analysis laboratory in the faculty of health. Picture of laboratory (A) and plan view of camera positions and force platforms (B)

some systems synchronize the opening and closing of the shutter on each camera ensuring that precisely the same image is captured by each camera and not just an image during the same frame. This is usually achieved using a charged coupled device on the cameras (Fig. 8.8). This allows for greater accuracy when reconstructing the individual camera images, especially when dealing with fast movements.

In addition to synchronizing the cameras with each other, it is important to synchronize with external devices such as force, pressure and electromyography (EMG) systems. This is achieved using transistor-transistor logic (TTL) signals which is usually in the form of a square wave pulse of about 2–5 V where the systems are triggered by the rising edge of the signal.

Most systems are able to either send or receive this signal (i.e. either to trigger or be triggered from an external source). There is usually a small associated delay between systems; however, this is usually constant (e.g. 20 ms), and can easily be allowed for in the analysis software. Once triggered, data can either be collected from two computers running different software or brought into the same software running the camera system via an analog-to-digital (A/D) board to convert the signal or, what is becoming more common, using direct input from a digital source through USB (Fig. 8.9). There are many technical challenges to the producers of motion analysis system and associated equipment regarding the precise synchronization using USB that we will not deal with here, but the benefits or not

FIGURE 8.8 ■ Oqus 7 cameras by Qualisys

FIGURE 8.9 ■ Analog-to-digital (A/D) board from Qualisys and direct digital source input from Delsys to Qualisys through USB

requiring an additional A/D board are many, including portability and setting up of software and hardware.

8.4.4 Calibrating Image Space

The process by which three-dimensional coordinates are extrapolated from two-dimensional images requires information from two sources: inside the camera and outside the camera. These are generally referred to as the intrinsic and extrinsic properties of the camera. The intrinsic parameters refer to information such as the focal length and the centre of the image in relation to the lens, and, importantly, the distortion parameters of the lens. The extrinsic parameters refer to information such as the position and orientation of the camera and image in the coordinate system of the measurement, which is generally the laboratory coordinate system or global coordinate system (GCS). The intrinsic and extrinsic parameters are generally acquired using calibration techniques; the calibrations used for each parameter differ significantly. Essentially, these calibrations are the calibration of the camera lens linearization and the calibration of the system.

Static Calibration

The image space, the area in which the movement is to be recorded, must be calibrated to allow for the calculation of the positional information with respect to a known frame of reference. To calibrate the image space, the location of fixed points within the area in which the movement is to be recorded must be known. These fixed points may be a calibration frame that is placed in the data collection area, or points suspended from the ceiling. This information is recorded and the frame removed so data from the activity may then be collected.

The accuracy of the data produced from motion analysis systems depends considerably on the accuracy of the calibration procedure. It is important that the calibration frame fills a significant proportion of the image space of each camera view. The coordinates of the calibration frame must also be extremely accurate: for laboratory-based work ± 0.1 mm is acceptable in all three planes. Any errors in the location of the calibration coordinates will affect the accuracy of the motion to be tracked.

The number of control points required depends on whether it is a two-dimensional or three-dimensional system. For a two-dimensional system, the position of at least four coplanar points must be known to define the measurements in one plane. When calibrating a two-dimensional system, care must be taken to position the calibration frame correctly. Two-dimensional systems suffer from perspective error, which occurs when markers or points of interest move closer to or further away from the camera recording the activity. Because of this limitation, it is extremely important that the calibration frame is in the same plane as the activity to be recorded. Even with this safeguard, the segments on the near side of the body will seem longer than those on the far side.

For the calibration of three-dimensional movement analysis systems, at least six non-coplanar control points are needed (Woltring, 1982) (Fig. 8.10). This means that there must be control points in all three planes. Many calibration frames for such systems have more than six calibration points so that they can cover a larger area for data collection and achieve greater accuracy. When setting up the calibration, the control points must be clearly visible from all the cameras. The accuracy of the calibration procedure can be seriously compromised if this is not the case, especially if the number of points visible falls below six. The area the control points cover should be approximately the same size as the event being filmed, as the accuracy of the measurements outside of the calibrated volume are compromised (Woltring, 1982).

Dabnichki and colleagues (1997) studied the accuracy and reliability of data collection with changes in

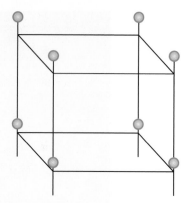

FIGURE 8.10 ■ Calibration frame from an early motion analysis corporation system

calibration setup. A series of tests were conducted using the Elite motion analysis system. Dabnichki systematically varied five different factors: camera–object distance, distance from calibrated field, size of calibration field, position in the calibration field and rotation speed of segment. The results showed that the error is sensitive to relatively small changes of the first four factors.

Dynamic Calibration

In order to define the extrinsic parameters of the camera, the position, and the orientation of the camera, the GCS must be defined. Dynamic calibration can be achieved in a number of ways; however, the most common and reliable way is to use a static frame to define the origin, or zero position, and the direction of the positive x and y axis.

In addition to the static frame, a wand is moved dynamically through the volume of the cameras. There are an incredibly large number of two-dimensional coordinates generated from the movement of the wand. To find the position and orientation of the cameras and the three-dimensional coordinates of the wand, a procedure known as bundle adjustment is used (Brown, 1966). From this, the position and orientation of the cameras and three-dimensional coordinates of the wand are calculated (Fig. 8.11A, B).

Norm of Residuals

The determination of the coordinates for each marker is an approximation with errors. This error in each marker may be reported as a 'norm of residuals'

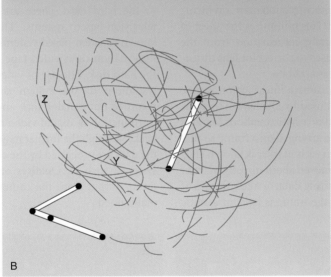

FIGURE 8.11 ■ (A and B) Dynamic calibration

(Nigg & Herzog, 1994), which is a summation of the errors present. This way of reporting on the errors involved in the calibration allows the user to determine whether there is any serious error. Typically, the norm of residuals is found for the calibration frame for each trial digitized. These residuals basically tell us what correction the movement analysis system will do; these are not the same as the errors in the data collection volume. The standard deviation of the wand length, or between static markers of the calibration frame, may also be reported to give an idea of the potential errors in the calculation of marker positions.

Lens Correction

The lens of all cameras is affected by distortions to some extent. These are caused by the material and imperfections produced during the manufacturing process. This introduces small errors into the system that, if not accounted for, can result in larger errors in the reconstruction of a two-dimensional image. If the error can be measured for the lens, it can be accounted for during or post acquisition. To study lens distortion, points of known position, relative to one another, are filmed. Tasi (1986) developed a method using 60 calibration points; Antonsson and Mann (1989) used over

12 000 points to obtain a far more detailed study. Ladin and colleagues (1990) studied lens distortion in two dimensions by filming an area with known positions of equally spaced points. They took measurements from film and compared them with the known values. The differences between these values were then plotted as vectors, the magnitude of the vector increasing as the lens distortion increased. In this way, the lens distortion could be mapped. Ladin and colleagues showed that substantial errors frequently occur due to lens distortion as the object being filmed moves away from the centre of the field of interest. To prevent such errors, the centre of the field of view should be used; however, this limits the user. With many movement analysis systems, the cameras are checked for lens distortion by the manufacturer and a lens correction matrix is incorporated into the software. This minimizes the effect of lens distortion. Alternatively, information from the calibration frame can be used to correct for lens distortion, although this is not as accurate.

Dynamic methods for camera linearization can be used, such as the method used by Qualisys. The same principle can also be used in a static method. Markers or a checkerboard grid are arranged on a frame so that the exact distance between each marker is known (Fig. 8.12). This is then either moved about directly in front of the camera whilst acquiring data, or a few frames of data are acquired whilst the frame is held in a static position. When the frame is moved, it provides information about the depth characteristics of two-dimensional space or aspect. From these data, the best-fit solution is then used to account for the errors associated with the camera lens.

8.4.5 Data Capture

Once the camera setup is calibrated and the subject has had the marker set attached, the movement can be recorded. After recording is complete, the video data are transferred to the computer hard disk. This process is called capture or video collection. Many movement analysis systems, such as VICON (Jarrett et al., 1974), Elite (Ferrigno & Pedotti, 1985) and Qualisys capture the video image information straight to the hard disk. These are known as camera- or television-based systems. The second category are known as video-based systems; these collect the video information onto video tape first, then transfer it to the hard disk. The use of video tape may decrease the resolution of the system slightly as the image is recorded onto analog tape and then re-digitized. The introduction of digital video no longer has this drawback, although the format may need to be altered when the data are processed by the computer. Camera-based systems, such as Qualisys and VICON, record the direct output from the cameras, so the resolution is maintained.

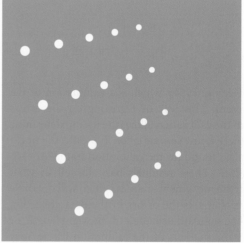

FIGURE 8.12 ■ Typical data used to correct for lens distortion. The image shows the frame in an orientation with and without roll introduced

The use of a video-based system does have an advantage as it keeps a valuable record on tape of the movement, which may be recollected should the transfer not be successful. However, the tracking of markers tends to be considerably more time consuming and file sizes tend to be larger than camera-based systems. Once the video information has been stored on the computer hard disk, the need for the video apparatus is eliminated and the information can be retrieved from the computer hard disk, one frame at a time, so that it may be digitized.

Clusters and Markers

For many movement analysis systems, it is necessary to have markers placed on various anatomical landmarks to represent body segments. These markers are either described as passive or active. The use of markers should not significantly modify the movement pattern being measured; if this were the case then movement analysis with markers would fail the criteria set by Brand and Crowninshield (1981): 'the measurement technique must not significantly alter the performance of the evaluated activity.'

Markers may be singular, to represent a joint, or in the form of clusters which are positioned on the segment itself. Much work has been carried out determining the optimal configuration of marker clusters and it is now widely accepted that a rigid shell with a cluster of four markers is a good practical solution (Cappello & Cappozzo, 1997; Manal et al., 2000) (Fig. 8.13A). Video-based systems tend to use either light-to-dark or dark-to-light contrast with the background, whereas markers for camera-based systems are generally made of a retroreflective material called Scotchlite. This material is used to reflect light emitted from around the camera back to the camera lens. Some camera-based systems use a stroboscopic light, whilst others use light from synchronized infra-red LEDs mounted around the camera lens. Whichever technique is used, the contrasting light markers on a dark background have the effect of showing the markers as bright spots (Fig. 8.13B).

In contrast, active markers produce light at a given frequency, so these systems do not require illumination and, as such, the markers are more easily identified and tracked (Chiari et al., 2005). The most frequently used active markers are those that emit an infra-red signal, such as LEDs (Woltring, 1976). LEDs are attached to a body segment in the same way as passive markers, but with the addition of a power source and a control unit for each LED. Active markers can have their own specific frequency which allows them to be automatically detected. This leads to very stable real-time three-dimensional motion tracking as

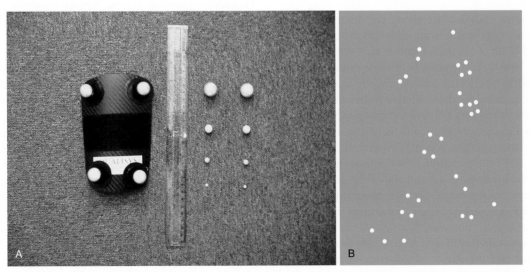

FIGURE 8.13 ▪ (A) Passive markers and marker clusters compared with a 30 cm ruler and (B) captured data from a camera-based system

no markers can be misidentified as adjacent markers. Active markers also have the advantage that they can be used outside (passive marker systems are usually confined to indoors as they are sensitive to incandescent light and sunlight). One example of an active system is CODA.

Errors Involved with Marker Placement

The positions of markers are susceptible to two types of error, generally referred to as relative and absolute errors. Relative errors are defined as the relative movement between two or more markers that define a rigid segment. Absolute errors are defined as the movement of a marker with respect to the bony landmark it is representing. Collectively, relative and absolute errors are often referred to as soft-tissue artefacts.

Markers are often placed on the skin over a specific anatomical landmark; however, this is not always acceptable to the subject being tested, especially on anatomical landmarks around the pelvis. Hazlewood and colleagues (1997) compared marker placement on skin and Lycra over the anterior superior iliac spine (ASIS). It was reported that there was significantly more movement when the marker was placed directly on the skin than on Lycra, which may lead to overestimation of the movement of the pelvis. Hazlewood and colleagues concluded that marker attachment may be preferable on close-fitting garments than directly on skin.

Relative and absolute errors are often caused by movement of the soft tissue on which the markers are placed (Lesh et al., 1979; Ladin et al., 1990; Cappozzo et al., 1996). The magnitude of these errors has been studied by using pins secured directly into the bone and comparing the data collected from skin-mounted markers to markers attached to bone pins. These data give a direct measure of soft-tissue movement with respect to the skeletal system (Levens et al., 1948; Cappozzo, 1991; Reinschmidt, 1996). Although this quantifies the errors involved and allows the development of corrective algorithms for skin movement, it is not ethically acceptable to use bone pins for routine motion analysis purposes.

A large quantity of data are available describing the amount and the effects of soft-tissue artefacts from skin markers on human lower-limb segments, but inconsistencies exist between the reported results. Differences can be accounted for by variation in marker placement and configuration, differences in techniques, intersubject differences and differences in the task performed (Leardini et al., 2005). Several studies (Cappello et al., 1997; Fuller et al., 1997; Lucchetti et al., 1998; Alexander & Andriacchi, 2001; Leardini et al., 2005) found skin movement to occur because of skin sliding due to adjacent joint rotations, or because of the transient response at impact, or because of muscle contraction such as pre-activation of the quadriceps prior to touchdown (Reinschmidt et al., 1997). Most of these studies assessed lower-limb motion and, in particular, knee kinematics, and largely concluded that unacceptable inaccuracies in extrasagittal motions are present mainly because of soft-tissue artefacts at the thigh segment. Conversely, soft-tissue movement recorded using non-invasive markers on the shank have only a small effect on three-dimensional kinematics and moment estimates at the knee (Holden et al., 1997; Manal et al., 2002).

A number of techniques for minimizing soft-tissue artefacts and compensating for their effects have been proposed. Again, these methods depend upon the marker configurations used in the analysis. Relative errors have been modelled using rigid body (Chéze et al., 1995) and non-rigid body (Ball & Pierrynowski, 1998) theory to best fit cluster marker trajectories during motion. However, as the cluster markers were fixed to a rigid plate, these methods were not able to address absolute errors. Surface modelling also includes a point cluster technique, where an array of markers is used to estimate the position of the centre of mass and reference system orientation of a segment (Andriacchi et al., 1998). Although an extended version of this method has reported improvements in estimation of the position of the underlying bones (Alexander & Andriacchi, 2001), it can only model skin deformations and has limited use in some applications due to the number of additional markers required.

A recent approach to compensate for skin sliding associated with joint flexion was proposed using an enhanced version of the calibrated anatomical system technique (CAST) (see Chapter 9), where static calibrations of the two extremes of motion of a specific task were recorded (Cappello et al., 1997). From these calibrations, a model was obtained that allowed for the change in relative position of each thigh marker between the flexed and extended positions. Improved femur orientation and position was achieved. The limitation of this method is that it should be designed specifically for the motor task under analysis; it may also be enhanced by using more sophisticated methods for characterizing skin deformation and sliding throughout the joint range of motion.

Another approach to this problem is known as global optimization. This involves simultaneous pose (position and orientation) estimation of multi-linked segmental models. These methods were originally developed by Kepple and colleagues (1994) and Lu and O'Connor (2000) to minimize global measurement errors by taking into account known joint constraints. With these methods, weighted sum of squares distances are minimized between measured marker locations and marker locations determined by a joint-constrained model. Similar methods to compensate for false identification and occlusion of markers were also extended to include computation of local marker displacement due to skin movement artefacts (Cerveri et al., 2005), but further validation is still required. Although advantages can be gained from global optimization, the constraints and/or complexity of the models limit their use for clinical assessment, particularly for patients with joint instabilities or deformities (Leardini et al., 2005).

Future studies that are able to model the three independent contributions of inertia, skin sliding near the joints and segment deformation due to muscle contractions may reduce relative and absolute errors. Another approach would be to collect a large series of measurements from several populations that compare the movement of the soft tissues directly with the movements of the underlying bone (Leardini et al., 2005). Until better compensation techniques become available, care must be taken in comparing movement data where different marker sets have been used.

8.4.6 Digitizing, Transformation and Filtering

Digitizing or tracking is the process of identifying points on the body using markers or a visual impression of the joint centres. There are two methods of digitizing, manual and automatic.

Manual Digitizing

Using a computer cursor, the location of each of the subject's body joints (e.g. ankle, knee, hip, shoulder and elbow joints) can be selected and entered into the computer (Ariel, 1974). Once this has been carried out for one frame, the information stored on the hard disk can then be advanced one frame and the same points on the body identified. This needs to be carried out on each frame, which builds up a stick figure of the movement. This allows movement with or without markers to be analysed, but it can be very slow, especially if three-dimensional movement is being digitized using a number of cameras.

Automatic Digitizing

Automatic digitizing uses markers that need to be identified once in the first frame but will be automatically tracked throughout all the remaining frames (Mann & Antonsson, 1983) and (Keemink et al., 1991). In all cases the centroid of the marker is calculated, producing a point representing each marker for each camera view (Fig. 8.14A, B). This technique is much faster than manual digitizing, but it does rely on the use of markers that can encumber some movement patterns.

Transformation

Transformation is the computation of the two- or three-dimensional coordinates of the markers on chosen points on the body. Computation is performed based on or adapted from a direct linear transformation (DLT) method developed by Abdel-Aziz and Karara (1971). The DLT provides a relationship between the two-dimensional coordinates of a marker from each camera view and its three-dimensional location in space. The requirement of DLT is that all the data recorded from the different camera views are synchronized (i.e. simultaneous camera views of the activity).

Transformation can be used for two-dimensional or three-dimensional movement analyses. The calculation of the coordinates is carried out by the computer from the calibration (see Section 8.4.4). The exact positions of the calibration points are stored in the computer, which can be used to calculate the exact position of a marker around the calibration area. The two-dimensional or three-dimensional coordinates of each marker can then be extracted from the system and used to represent the segment as a rigid body.

Data Filtering

A smoothing or filtering operation is performed on the coordinates of each marker or position to remove small random digitizing errors. There are several smoothing algorithms or filters available; the most common are digital filters and spline techniques.

Spline techniques are described as piece-wise polynomials that are joined together at points called knots (Wood & Jennings, 1979; Woltring et al., 1985). These are usually based on cubic or quintic polynomials and are known as cubic spline or quintic spline techniques. Cubic spline is the most common smoothing technique in motion analysis as it balances closeness of fit with speed of calculation. Some motion analysis system's software allows the operator to choose the smoothing algorithm and the level of smoothing,

whereas others smooth automatically. The least squares spline approach requires the operator to input a smoothing parameter, which controls the smoothness and closeness of the fit. The general cross-validation spline technique developed by Woltring et al. (1985) involves an estimation of the smoothing parameter based on all the data points and prediction of a best-fit curve, which subsequently does not require any input from the operator. The general cross-validation is the technique favoured by many motion analysis systems' software.

Low-pass filters are also commonly used, typically second- or fourth-order Butterworth filters with the

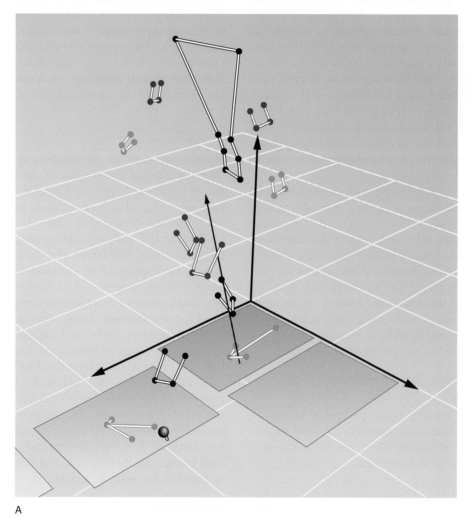

A

FIGURE 8.14 ■ (A and B) Automatic digitizing using Qualisys Track Manager (QTM)

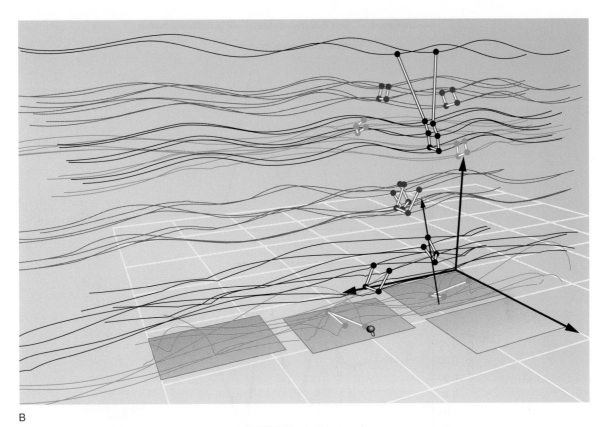

B

FIGURE 8.14, Continued

cut-off frequency set to 6 or 7 Hz for walking data. For faster activities, such as running, higher cut-off frequencies are required, typically in the range of 10–15 Hz. Low-pass filters allow the low-frequency data through, but prevent the high-frequency data. The practical upshot of this is that the small random digitizing errors and some errors of soft-tissue artefacts are removed. However, much care needs to be taken so that the filtering does not change the movement data itself.

Figs 8.15A, B and C shows an example of the vertical toe displacement, velocity and acceleration, and the effect of no filtering, a 6 Hz fourth-order Butterworth filter and a 4 Hz fourth-order Butterworth filter. The unsmoothed data show the higher-frequency 'noise' from the random digitizing errors causes a disproportionate error. The 6 Hz fourth-order low-pass Butterworth filter removes the peaks of the displacement graph, but allows a useable signal for velocity and acceleration. The 4 Hz fourth-order Butterworth filter

significantly affects the displacement data and causes an underestimation of the velocity and acceleration data. Therefore, not filtering will produce unsmooth displacement graphs and leave much of the velocity and all the acceleration data unusable. Filtering too aggressively with a 4 Hz cut-off frequency will distort and falsify the data, but filtering at a cut-off frequency of 6–7 Hz will produce useable data for displacements, velocities and accelerations for walking. It is always worth looking at the marker velocity and acceleration data to check that the data are not over- or under-filtered.

Different time and frequency domain filtering techniques have been developed to remove these high-frequency error components. Fourier series filtering and digital filtering are currently popular. However, these filters do not account for higher frequencies generated by impacts or rapid energy transfers. Alternative methods have been proposed and results so far are encouraging (Chiari et al., 2005), but irrespective of

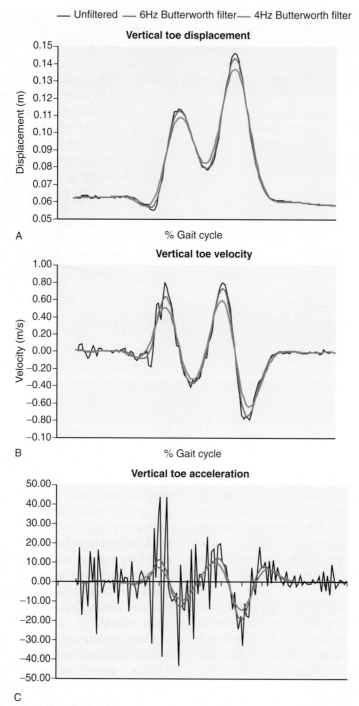

FIGURE 8.15 ■ (A) Filtering of vertical toe displacement, (B) filtering of vertical toe velocity and (C) filtering of vertical toe acceleration

the processing technique, random errors must be minimized as data differentiation amplifies the noise to signal ratio, which results in considerable inaccuracies in derivative data.

8.4.7 Errors Due to Digitizing

We previously covered errors involved with marker placement; however, other sources include random errors in the form of electronic noise, digitizing errors, marker flickering or marker distortion (Chiari et al., 2005). Two-dimensional video images using standard video cameras commonly suffer from marker distortion from higher-speed distal segment movement. Pattern recognition algorithms and more advanced digitizing palettes or skilled manual digitizing can reduce errors in locating the centre of distorted markers, but inevitably notable errors may still remain. Camera-based systems use threshold detection algorithms to detect the two-dimensional coordinates of the brighter pixels covered by a marker. Detection may then be enhanced with a bi-dimensional cross-correlation template matching method provided that the markers are spherical (Chiari et al., 2005). Errors in estimating the centre of a marker vary with technique, but can be improved with circle fitting using a least squares approach. Marker size can also influence centroid estimation as larger markers can merge with neighbouring markers and small markers may not be detected by a sufficient number of cameras, which will result in tracking errors. Optimal camera setup and calibration procedures may reduce errors resulting from marker merging and distortion, but further developments are still required to minimize these errors from the movement data.

8.5 CONFIGURATIONS FOR CAMERA-BASED MOTION CAPTURE

Motion capture used in movement analysis laboratories can generally be grouped into two-dimensional and three-dimensional systems. The type of system used varies dependent on the requirements of the depth of analysis required. Two-dimensional systems are normally based on the use of simple video cameras positioned orthogonally to capture either the coronal or sagittal plane movement. Three-dimensional systems are normally made up of many cameras (generally

between 4 and 10): these types of systems allow for a detailed three-dimensional analysis of human movement. Three-dimensional systems are generally much more expensive than two-dimensional systems. When purchasing a three-dimensional system one must be aware of the amount of cabling associated, so, if you have the opportunity to design your laboratory, try to incorporate cabling into the walls and floor wherever possible. This will allow for a much safer working environment for both you and any research participants/patients.

8.5.1 Configuration of Two-Dimensional Motion Analysis Systems

Two-dimensional analysis configurations consist of single or multiple video camera. These can give valuable information about single plane movements. A standard video camera will usually suffice, but this will not be able to pick up more complex multiplanar movements. The only additional cost associated with two-dimensional systems is the software to digitize and process the data. This software is produced by a number of manufacturers; these include HU-M-AN (HMA Technologies Inc.), APAS (Ariel Performance Analysis System) and Silicon coach. Some of these systems work on the basis of manual digitizing of markers, whereas others use colour and shape recognition to identify markers. When using systems of this type there are always a number of risks associated with the quality of the data. These include parallax error, perspective error, cross-planar errors and digitizing errors. The use of two-dimensional systems relies on the placement of markers on the lateral aspect of the joint, which is used to identify the joint. This is not, in fact, the joint; however, the relative movements between the markers can provide some information with regards to the joint movement.

8.5.2 Video Camera Configuration

Setting up video cameras directly in the sagittal plane and coronal planes can give a useful record. This allows for analysis of the joints of the lower limbs in the sagittal plane and can give an indication of varus/valgus knee and hip joint movement. Care must be taken to ensure the camera is exactly in the plane of movement and to not incorrectly identify internal rotation with flexion as valgus. In addition, a posterior camera can

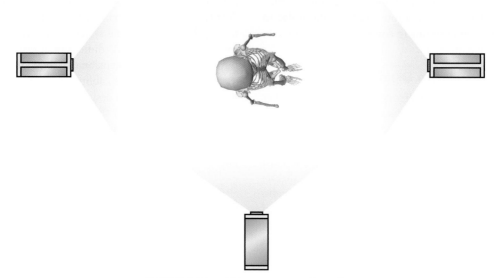

FIGURE 8.16 ■ Video camera configuration

be used, which can be useful when looking at rearfoot movement in the prescription of orthoses. Again, care must be taken not to misinterpret cross-planar motions in a single plane (Fig. 8.16).

8.5.3 Three-Dimensional Motion Analysis Systems

Three-dimensional systems remove much of the risk associated with two-dimensional systems; however, this can be directly related to the cost of the system. In general, the camera value can range from between £1000 and £12 000 per camera. This can bring the overall cost of the hardware to well in excess of £100 000. One must then account for support and software, which may be as much as another £2000 per year. However, the cost of such systems is generally matched by their performance. A three-dimensional system allows the user to take advantage of more advanced marker configurations, as the system will generally automatically identify the markers and will not be affected by errors such as perspective and cross-planar errors. As with two-dimensional systems, care must be taken when setting up three-dimensional systems. There are two main types of configuration, linear and umbrella. Linear refers to the cameras being positioned at set distances apart, all running in two parallel lines. Note that this is only possible with certain systems which do not require all the cameras to track

the calibration frame. Umbrella setups refer to those which use a bank of cameras round the front and rear.

Linear Camera Configuration

The linear camera configuration allows much greater data collection volume as not all the cameras have to 'see' the reference frame. However, there is an increased risk of the cameras identifying one another as a camera due to cameras tracking the opposite camera. And there is increased risk of marker occlusion due to the limited number of cameras in a certain area (Fig. 8.17).

Umbrella Camera Configuration

This type of configuration is more suited to more advanced marker sets that exploit marker sets for multiple segment feet. The cameras are positioned in such a way that ensures at least three cameras are always tracking the data for each marker. This configuration will result in a slightly smaller calibrated volume (Fig. 8.18).

SUMMARY: METHODS OF ANALYSIS OF MOVEMENT

- Movement analysis systems can vary from the study of foot contact distances using ink pads to advance multicamera three-dimensional movement analysis.

■ Each type of system can give very useful clinically relevant information about changes through rehabilitation and treatment.

■ As the complexity of movement analysis systems increases, so does the ability to look at the interactive effects between joints and body segments and the interaction between the different planes of the body.

■ With an increase in complexity comes longer setup times and the requirement for training for the users. However, with many systems now on the market, the training requirements are reduced with improvements in the software.

■ There are many systems currently on the market. However, the most important aspect for the user is having a clear idea of what measures will produce clinically useful data; without this, investment in equipment is a folly.

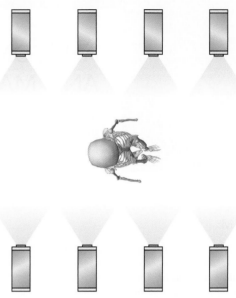

FIGURE 8.17 ▪ Linear camera configuration

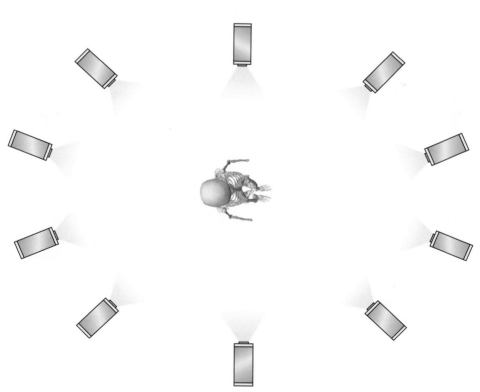

FIGURE 8.18 ▪ Umbrella camera configuration

9

ANATOMICAL MODELS AND MARKER SETS

JIM RICHARDS ■ DOMINIC THEWLIS ■ RICARDO MATIAS ■
ROBERT NEEDHAM ■ NACHIAPPAN CHOCKALINGAM

AIM

This chapter covers different marker sets commonly used in movement analysis. This includes modelling of the foot, lower limb, spine and shoulder. The nature of six degrees of freedom measurement is considered and the associated errors encountered when considering different coordinate systems.

OBJECTIVES

■ To describe the different methods of marker placement in movement analysis of the lower limb, upper limb and spine

■ To summarize the nature of six degrees of freedom analysis

■ To contrast and compare the effect of using different anatomical landmarks on normal and pathological gait data

■ To contrast and compare the effect of using different coordinate systems on normal and pathological gait data.

9.1 LOWER-LIMB MARKER SETS

9.1.1 The Simple Marker Set

The simplest marker set involves directly fixing markers on the skin over a bony anatomical landmark close to the centre of rotation of a joint. The position of the limb segment is then defined by the straight line between the two markers.

The first method requires less markers and, so, theoretically, has less interference with the movement pattern, but does not allow the calculation of axial rotation of the body segment. The anatomical landmarks used are the head of the fifth metatarsal, the lateral malleolus, the lateral condyle of the femur, the greater trochanter, the anterior superior iliac spine (ASIS), the acromion process, the lateral condyle of the humerus and the styloid process at the wrist (Fig. 9.1).

9.1.2 Vaughan Marker Set

The Vaughan marker set consists of 15 markers on the lower limb and pelvis. This allows for more detail on the location of the knee joint centre by including a marker in the coronal plane on the tibial tuberosity. The inclusion of the heel marker allows a more appropriate functional reference for the long axis of the foot. The inclusion of the sacral marker also allows for a more functional reference for pelvis inclination in the sagittal plane and a meaningful measurement of pelvic tilt. The anatomical landmarks used are the head of the fifth metatarsal, the lateral malleolus, the heel, the tibial tuberosity, the femoral epicondyle, the greater trochanter, the anterior superior iliac spine and the sacrum (Fig. 9.2).

9.1.3 Helen Hayes Marker Set

As with the Vaughan marker set, the inclusion of the heel marker and the sacral marker allows for a more appropriate functional reference for the foot and pelvis. However, the Helen Hayes marker set also includes tibial and femoral wands. These are markers on sticks that are attached to a pad fixed to the segment using tape or bandage. They are not placed on any anatomical position as such, and variations on the length of wand and positioning, anterior versus lateral, have been used. The inclusion of these wands allowed the femoral and tibial rotations to be quantified for the

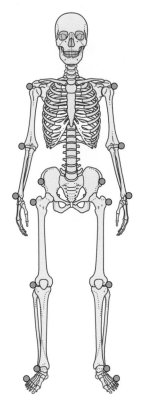

FIGURE 9.1 ∎ The simple marker set

first time. The anatomical landmarks used are the head of the second metatarsal, the lateral malleolus, the heel, the tibial wand, the femoral epicondyle, the femoral wand, the greater trochanter, the anterior superior iliac spine and the sacrum (Fig. 9.3).

9.1.4 The CAST Marker Set

The calibrated anatomical system technique (CAST) was first proposed by Cappozzo and colleagues (1995) to contribute towards standardizing movement description in research labs and clinical centres for the pelvis and lower-limb segments. This method involves identifying an anatomical frame for each segment through the identification of anatomical landmarks and segment tracking markers, or marker clusters. Marker clusters can be directly attached to the skin or mounted on rigid fixtures, which are dependent upon the anatomy, the activity and the nature of the study. These markers may be anatomical or arbitrary, individual or clusters, and mounted on the skin, on wands, or on rigid plates (Manal et al., 2000).

'Anatomical Calibration' Markers

Markers are placed on lateral and medial aspects of joints on anatomical landmarks at the proximal and distal ends of the segment. This is similar to previous marker sets; however, with CAST an additional cluster of markers is also placed on each segment (Fig. 9.4). The anatomical landmark markers enable the proximal and distal ends of the segment to be identified in relation to the cluster of markers. For techniques using three-dimensional kinematics, the coordinates in the laboratory or global coordinate system (GCS) defining the segment are transformed into a local coordinate system (LCS) or segment coordinate system (SCS) using coordinate transformation. Different anatomical markers may be used to define the proximal and distal ends of a segment. For example, the segments of the lower limb may be defined by:

- Foot segment: first and fifth metatarsal heads and the medial and lateral malleoli;
- Tibial segment: medial and lateral malleoli and the femoral epicondyles;
- Femoral segment: femoral epicondyles and the greater trochanter; and
- Pelvis segment: posterior superior iliac spines (PSIS) and anterior superior iliac spines.

Dynamic Tracking Markers

Marker clusters are placed on each body segment. These must be in place during the static 'anatomical calibration'. The exact placement of the clusters does not matter as the CAST technique uses the relative positions to the anatomical landmarks used in the static 'calibration'. These markers may be anatomical or arbitrary, individual or clusters, and mounted on the skin, on wands, or on rigid plates (Manal et al., 2000). During placement, make sure that all markers on the clusters are positioned so they can be tracked effectively. To get the most effective tracking this usually requires the markers be placed at an angle between the coronal and sagittal planes.

At least three non-collinear markers (markers that do not lie in a straight line) are required to track the segment position and orientation (the ensemble of position and orientation of a rigid body from any one frame relative to another in six degrees of freedom [Cappozzo et al., 2005]); however, up to nine have

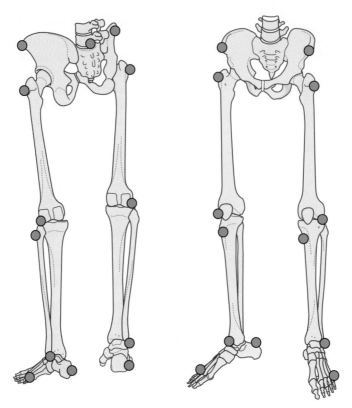

FIGURE 9.2 ■ Vaughan marker set

been used. The usually accepted number is four or five markers per segment or cluster, allowing for one or two markers to be lost at some stage during the movement tasks and still allow data in six degrees of freedom to be found (Fig. 9.5). This is sometimes referred to as marker redundancy (i.e. if you lose a marker during tracking, the model will still work).

9.1.5 What is the Benefit of Using CAST Compared with Other Marker Sets?

CAST offers the ability to model each body segment in six degrees of freedom. As long as a segment has got an anatomical frame by using the 'static' markers and a cluster of 'dynamic' tracking markers, this further allows interactions and movements between the body segments, again in six degrees of freedom, along the anatomical axis defined by either proximal or distal segment.

9.1.6 What Do We Mean by 'Six Degrees of Freedom'?

Any body segment may move in six ways, independently, although in human movement all six ways often happen at the same time, which is good functionally as it allows a very adaptive mechanism; however, it makes understanding the function of each of these movements quite challenging.

The six ways a body segment can move are three linear or translational movements, vertically, medial–laterally and anterior–posteriorly, and three rotational or angular movements in the sagittal, coronal and transverse planes. Figs 9.6 and 9.7 show these different movements on the foot, but these can be equally applied to any body segment.

However, if we are trying to look at the interaction of two segments about a joint, then there are 12 degrees of freedom, six for each body segment. We can now

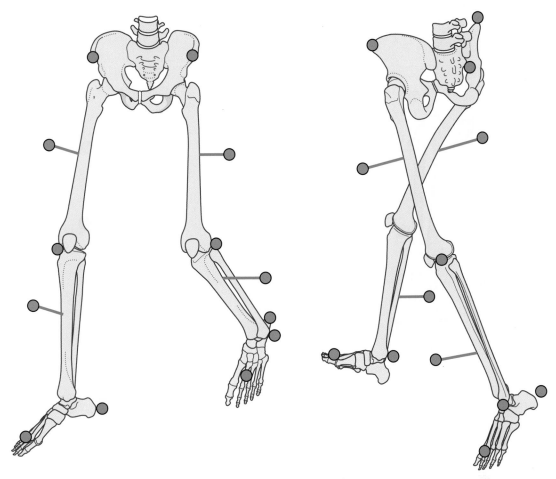

FIGURE 9.3 ■ Helen Hayes marker set

start to get really complex when considering all the segments within, say, the foot, and this is where we need to focus on the clinical sense of what we are trying to quantify in terms of functional anatomy before we tie ourselves in potentially unnecessary and unintelligible knots.

9.1.7 Why Do We Need 'Six Degrees of Freedom'?

For many joints the use of six degrees of freedom is not strictly necessary to gain clinically useful information; therefore, simpler anatomical models are extremely useful for determining changes in function due to surgical and conservative management, such as physiotherapy and orthotic management. However, simpler anatomical models do not give a full picture of the functioning and interaction of different joints in the body, and on occasions could give inaccurate measurements and misleading information.

Consider the measurement of knee valgus. With a simple marker setup, what may appear as a valgus knee may, in fact, be nothing of the sort. If you were to stand, for instance with your knee flexed and your hip internally rotated, then this would be presented as an extremely valgus knee, far beyond what the anatomy would in fact be capable of. If we now consider the same position using CAST and six degrees of freedom, then we are able to isolate the angular movements in

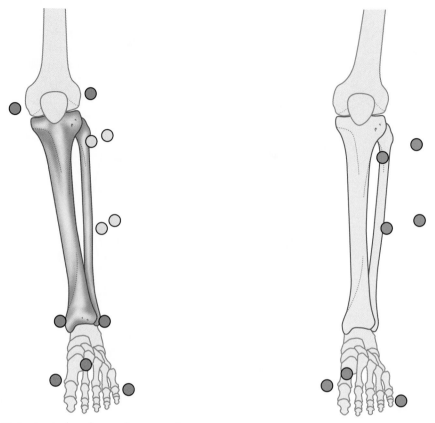

FIGURE 9.4 ■ Anatomical markers and segment cluster

FIGURE 9.5 ■ Dynamic tracking markers

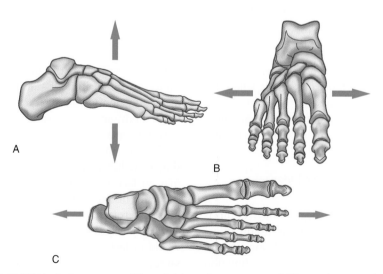

FIGURE 9.6 ■ (A) Vertical movement, (B) medial–lateral movement and (C) anterior–posterior movement

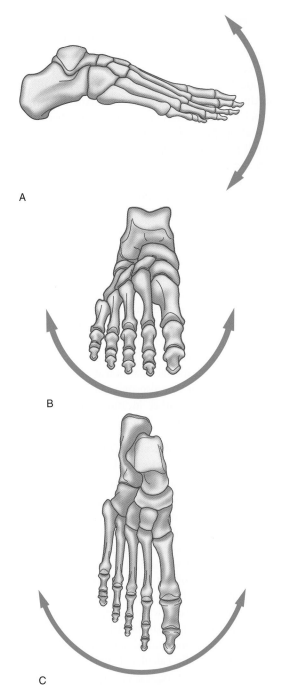

FIGURE 9.7 ■ (A) Sagittal plane angular movement, plantar–dorsiflexion of the foot (engineers call pitch), (B) coronal (frontal) plane angular movement, inversion-eversion of the foot (engineers call roll) and (C) transverse plane angular movement, Internal-external rotation of the foot (engineers call yaw)

the three rotational planes independently and, therefore, gain the correct interpretation of the joint and segment positions and movement.

Much has been published on sagittal plane movements, but there has been far less attention on the interaction of coronal, transverse and sagittal movements. The question is how important are the coronal and transverse movements; the answer is clear, VERY.

If we consider pretty much any foot orthotic management, clinicians will talk about pronation–supination, inversion–eversion and plantar–dorsiflexion in terms of triplanar movement of the joints in the ankle foot complex. Therefore, any functional foot orthotic management is likely to be changing the foot mechanics in six degrees of freedom.

If we consider patellofemoral bracing and taping in individuals with instability or pain in the patellofemoral joint, most of the previous work has focused on the sagittal plane, although both these interventions (i.e. bracing and taping) are actually trying to change the coronal translational position of the patella, which in fact will change both the coronal and transverse plane rotational movements and probably have only a secondary effect in the sagittal plane. The clinicians involved are very focused on what they were trying to achieve. Had the body segments been modelled in a more complex way, data would have been available to determine the nature of these changes.

9.2 METHODS OF IDENTIFYING ANATOMICAL LANDMARKS

9.2.1 The CAST Marker Set with the Davis Dynamic Pointer (or Pointy Stick Method)

CAST requires a medial and lateral, and proximal and distal segment reference. Previously, we talked about the use of anatomical markers positioned on bony landmarks on the body. However, these anatomical landmarks may also be determined by using a wand of at least two coincident markers to point to the target anatomical landmark (Davis et al., 1991). If the distance between these markers is known in relation to the end of the point of the wand, or stick, then the position of the end of the stick can be determined.

This means that anatomical markers are not required to be stuck to the individual being tested. This method is now being incorporated into movement analysis software, such as Visual 3D Motion Analysis Software by C-Motion, which allows this method of anatomical referencing to be applied simply and quickly. This has certain advantages; for example, once the anatomical frame is defined and the cluster moves during a movement or becomes uncomfortable, it is possible to redefine the anatomical frame quickly without having to reapply markers. In addition, a single anatomical frame, for example, the shank, can be redefined. This system may also be more accurate at defining the anatomical landmark as the pointer end is positioned directly on the location. This removes the need to account for marker diameter and marker centroid calculation.

9.2.2 The CAST Marker Set with Functional Joint Centre Identification

Locating the centre of the hip is required to accurately quantify hip and knee joint rotations. Functional and projection methods are currently available; however, modelling of the thigh segment remains difficult, particularly at its proximal insertion with the pelvis.

Arguably the most reliable method for the calculation of the position of a joint centre, and the International Society of Biomechanics (ISB) recommendation for identifying hip joint centre, is the 'functional' approach (Leardini et al., 1999; Schwartz & Rozumalski, 2005; Camomilla et al., 2006). The functional joint centre calculates the position between two moving body segments and defines the common point of rotation segments. Essentially this method examines the six degrees of freedom of motion of two adjacent body segments defined using clusters. For the pair of body segments the axis of rotation is computed. From this information, the most likely intersection of all axes for the two segments is found, providing the position for the joint centre. This requires the subject to perform a number of movements, such as flexion–extension, internal–external rotation and abduction–adduction and circumduction on each leg. Other movements such as a 'hula hoop' movement have also been used. This method can be used on all joints of the body; however, it is recommended that the shoulder joint is not attempted due to the complexity of the shoulder girdle.

For some subjects, the previously mentioned movements may be challenging. The position of the hip joint centre may also be projected from static markers using projection methods. Projection methods use regression equations to estimate the location of the hip joint centre with respect to the pelvis LCS. A number of models exist that estimate the origin and pelvis coordinate systems. In the coronal plane, the origin of the pelvis is defined as the point midway between the left and right anterior superior iliac spines, and in the sagittal plane, as a proportion of the depth measured from the anterior superior iliac spines and posterior superior iliac spines (Bell et al., 1989; Davis et al., 1991; Harrington et al., 2007). A more recent study by Kiernan and colleagues (2015) concluded that although the different models produce different hip joint locations, these do not show any clinically important differences and it is unlikely any error would be incorrectly considered clinically meaningful.

Once the pelvis coordinate system and the hip joint centre have been defined, the thigh coordinate system can also be defined. This commonly uses the medial and lateral femoral epicondyles at the distal end, and the hip joint centre defined by the pelvis at the proximal end; an additional anatomical reference of the greater trochanter can also be used, therefore giving three anatomical marker locations and one virtual marker defined from the hip joint centre (Fig. 9.8).

9.2.3 The Effect of Using Different Anatomical Landmarks on Gait Data

In order to calculate accurate joint kinetics, it is essential to locate the centre of rotation in a repeatable manner through the definition of an anatomical frame. This issue was highlighted by della Croce and colleagues (1999), who identified that the errors associated with incorrect anatomical frame definition are as, if not more, important than those caused by skin movement. Clinically, the definition of the joint centre is generally achieved by using palpable anatomical landmarks to define the medial–lateral axis of the joint. From these anatomical landmarks the centre of rotation is generally calculated in one of two ways: through the use of regression equations based on standard radiographic evidence, or simply calculated as a percentage offset from the anatomical marker based on some kind of anatomical landmarks (Bell et al., 1990; Cappozzo et al., 1995; Davis et al., 1991; Kadaba et al., 1989).

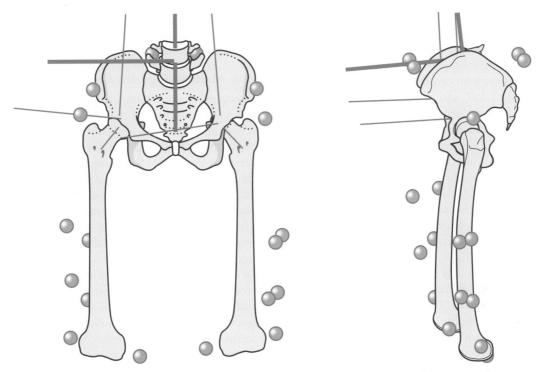

FIGURE 9.8 ■ Pelvis and thigh coordinate systems defining hip joint centre and hip joint angle

The CAMARC (computer-aided movement analysis in a rehabilitation context) consortium proposed a standardized list of palpable anatomical landmarks in order to define the anatomical frame of the lower limbs; these were largely based on the work conducted by Cappozzo and colleagues (1995). This work was designed to standardize methodology and resolve many of the historical issues found with modelling the segments of the lower limbs. The issue of hip joint identification is one that has been covered in much depth and there are still many debates around this area. However, the errors associated with knee joint location have received much less attention. Initially, it should appear a simple choice as to the correct landmark to use based on the CAMARC suggestions. Traditionally there have been a number of landmarks used around the knee, which may not be a true representation of the knee joint centre of rotation. Some of the anatomical landmarks may have an effect on anterior–posterior position of the centre of rotation, whereas others may affect the relative angle of the anatomical frame through a change in orientation.

Holden and Stanhope (1998) examined the effect of changes in the anterior–posterior position of the knee joint centre on sagittal plane knee joint moments. Importantly, they identified that a displacement as small as 10 mm can result in a change in the functional interpretation of the moment. The methods used by Holden and Stanhope (1998) were to establish the anatomical frame based on a standard model, and then to virtually reposition the anatomical landmarks to represent an anterior–posterior shift, rather than considering the effect of incorrect anatomical landmark identification. Manal and colleagues (2002) examined the effect of expressing knee joint moments in two different orthogonal anatomical frames. The anatomical frames differed by approximately 15° throughout the transverse plane. It was noted that large differences were found in sagittal and coronal plane knee moments for what was deemed to be only a small change in the orientation of the anatomical frame.

Thewlis and colleagues (2008) examined the common methods of identifying the knee with different anatomical landmarks. The different landmarks considered were the femoral epicondyles, femoral condyles and tibial ridges. Each method was used to define the anatomical frame; thus the segment end points of the distal femur and proximal tibia represented the knee joint centre. Gait data were then collected using CAST and the external net knee joint moments were calculated based upon the three different anatomical frames. The moments were analysed in the sagittal, coronal and transverse planes. The maximum deviation about the femoral epicondyle frame was found when using the femoral condyle frame with mean deviations of 30 mm anterior and 20 mm medial. This was found to have the effect of changing the peak knee moments by approximately 25% and 8% in the sagittal and coronal planes, respectively (Fig. 9.9A, B, C).

From this work, it is clear that anatomical frames need to be well defined and clearly reported to ensure clinically useable data. This study has identified that Holden and Stanhope (1998) and Manal and colleagues (2002) both identified independent sources of discrepancy by changing the position and orientation of the segment, respectively. Because of the relatively small distances between anatomical landmarks about the knee, there is potential for misidentification. Although these errors may be small in terms of the distances between landmarks, the error is carried forward through all the calculations, resulting in a much larger systematic error in the moment calculations. Therefore, the correct marker placement is of paramount importance in the reliability and repeatability of joint moment data.

9.3 FOOT MODELS

9.3.1 Single Segment Foot Models

In gait analysis, the foot is nearly always considered as a single segment system with the ankle–joint complex considered as a simple hinge. In some of the early models, markers were placed on the heel, the metatarsal heads and on the malleoli, which is sufficient to measure plantar–dorsiflexion; however, this was not detailed enough to measure foot-to-tibial inversion–eversion or internal–external rotation. The addition of markers that define the medial and lateral metatarsal heads and the medial and lateral malleoli with additional tracking markers allow a more detail six degrees

of freedom that is able to measure all three planes. However, this is not sufficient to measure the relative motion of the different parts of the foot.

9.3.2 Multiple Segment Foot Models

It is becoming more widely accepted that the foot and ankle joint complex is better described as rearfoot, midfoot and toes. This, interestingly enough, is what podiatrists have been telling us for years! It is also very interesting to note that modelling the foot as a deformable segment, rather than a rigid single segment, has a significant effect on the accuracy of instantaneous power calculations between segments.

It has also been suggested that modelling the metatarsophalangeal joint would improve the instantaneous power calculations. How the single segment foot came about was due to restrictions in the capability of biomechanical measurement to identify many small markers. However, now with careful marker placement and camera positioning, and with camera systems containing 8–12 cameras, it is possible to track the movement of the tibia, the rearfoot, the midfoot and the forefoot as separate segments through multiple gait cycles in six degrees of freedom.

There are now many foot models emerging from the literature (Carson et al., 2001; MacWilliams et al., 2003) where aspects of the rearfoot, midfoot and forefoot are modelled as separate segments, and, more recently, the Oxford foot model (Stebbins et al., 2006) (Fig. 9.10). Woodburn and colleagues (2004) considered the use of multi-segment foot motion during gait in rheumatoid arthritis and concluded 'this technique may be useful to evaluate functional changes in the foot and to help plan and assess logical, structurally based corrective interventions'. Since 2005, multiple segment foot models have been used routinely in the assessment of patients with foot-related musculoskeletal and neurological disorders. However, almost all studies so far have considered barefoot walking due to the nature of the marker placements on the feet, which makes the study of foot orthoses and the effect of footwear very difficult, if not impossible.

The issue of putting markers on the shoe is an interesting one. It is questionable whether this gives us the same data as on the foot itself; however, this is currently the only way we can look at shod walking and

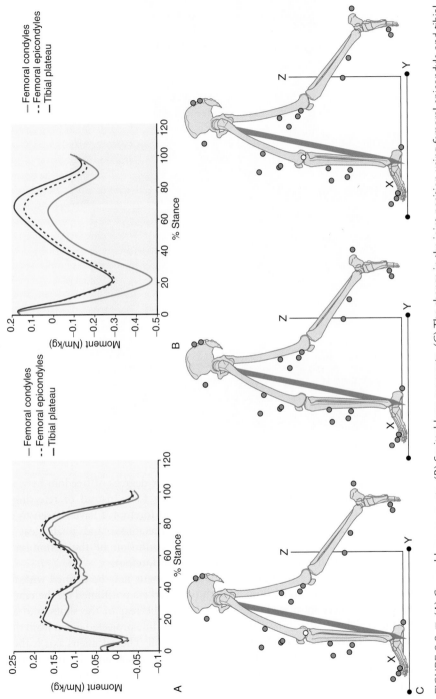

FIGURE 9.9 ■ (A) Coronal knee moments. (B) Sagittal knee moments. (C) The change in the joint position using femoral epicondyle and tibial plateau markers

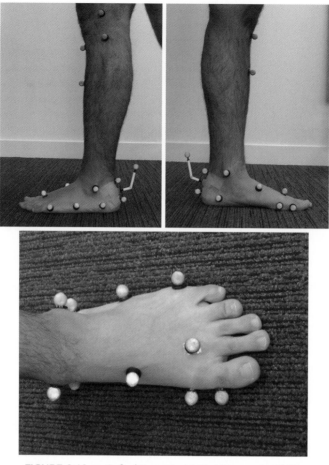

FIGURE 9.10 ■ Oxford Foot Model (Stebbins et al., 2006)

the effect of foot orthoses. It is possible to modify the shoes to allow markers on the foot to be seen by cutting holes in the upper of the shoe. However, this involves the structure of the footwear being modified, which could be considered as an integral part of the foot orthotic management. Currently, the practice in our laboratory at UCLan, Preston, is to look at orthotic management with a three-segment foot model, but being mindful of the fact that markers are placed on the shoe and not the foot itself. The model shown in Figs 9.11 and 9.12 is an adaptation of the Carson foot model that may be appropriate to also attach to footwear, although care is required in the analysis as shoe-to-foot movement could produce substantial artefacts due to the small nature of some of the relative movements between foot segments. This adaptation allows

analysis in six degrees of freedom between three segments of the foot. Instead of referring to rearfoot, midfoot and forefoot, I have chosen to call these the calcaneus, metatarsal and phalangeal segments to avoid any confusion of the definition of rearfoot, midfoot and forefoot.

The calcaneus may be defined with four markers, with two markers positioned on the rear of the calcaneus or on the rear of the shoes, and a further two markers on a line projected down from the medial and lateral malleoli. The calcaneus axis is then defined by the malleoli markers and the two markers projected down. All four markers are then used to track the movement of the calcaneus. The markers on the medial and lateral malleoli define the proximal end of the calcaneus, and the markers projected down define

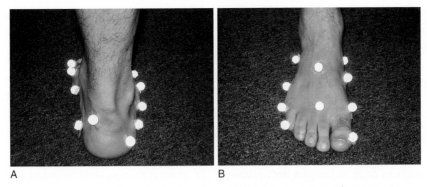

FIGURE 9.11 ■ (A and B) Marker placement for a six degrees of freedom three-segment foot model

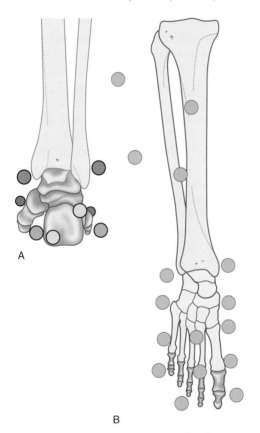

FIGURE 9.12 ■ (A and B) Anatomical calibration and tracking markers for a six degrees of freedom foot

the distal end. In this way, the SCS of the calcaneus and tibia are similar, and allow data analysis with minimal cross planar measurement errors. The metatarsals may also be defined with four markers. To reduce the number of makers required on the foot or shoe, the two markers projected down from the medial and lateral malleoli define the proximal segment end, and markers on the lateral aspect of the fifth metatarsal, and medial aspect of the first metatarsal define the distal end. An additional tracking marker may be added on the dorsal surface between these proximal and distal anatomical landmarks. Again, this allows an SCS similar to that of the rearfoot. The phalanges may be defined by the proximal markers on the lateral aspect of the fifth metatarsal and medial aspect of the first metatarsal, and the distal markers on the hallux or medial distal part of the shoe on one of the lateral phalanges or the lateral distal part of the shoe. An additional tracking marker may also be placed just anterior to the second or third metatarsal on the dorsal surface (Fig. 9.11A, B). This model gives the ability to consider each section of the foot in six degrees of freedom as each segment is defined by at least four markers (Fig. 9.12A, B).

A more extreme foot marker set involves putting pins in the different bones of the foot with trihedron marker clusters attached, a trihedron being a configuration of markers in all three planes and meeting at a single point (Lundgren et al., 2008). This enables the quantification of the relative movement between the different bones of the foot. This may well give very interesting data on the relative movement; however, it is questionable whether the person will walk exactly the same with pins fixed in the bones of the foot. It is also not ethically acceptable to use bone pins for routine motion analysis purposes.

It is very important when considering which foot model to use to consider the area of interest and the outcome measures you wish to take. It is possible that the foot model chosen is either too simplistic or unnecessarily complicated for the research/clinical question being asked.

9.4 MODELS FOR THE TRUNK AND SPINE

Various types of imaging modalities are commonly used for the assessment of spine. These images can produce a three-dimensional representation of the spine. However, some of these techniques are invasive in nature and expose the patient to unnecessary radiation and will not help to complete a dynamic assessment of the trunk and spine.

A dynamic assessment is important to understand the segmental coordination, variability in movement and the control strategies adopted by patients to overcome difficulties in mobility. This assessment has to be non-invasive and should be able to measure the function of the spine and back. Given that the trunk, head and upper limbs normally account for two-thirds of an individual's body mass, any changes in gait pattern due to specific clinical conditions can result in compensatory movements of the trunk.

Some of the published research indicate that the interaction between the spine, pelvis and lower limbs is essential in maintaining balance and to achieve a smooth and efficient gait (MacWilliams et al., 2013; Cappozzo, 1983; Van Emmerik et al., 2005). With this background, it is important to review and understand the models which could be used for the non-invasive dynamic assessment of the back.

9.4.1 Modelling of the Trunk

The trunk can be defined into two components, the thorax and abdomen. A line between the seventh cervical spinous process (C7) and sacrum models the trunk as a rigid segment (Frigo et al., 2003). This approach offers an assessment of trunk inclination in the sagittal and frontal plane. A projection of the line between left and right acromion process (outlines shoulder girdle) and a line between left and right PSIS onto a horizontal plane can define trunk axial rotation (Frigo et al., 2003). Markers on the shoulder girdle, C7, and sternum can define a three-dimensional trunk model (Rab et al., 2002), although a confounding influence of upper limb movement is a potential

limitation of using acromion landmarks. Alternatively, markers fixed to the spinous process of the second thoracic vertebrae (T2), deepest point of incisura jugularis (SJN), and midpoint of the PSIS defines the centre line of the trunk and a is a non-linear marker configuration that can represent three-dimensional movement (Baker, 2013).

9.4.2 Modelling of the Thorax

The trunk is commonly characterized by the analysis of thorax motion and is advocated as a suitable approach for clinical gait analysis (Armand et al., 2014; Leardini et al., 2011b). This is apparent, for example, in the study of cerebral palsy (CP) gait. In a recent systematic review, Swinnen and colleagues (2016) noted that children with bilateral CP displayed greater range of motion of the thorax and pelvis during walking in comparison to typically developing children. The increase in thorax displacement and mechanical work of the passenger unit were considered the main contributors to decreased gait efficiency in children with bilateral CP (Van de Walle et al., 2012). Knowledge of the interaction between the pelvis and trunk has also provided a valuable insight into the differences in segmental coordination strategies between healthy individuals and those suffering from chronic low back pain (LBP). For instance, Lamoth and colleagues (2002) reported pelvic–trunk coordination is generally in-phase at lower walking speeds with transition to anti-phase at higher speeds. In contrast, individuals with chronic LBP have a reduced ability to transition pelvic–trunk coordination from in-phase to anti-phase walking as speed increases (Lamoth et al., 2006; Selles et al., 2001).

Whereas various protocols for kinematic modelling of the thorax have been described and documented (Wu et al., 2005; Thummerer et al., 2012; Armand et al., 2014; Leardini et al., 2011b), there is a lack of consensus on which thorax model offers greater clinical benefit over another. Based on the ISB recommendations (Wu et al., 2005), the thorax segment is defined using anatomical landmarks on the sternum and the spinous process of C7 and T8 (eighth thoracic vertebra) (Fig. 9.13B). A similar method is noted with regards to the Plug-in-Gait model (Vicon, OMG, UK), whereby the only difference being the use of T10 (tenth thoracic vertebra) instead of T8. Quantifying

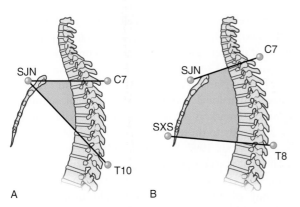

FIGURE 9.13 ■ (A) Marker placements for thorax model according to Baker (2013) and Wu et al. (2005). (B) Anatomical landmarks. SJN, Deepest point of incisura jugularis; SXS, xiphoid process; C7, 7th cervical vertebra; T2, 2nd thoracic vertebra; T8, 8th thoracic vertebra; T10, 10th thoracic vertebra

thorax movement using C7 is questionable, as the C7 vertebra is not representative of the thoracic region, and movement of the head and neck can influence C7 position (Armand et al., 2014). The use of T2 as an alternative to C7 has been proposed (Armand et al., 2014; Baker, 2013; Leardini et al., 2011).

Leardini and colleagues (2011) recommended that the posterior-distal reference position of the thorax segment should be aligned with the distal apex of each scapula (defined as MAI). This approach was based on reported difficulties in identifying single thoracic vertebrae. The MAI landmark is approximately at a spinal level of T9, although this can range between T7 and T10 (Haneline et al., 2008). This range could be partially attributed to differences in scapula position and spinal postures between individuals. Armand and colleagues (2014) found similar results for thorax movement during gait when either T8 or T10 was included as part of the thorax pose estimation, suggesting that the MAI marker could be a suitable landmark. Armand and colleagues (2014) also noted a consistency between movements performed and estimated range of motion when a thorax segment was defined by one marker on the manubrium and two markers on the spine (T2 and T8 or T10). Baker (2013) advocates the use of T2, T10, and the manubrium as a minimal marker configuration to define a three-dimensional thorax segment (Fig. 9.13A). The placement of one marker

on the superior aspect of the sternum would remove the practical concerns during the assessment of female participants.

Thorax models thus far do not assess movement beyond T10. Armand and colleagues (2014) reported the highest error for axial rotation of the thorax and for all movement of the trunk when T12 was included as part of the marker set. Defining the posterior-distal position of a rigid thorax segment beyond the sixth to eighth thoracic vertebrae is debatable because the degree of displacement and pattern of movement of the thoracic spine differ above and below the level of T6 (sixth thoracic vertebrae) during gait (Crosbie et al., 1997; Needham et al., 2015; Syczewska et al., 1999). Such differences could be partially attributed to variation in spinal anatomy between the upper and lower region of the thoracic spine (i.e. facet joint orientation).

As covered earlier, reflective markers are attached over anatomical landmarks in accordance with the kinematic model guidelines and are used to define the position and orientation of a segment and respective coordinate system. The proximal and distal end of a segment usually defines the primary axis. For example, the line connecting the midpoint between C7 and SJN and the midpoint between T10 and xiphoid process (SXS) is one approach to define the proximal and distal end of the thorax (Plug-in-Gait, Vicon, OMG, UK). A similar approach is adopted by the ISB thorax model, although T8 defines the posterior-distal aspect (Wu et al., 2005). Virtual markers are used to create these midpoint locations. The proximal and distal end of a thorax segment can also be defined by the physical markers attached to the spine (Armand et al., 2014; Leardini et al., 2011a). The ISB thorax model defines the secondary axis (medial–lateral) by a line pointing to the right that is perpendicular to the plane formed by the SJN, C7, and the midpoint between SXS and T8. In contrast, the Plug-in-Gait thorax model defines the secondary axis (anterior–posterior) by a line connecting the midpoint between C7 and T10 to the midpoint of SJN and SXS. The final axis is created based on the cross-product of the primary and secondary axis.

It is important to note that defining the proximal and distal end of a thorax using the approaches previously described will produce different kinematic waveform profiles. This is due to differences in the sagittal plane offset between the respective coordinate systems. For example, Fig. 9.14 represents thorax movement relative to the pelvis in the sagittal (A), frontal (B) and transverse plane (C) during gait. The black and blue lines outline data from the Plug-in-Gait thorax model. The black line represents the conventional definition of the thorax segment, whereas the blue line outlines data from a thorax that defines the proximal and distal end by physical markers on the spine. For both modelling approaches all four markers were used to track movement. Differences between individuals in the offset of sagittal plane posture has been suggested to be the reason for large variability in reported spine movement during dynamic tasks (Leardini et al., 2011a).

9.4.3 Segmental and Intersegmental Movement of the Spine

Primarily focusing on trunk or thorax movement relative to the pelvis or to a laboratory location is dependent upon the research question and/or clinical application. An approach to assess intersegmental movement within the thoracic and lumbar region would provide an understanding of the compensatory strategies at a spinal level as a result of various clinical conditions: for example, scoliosis, LBP, or leg length discrepancy.

One simple method is to attach single markers on several spinous processes of the vertebrae. A number of configurations have been proposed (Chockalingam et al., 2002, 2008; Frigo et al., 2003; Syczewska et al., 1999). Segmental angle can be estimated by projecting a direct line between two markers onto the planes of the global reference frame, i.e. the vertical (Fig. 9.15). A planar projection using three or four markers can provide an angle between adjacent segments. This approach, for example, allows for the static and dynamic assessment of spine posture: that is, the angle of kyphosis and lordosis in the sagittal plane (Frigo et al., 2003; Heyrman et al., 2013, 2014) (Fig. 9.15B, C).

As stated previously, markers attached to the spine in addition to the sternum provide a non-linear configuration that can track movement of a thorax segment in three dimensions. Due to difficulty in

finding relevant anatomical landmarks to effectively define axial rotation in the transverse plane, the placement of markers solely on the spinous processes limits the analysis of movement of the spine to the sagittal and frontal plane (Leardini et al., 2011b). To define a three-dimensional thoracic and/or lumbar segment, additional markers can be applied on the surface of the back laterally to those attached to the spinous process (Crosbie et al., 1997; Seay et al., 2008). Whereas these additional markers provide transverse plane information, skin elasticity, soft-tissue artefact and the morphology of the back surface could influence the independent movement between markers. Applying markers on Elastikon elastic tape that has been wrapped around the lumbar region is an approach that has been used to minimize the potential influence of soft tissue artefact (STA) (Seay et al., 2008; Mason et al., 2016).

9.4.4 Intersegmental Movement of the Spine Using Marker Clusters

One alternative approach is the utilization of a three-dimensional cluster, which consists of a cluster of at least three markers that are positioned in a fixed non-linear configuration, and attached to a rigid or semi-rigid base that can be placed onto the spinous process of a vertebra (Fig. 9.16). If an appropriate number of three-dimensional clusters are placed on relevant spinous processes as described earlier, it is possible to assess static and dynamic posture using this technique. Whilst it is also possible to measure axial rotation using this technique, the potential limitations (Pearcy et al., 1987) and the lack of information provided by authors on the design and construction of the three-dimensional cluster could be a reason for scarce evidence to support the practical application of this technique. Nevertheless, consistent and reliable measurements of lumbar movement during gait have been shown within the same gait laboratory which employed the three-dimensional cluster technique (Schache et al., 2002; Needham et al., 2015; Taylor et al., 1996; Thurston, 1982).

Further methodological considerations and recommendations on the use of three-dimensional clusters are reported elsewhere (Needham et al., 2015, 2016). The three-dimensional cluster technique has been

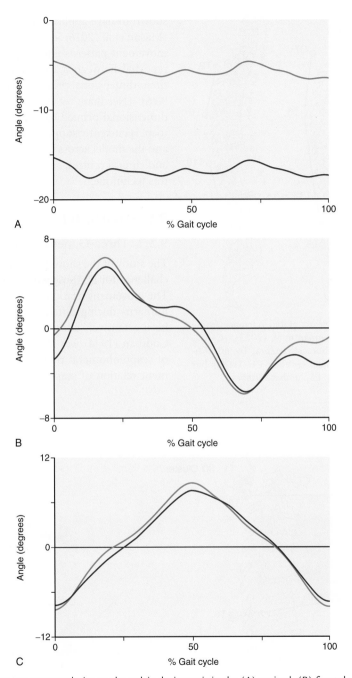

FIGURE 9.14 ■ Thorax movement relative to the pelvis during gait in the (A) sagittal, (B) frontal and (C) transverse plane

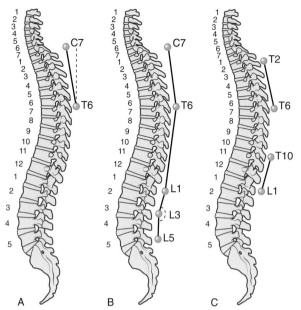

FIGURE 9.15 ▪ Segment angle to global reference frame (A) Frigo et al. (2003). (B) Heyrman et al. (2013). (C) C7, 7th cervical vertebra; T2, 2nd thoracic vertebra; T6, 6th thoracic vertebra; T10, 10th thoracic vertebra; L1, 1st lumbar vertebrae; L3, lumbar vertebrae; L5, lumbar vertebrae

shown to have application to track thorax movement (Mason et al., 2016; Seay et al., 2011). Recently, similar movement patterns and ranges of motion have been reported between a traditional thorax model and a three-dimensional printed cluster attached to T1 (Fig. 9.16) (Needham et al., 2016). The use of three-dimensional printed clusters and a standardized protocol is currently supporting the use of this marker set and the model across various gait analysis laboratories and clinical centres, enabling an external validation of this technique.

9.5 SHOULDER MODELLING

9.5.1 Three-Dimensional Shoulder Models

The study of the shoulder mechanism has been a great challenge for all those who have been interested in it. The movement relation of the shoulder girdle and the humerus during different upper extremity tasks has been extensively addressed. In the early 20th century, Codman (1934) was the first to propose the concept of scapulohumeral rhythm as the continuous movement relation of scapula on thorax and humerus on

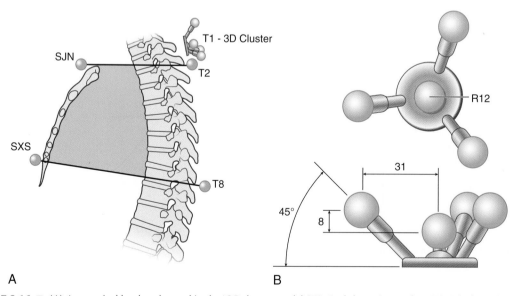

FIGURE 9.16 ▪ (A) Anatomical landmarks used in the IOR thorax model (T2, 2nd thoracic vertebra; T8, 8th thoracic vertebra; SJN, deepest point of incisura jugularis; SXS, xiphoid process) and 3D cluster on T1 (1st thoracic vertebra). (B) 3D cluster structural dimensions. Unless stated measurements are in millimetres

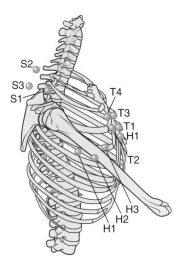

FIGURE 9.17 ■ An example of a shoulder flexion activity reconstructed in visual 3D with 3 clusters skin-mounted over the sternum, scapula and humerus.

scapula during the elevation of the arm. Ten years later, Inman and colleagues (1944) quantified this relation as a 2:1 ratio of arm elevation angle to glenohumeral elevation angle. This ratio has been undergoing changes that prove to vary among individuals, depending on the humerus plane of elevation, on the arm load magnitude, etc. (e.g. Bagg & Forrest, 1988; Hogfors et al., 1991; de Groot et al., 1999; Pascoal et al., 2000).

Mathematical models have been developed to further understand the complex mechanisms of the shoulder: two-dimensional models like those from DeLuca and Forrest (1973) and Poppen and Walker (1978) to determine the forces in the glenohumeral joint; three-dimensional models of the shoulder (e.g. Hogfors et al., 1991; Karlsson & Peterson, 1992; van der Helm, 1994a; Maurel & Thalmann, 1999; Charlton & Johnson, 2006; Dickerson et al., 2007) and complete upper extremity models (e.g. Garner & Pandy, 1999, 2001; Holzbaur et al., 2005).

In biomechanics research, as in clinical research, the importance of the scapulothoracic joint in upper extremity motion and, consequently, in the function and participation of individuals, is increasingly evident. The scientific body of knowledge of scapulothoracic altered kinematics in patients with shoulder dysfunctions is increasing (Ozaki, 1989; Warner et al., 1992; Deutsch et al., 1996; Paletta et al., 1997; Lukasiewicz

et al., 1999; Ludewig & Cook, 2000; Endo et al., 2001; Hebert et al., 2002; Vermeulen et al., 2002; Lin et al., 2005, 2006; von Eisenhart-Rothe et al., 2005; Mell et al., 2005; Illyés and Kiss, 2006; Laudner et al., 2006; Matias & Pascoal, 2006; McClure et al., 2006; Ogston & Ludewig, 2007; Rundquist, 2007; Fayad et al., 2008).

From two reviews (Ludewig & Reynolds, 2009; Struyf et al., 2011) of scapular kinematics in patients with shoulder dysfunctions, it is possible to conclude that, when compared to healthy control groups, patients with shoulder dysfunctions have a consistent pattern of decreased scapula lateral rotation, increased protraction and decreased posterior tilting. Some of the previously mentioned models have addressed the shoulder girdle. From an anatomical point of view, the scapulothoracic is not composed of articular structures, but, due to the surrounding soft tissues, it is usually assumed that the scapula is constrained to glide on the thorax surface (e.g. van der Helm, 1994a; Maurel & Thalmann, 1999; Garner & Pandy, 2001), in agreement with the motion description by Dvir and Berme (1978). This constraint has been represented by one or two fixed scapula points constrained to an ellipsoid gliding surface matched to the thorax dimensions, leading to a four or five degrees of freedom shoulder girdle, respectively. A more simplistic approach is determining the orientation of the scapula and clavicle bones by means of the regression equations based on the humeral elevation angle and/or elevation angle (e.g. Karlsson & Peterson, 1992; Holzbaur et al., 2005; Charlton & Johnson, 2006; Blana et al., 2008).

Recently, Bolsterlee and colleagues (2011) demonstrated that the kinematic optimization solution implemented in the Delft Shoulder and Elbow Model (van der Helm, 1994a) can lead to unrealistic positions, and can cause inaccuracies in model predictions, if one of the shoulder girdle close-chain constraints (i.e. the distance between scapula *trigonum spinae*, *angulus inferior* and the thorax surface) has to be fulfilled, especially in higher humeral elevation angles.

Neither the rigid constrained scapulothoracic nor the regression solutions seem to completely reflect the three-dimensional scapulothoracic movement. More recently, Seth and colleagues (2016) published a three-dimensional biomechanical shoulder model (Fig. 9.18) developed in OpenSim (Delp et al., 2007) that includes the thorax/spine, clavicle, scapula, humerus,

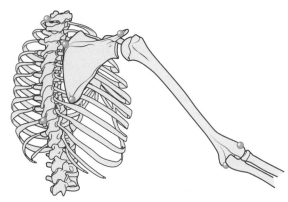

FIGURE 9.18 ■ An example of a shoulder model (Seth et al., 2016) with the International Society of Biomechanics recommended thorax, scapula and humerus anatomical landmarks (green spheres). The thorax anterior markers and the glenohumeral rotation centre are not shown

TABLE 9.1			
Summary of Three-Dimensional Shoulder Models, Their Number of Degrees of Freedom, Modelling Approach for Scapulothoracic Kinematics and Model's Accessibility for Others to Reuse			
Model	Degrees of Freedom	Scapula Kinematics	Accessible
van der Helm (1994)	7	Gliding plane	No
Garner & Pandy (1999)	10	Gliding plane	No
Maurel & Thalman (2000)	10	Gliding point	No
Holzbaur et al. (2005)	15	Regression	Yes
Dickerson et al. (2007)	8	Regression	No
Blana et al. (2008)	9	Gliding plane	No
Chadwick et al. (2014)	11	No	No
Saul et al. (2015)	15	Regression	Yes
Seth et al. (2016)	11	Gliding plane	Yes

radius, ulna, and joints that allow for 11 degrees of freedom relative between these segments. This model produced the three-dimensional kinematics of the scapula from surface markers to within the accuracy of bone-pin measurements. Inverse kinematic analyses run-to-run 12 000 trials showed the model's ability to reduce the effects of experimental errors and noise in marker locations when describing scapular Euler angles, reaching one-half to one-third of the variability of those angles calculated directly from noisy markers. See Table 9.1 for a summary of three dimensional models.

Normally, the glenohumeral is modelled as a three degrees of freedom ball-and-socket joint (e.g. Garner & Pandy, 1999); however this is an oversimplification of a complex six degrees of freedom joint. If dynamic stabilization is to be addressed in future research, then the three translations need to be considered in future models, as these will have a significant contribution to glenohumeral stability and the distribution of the forces in the muscles.

9.5.2 Shoulder Motion Reconstruction

Scapulothoracic and glenohumeral motion is difficult to measure objectively, due mostly to the scapula movement under the skin and muscles. Bone-pin kinematics has been used as the gold standard procedure and involves fixing cortical pins into the anatomical

segments. Although accurate, it is not a procedure that can be commonly adopted in the clinical or research practices to study shoulder movements. Skin-mounted (markers, cluster of markers or sensor) kinematics has been the most commonly used technique to study in-vivo scapula and humerus kinematics using electromagnetic or optoelectronic systems. This technique has been applied by covering the scapula with a grid of markers, or by placing a sensor, a device or a cluster of markers mounted over the skin. The former (grid of markers) uses soft-tissue deformation caused by scapular orientation to extrapolate the position of the scapula beneath the overlying soft tissue, while the coordinates of the latter (sensor or a cluster of markers) are then converted to anatomical axes calculated from digitized bony landmarks. Fig. 9.18 shows the use of 3 clusters of markers mounted: (a) on a rigid plate over the body of the sternum; (b) on rigid fixture over the flat upper surface of the acromion, and (c) on a rigid plate over lateral aspect of the humerus.

Skin deformation and displacement challenge the accuracy of the skin-mounted technique due to the marker or sensor movement with respect to the underlying scapula segment (the so-called soft-tissue artefacts), especially during large ranges of arm motion. Different methods have been developed with the aim of improving the accuracy when describing scapula kinematics: (a) single calibration at rest (CAST, see Section 9.1.4) and (b) the double or multiple calibration which combines calibrations at different arm elevations.

The Standardisation and Technology Committee of the ISB proposed joint coordinate system (JCS) standards for human joints in 2005 (Wu et al., 2005). These standards recommend that segments' LCSs should be derived from digitized bony landmarks. The thorax LCS is calculated using the following bony landmarks: spinous process of the seventh cervical vertebra, spinal process of the eighth thoracic vertebra, deepest point of the suprasternal notch, and the most caudal point on the sternum. The original recommendation suggested for scapula bony landmarks included the root of the scapular spine (*trigonum spinae*), the posterior acromioclavicular joint, and the inferior angle of the scapula (*angulus inferior*). More recently, a modification of the original recommendation was proposed aiming to substitute the acromioclavicular landmark with the posterolateral acromion (*angulus acromialis*) with the support reasoning that it would reduce the potential for singular positions (gimbal lock). Using the current standard, the same scapula motion is described with less internal rotation and upward rotation, and more posterior tilting than the original (Ludewig et al., 2010). For the glenohumeral, Euler sequence XZ′Y″ (angle of elevation, angle of horizontal adduction/abduction [or flexion/ extension] and axial rotation) seems a preferable choice when compared to the YX′Y″ Euler sequence, not only because it captures the glenohumeral motion in a more clinically meaningful way but also because it is less sensitive to discontinuities (due to gimbal locks) (Phadke et al., 2011). Recommended humerus bony landmarks include the glenohumeral rotation centre (estimated by regression or by computing the pivot point of instantaneous helical axes), the most caudal point on the lateral epicondyle, and on the medial epicondyle.

There has been significant development of Inertial Measurement Units (IMUs) in the last decade. These are showing great promise when the data are merged with biomechanical models. IMUs are now lightweight, small, wireless sensors and are not restricted to a certain measurement volume, which make them very practical both in clinical and laboratory scenarios. Despite some limitations of the IMUs, preliminary studies have demonstrated satisfactory results to their use in shoulder analysis (e.g. Cutti et al., 2008). However, further research is needed to address clinometric characteristics and clinical feasibility when studying patients with shoulder dysfunctions.

9.6 BIOMECHANICAL MODELS USING INERTIAL MEASUREMENT UNITS

9.6.1 Anatomical Calibration and Joint Angles Using IMUs

An area of considerable development in the last 10 years is the use of IMUs. IMUs usually include a combination of gyroscopes and accelerometers, but may also include magnetometers and sometimes a global positional system (GPS). These have been used in a number of ways, from getting estimates of spatial and temporal parameters of gait outside of the laboratory through to the estimation of three-dimensional joint angles.

A fundamental issue with using IMUs to assess human motion is that they require an LCS. Each IMU sensor will have a global orientation which is not aligned with any physiologically meaningful axis (Seel et al., 2014). This orientation can be represented by a quaternion, a rotation matrix, or Euler angles. A number of algorithms have been proposed for sensor orientation estimation. Typically, these use the integration of the angular velocity to estimate angular orientation; however, the signal can 'drift'. This can in part be reduced using acceleration and/or magnetometer measurements. although the presence of magnetic disturbances can affect the accuracy of the orientation estimates.

Through a fusion and integration of these different measurements, the IMUs are able to estimate the orientation (roll, pitch and yaw), the position, and the direction of movement of the sensor in a GCS. From this, the relative positioning of two or more sensors can be found. These can then be used to calculate the angle between sensors, which can be used to calculate joint angles in the different planes, providing an anatomical frame can be obtained (Fig. 9.19).

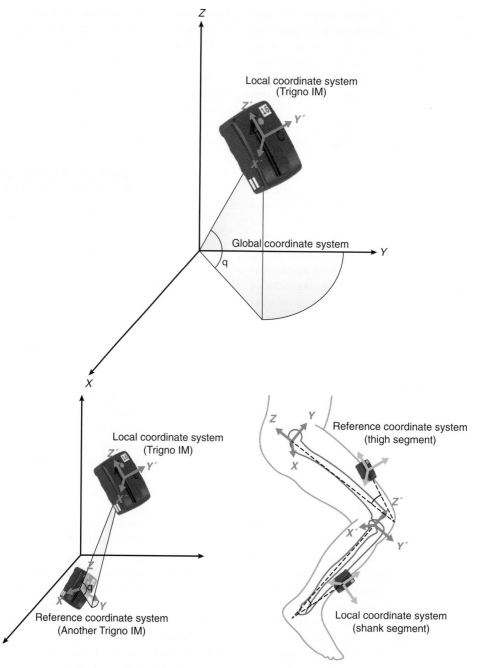

FIGURE 9.19 ■ IMU local coordinate system

All IMU systems require some kind of static anatomical calibration of the person if joint angles are being measured. This requires a reference static posture to create a 'neutral orientation' of the joints from which the three-dimensional joint angles can be projected into specific planes. This remains one of the major challenges in biomechanical models using IMUs, as patients may not be able to hold a neutral orientation. The identification of a JCS without any true positional data is far from trivial and is a major challenge in IMU-based joint angle measurement, in particular when joints move with three or more degrees of freedom.

The majority of papers have focused on sagittal plane joint angles and there is good agreement that the sagittal plane can be estimated with high levels of accuracy in comparison to camera-based systems. Those that have looked at the coronal and transverse plane have found differences due to the biomechanical models, as these can suffer from considerable crosstalk, which makes this data difficult to use. In saying this, the development of new IMU hardware and software is moving at a considerable pace and it may be possible in the near future to be able to measure body segment and joint rotations reliably in all three planes.

9.7 COORDINATE SYSTEMS AND JOINT ANGLES

Whichever system is used, one of the key measurements reported are joint angles. There are different ways in which we can define joints and segments. These vary from the simple method of using the GCS and simple trigonometry to more advanced methods. Rather than delve into the mathematical manipulation we will look at what the terms mean and what effect these can have on the data. The most commonly found methods include the laboratory coordinate system or GCS, the SCS, and the JCS.

A GCS is where the segment angles are calculated from the x, y, z axis of the laboratory or global frame. An SCS uses the proximal and distal endpoints of the segment to determine an orientation of the x, y, z axis of the joint. A JCS is where the axis of two body segments (proximal and distal) is used to create a third floating axis, or agreeable axis, to the proximal and distal segments. Both the SCS and JCS are ways of identifying an LCS at the joint, which is more meaningful anatomically than the GCS. A third way of considering three-dimensional movement is with helical analysis. This does have some advantages over SCS and JCS; however, it does not consider rotations about the anatomical frame axes.

9.7.1 Calculation of Joint Angles in the Global Coordinate System

The segment angles can be calculated in GCS by knowing the coordinates of the proximal and distal end of a body segment in a particular plane. The segment angles can then be found by using simple trigonometry. If we assume that two segments are rigid in all other planes than the one we are interested in (i.e. do not move in the coronal and transverse when we are looking at the sagittal plane), and that they are aligned perfectly with the plane we are interested in, then the calculation of the joint angles is a simple matter of subtracting the two angles. This is the method relied upon for all two-dimensional movement analysis and this has been shown to have considerable errors.

Work conducted by McClay and Manal (1998) suggested that two-dimensional data are adequate to indicate peak rearfoot eversion displacement and velocity if foot placement is within normal limits. However, when the foot was not aligned in the sagittal plane, i.e. at heel strike or at toe off when the foot was plantarflexing, the two-dimensional and three-dimensional data did not agree. This was also apparent in the transverse plane; if the foot placement angle was excessively abducted, it magnified any differences between the two-dimensional and three-dimensional data. Areblad and colleagues (1990) also quantified the magnitude of this error. They compared data when the foot was abducted from 10° to 30° from the axis of the camera, and found that for every 2° of change in the alignment angle, there was 1° of error introduced into the computed angle.

9.7.2 Errors between Global and Segment Coordinate Systems

The nature of these potential errors should be considered very carefully when interpreting any data

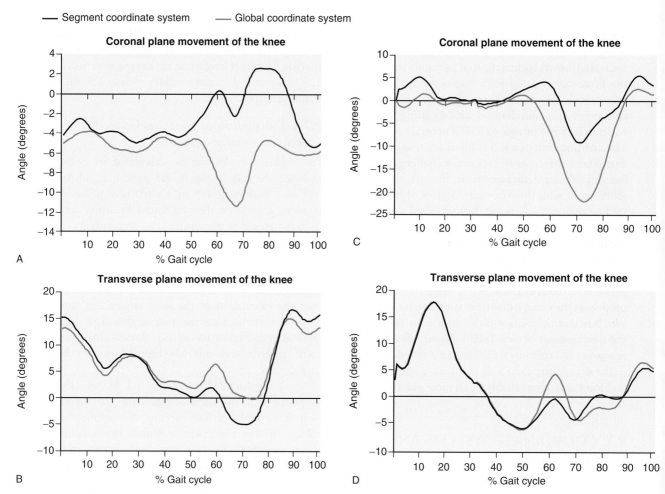

FIGURE 9.20 ■ Errors between global and segment coordinate systems: (A, B) normal walking and (C, D) in an individual with cerebral palsy

where the GCS has been used. The following examples show the errors involved first for normal walking, and then for an individual with CP who has significant internal rotation of the femoral segment (thigh). In introducing internal rotation of the thigh, the GCS gives an enormous error in the coronal plane, suggesting the knee reaches 22° abduction or valgus deformity, when in fact this is mostly due to cross-planar error from the knee flexion angle, and the true abduction or valgus angle is only 8°. Such a large error could easily lead to incorrect management of this individual.

The nature of these errors means that if we are interested in movement in the extra-sagittal planes (i.e. coronal and transverse planes), we have to think about whether what we are measuring has been forced into a false anatomical frame rather than a true anatomical frame (Fig. 9.20A–D).

9.7.3 Cardan Sequences and Their Effect on Gait Data

Girolamo Cardan (1501–1576) lectured and wrote on mathematics, medicine, astronomy, astrology, alchemy and physics. So, it is in some ways appropriate that his contributions should be considered in a book on clinical biomechanics. Cardan's fame rests on his work in mathematics, and especially in algebra. In 1545 he published his 'Ars Magna', which was the first Latin treatise devoted solely to algebra and contained the solution of the cubic equation.

The Cardan sequence itself is a method where a series of three rotations, one about each of the coordinate axes, is calculated that would place the joint in the same final orientation as the true movement. In other words, they describe one LCS or SCS relative to another.

The Cardan sequences are characterized by the rotations about all three axes: xyz, xzy, yzx, yxz, zxy, zyx. Anatomical meaning has been given to one of the Cardan sequences, which is referred to as the JCS, by Grood and Suntay (1983). Assuming an SCS with z – up, y – anterior and x – lateral, the xyz sequence is the JCS; however, if a different definition of x,y,z directions is used, the Cardan sequence for the JCS will change.

If we consider the Cardan sequence that assumes that the x axis is in the medial–lateral direction, then the y axis is anterior–posterior (or the direction of travel), and the z axis is in the up and down or axial direction.

Therefore, we can describe the JCS as x, y and z as:

x = flexion–extension,
y = abduction–adduction,
z = longitudinal internal–external rotation

The JCS proposed by Grood and Suntay, for instance, usually relates to the Cardan sequence xyz, or flexion–extension, abduction–adduction, axial rotation. The JCS does have the advantage of relating to the commonly used clinical terms for lower-limb joint motion; however, the drawback is that an orthogonal coordinate system is not guaranteed, i.e. the coordinate systems of the different segments may not be correctly aligned, which can lead to crosstalk between the different planes of movement.

So, what effect does choosing different Cardan sequences actually have on joint kinematics? In the sagittal plane, there is little or no effect in picking different sequences and the best would be xyz, where x is the flexion–extension axis for the reference segment,

which would also be the default for most systems. This is due to the relatively large amount of movement that occurs in the sagittal plane. The following graphs show the knee flexion–extension patterns for a normal individual and an individual with CP (Fig. 9.21A and B).

However, when we look in the coronal and transverse planes where far less movement occurs, we should be far more careful (Fig. 9.22A–D). Much of the published research does not state which Cardan sequence is used; therefore, we must assume that the 'default' xyz has been used.

The black lines for all graphs show the Cardan sequence xyz (flexion–extension, abduction–adduction, internal–external rotation) and xzy (flexion–extension, internal–external rotation, abduction–adduction). It is very interesting to note that both these sequences produce very similar results. It is also interesting to note that the peaks occur at peak knee flexion, which would imply that crosstalk between the planes is occurring. When a flexion–extension is placed second in the order of rotations yxz and zxy, two very different patterns are produced, which seem to increase the crosstalk between planes. When x is placed last in the order of rotations (yzx, zyx), then x or the flexion–extension has least effect and the patterns for the coronal and transverse planes look independent to the movement in the sagittal plane (i.e. no planar crosstalk) and show near identical results. The reason for the difference is an effect due to the dependence in the sequence order, which is orientated to the movement in the anatomical plane being considered.

We now think about the implications of this to clinical research. Much has been talked about regarding the three-dimensional movement of the foot, ankle and knee joints, and the effect of orthoses. It is quite likely that all, or most, of the results found with research to date are true effects. However, when we are trying to determine very small changes in the coronal and transverse planes that may still be clinically significant, we need to be clear in our reporting to allow comparison and replication of appropriate data analysis.

Schache and colleagues (2001) studied the effect of the different Cardan angle sequences on the three-dimensional lumbo-pelvic angular kinematics during running. They concluded that different Cardan angle

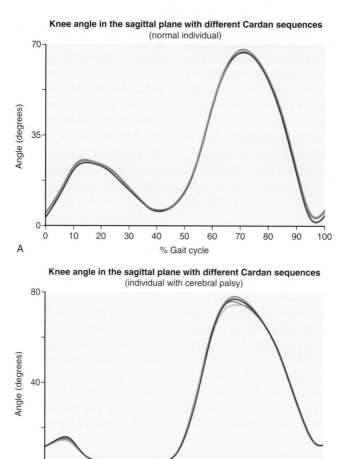

FIGURE 9.21 ■ (A) Knee motion with the different Cardan sequences for the same gait trial for normal walking and (B) knee motion with the different Cardan sequences for the same gait trial for an individual with cerebral palsy

sequences were not found to substantially effect typical three-dimensional lumbo-pelvic angular kinematic patterns during running. However, Nguyen and Baker (2004) found clinically significant differences between the different sequences of rotation for subjects with pathological thoracic motion, and concluded that the conventional sequence (flexion, lateral bending, axial rotation) is preferable for the thorax.

Clearly the issue of the order of Cardan sequence is coming to the fore. One suggestion for a possible way forward is to look at the 'principal axis' where a Cardan sequence is picked based on the plane being

considered; for example, to analyse the transverse plane (z) the Cardan sequence will be zyx, and to analyse the coronal plane (y) the sequence will be yzx. In both these cases I have placed the sagittal plane (x) last to reduce its weighting to avoid possible crosstalk, as the knee flexion contained by far the most significant movement during walking. However, if we are dealing with a joint where this is not the case, different Cardan sequences may be required.

These are far from being recommendations, as there are currently very few published papers on the effect of Cardan sequences, and the best mathematical conventions have yet to be universally agreed for all

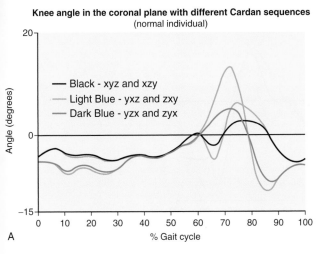

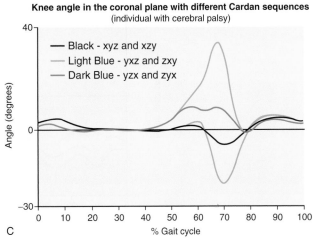

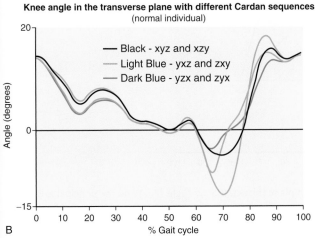

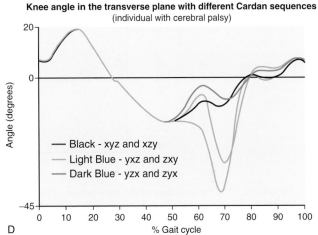

FIGURE 9.22 ■ Knee motion in the coronal and transverse planes when using the different Cardan sequences. (A, B) Normal walking and (C, D) in an individual with cerebral palsy

joints. However, it is clear that the different Cardan sequences do yield significantly different data.

9.7.4 Helical Angles

Helical angles are often included in movement analysis software. A helical angle, finite helical axes, or screw axis, can be described as a three-dimensional angle of one segment to another in terms of a rotation about and translation along a single axis; this was first suggested by Woltring et al. (1985). This does have some advantages over SCS and JCS as it allows for a combined effect of all three planes and, therefore, does not suffer from planar crosstalk. However, helical angles do not consider rotations about the anatomical frame, so the clinical interpretation of the results is difficult. One further development is that helical angles may be projected into the different anatomical frames. In Fig. 9.23A, B we consider the effect of using the helical angle of the knee projected into the coronal plane (black) compared with the JCS xyz (light blue)

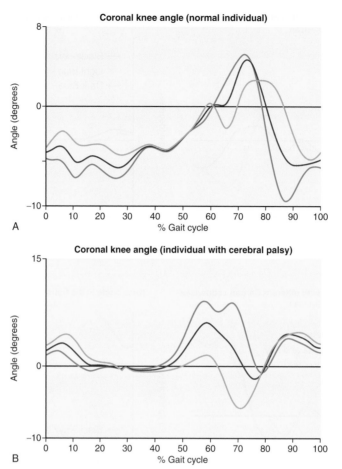

FIGURE 9.23 ■ Helical angle of the knee projected into the coronal plane (black), JCS xyz (light blue) and the principal axis yzx (dark blue) in normal (A) and cerebral palsy (B) gait

and the principal axis yzx (dark blue) in normal and cerebral palsy gait.

9.7.5 Recommendations

In the literature, there are many reported techniques: most notable are the original JCS proposed by Grood and Suntay (1983), finite helical axes (Woltring et al., 1985) and, more recently, a comparison of different techniques by Cappozzo and colleagues (2005).

Although it may appear a daunting task to pick the correct mathematical technique, it should be noted that the majority of work, unless stated, has used the JCS, which is equivalent to the Cardan sequence (xyz), and is recommended by the ISB for the lower limb

(Wu et al., 2002). Therefore, to enable comparison to previously published data, the JCS (xyz) should be used unless there is a VERY good reason not to. And whichever sequence or method is used, make sure it is reported clearly to allow comparison of data, as without comparison of data we have no way of sharing knowledge.

One other recommendation comes from the International Shoulder Group who state that the shoulder angle (upper arm relative to torso) should be described by Euler sequences (zyz, zxz, yxy, yzy). *The ISB recommendation on definitions of joint coordinate systems of various joints for the reporting of human joint motion – Part II: shoulder, elbow, wrist and hand* (Wu et al., 2005) considered the use of different sequences for the different relative movements between segments. This produced the recommendation of the sequence yxy for the movement between the humerus and the scapula, and between the humerus and the thorax, which are thought to be more appropriate for shoulder movement than the 'traditional' Cardan sequences. However, this does not seem to work for all movements during different tasks and, therefore, great care must be taken to ensure that the sequence gives clinically and anatomically meaningful data.

SUMMARY: ANATOMICAL MODELS AND MARKER SETS

- There are many anatomical models and marker sets reported in the literature. From these it is possible to plot an increase in complexity over time.
- The increase in complexity in the models relates to the ability of movement analysis systems to track more and more markers, but also the increase in the knowledge of modelling human movement.
- Simple marker sets or models have the advantage of requiring only a few cameras, whereas the more advanced models often require more than eight cameras to track the markers. However, simple models are not able to look at the more complex movements between joints and the different planes of the body.
- With more advanced models, we are now able to quantify foot movement by considering the foot in multiple segments. This will lead to more clinically relevant research to find the nature of the interactions between the foot segments and the efficacy of foot orthotic management.
- The three-dimensional cluster technique has been shown to have application to track segments of the spine in six degrees of freedom which will further our understanding of multiplanar movements of the shoulder and spine.
- The use of IMU sensors is now commonplace within biomechanics research, and clearly these have significant potential to be used in clinical and sports practice environments, especially when combined with EMG data.
- With an increase in the complexity also come questions on how we define the orientation of body segments and joints. The use of simple joint geometry can lead to misinterpretation of data, especially when the movement of the joints occurs in multiple planes of body simultaneously. Therefore, great care must be taken when describing and reporting clinical data.

10

ELECTROMYOGRAPHY

PAOLA CONTESSA ■ CARLO J. DE LUCA ■ SERGE H. ROY ■ JIM RICHARDS

This chapter covers the methods commonly used to assess muscle activity and function. This includes the techniques used to record electromyographic (EMG) signals, the best practices for obtaining good-quality EMG signals, and methods for the analysis and interpretation of the EMG recording.

AIM

To learn the bases of EMG: its use, recording and processing techniques; the methods of assessing muscle activity and the information that can be drawn from their use.

OBJECTIVES

- To describe the nature of EMG signals
- To explain the different methods of detecting EMG
- To describe the factors that influence the EMG recording
- To present the best practices to obtain good-quality EMG signals
- To describe different methods of processing the EMG signal
- To describe the process of decomposition of the EMG signal and the information that can be obtained.

10.1 BACKGROUND TO ELECTROMYOGRAPHY

10.1.1 What is the Link between Electricity and Muscle Activation?

Electromyography is the study of the electrical signal generated by contracting skeletal muscles. The early work on the relationship between muscle contraction and electricity was conducted in the 1600–1700s.

Before the 17th century, it was believed that movements were produced by 'moving spirits' that travelled from the brain down hollow nerves to the muscles and that caused the muscles to increase in volume during a contraction. In contrast to this belief, in 1664 Jan Swammerdam (1637–1680) demonstrated that stroking the innervating nerve of the frog's gastrocnemius muscle generated a contraction even on an isolated muscle-nerve preparation when the connection to the brain was cut. The first intuition of the relationship between muscles and electricity is documented in the work of Francesco Redi (1626–1697) in 1666. Redi showed that the shock of the electric ray fish was muscular in origin, highlighting the relationship between voluntary contractions and the production of an electrical signal. A decisive breakthrough came in 1791, when Luigi Galvani (1737–1798), who is credited as the father of neurophysiology, showed that electrical stimulation with metal rods of muscular tissue in frog's legs produced a contraction which generated a force. He reasoned that the factor responsible for movement was probably of electrical origin. Carlo Matteucci (1811–1868) definitively proved that electrical currents originated in muscles by showing that injured excitable biological tissues generated direct electrical currents, and that they could be summed up by connecting elements in series, as in an electric pile.

208

In 1849, DuBois-Reymond (1818–1896) was the first to report the detection of voluntarily elicited electrical signals from human muscles. He constructed an apparatus that consisted of a surface electrode (a wire attached to a blotting paper immersed in a jar of saline solution) that caused a small deflection on the galvanometer when the fingers immersed in the saline solution were contracted. He understood that the impedance of the skin reduced the current that drove the galvanometer, and that a greater deflection could be produced by induced blisters so that the open wounds were in direct contact with the saline solution. However, methods to detect the electromyographic signal from a human improved slowly in the 19th century, when electricity started to be used for clinical applications. For instance, Guillaume Duchenne (1806–1875) applied electric stimulation to intact skeletal muscles to investigate muscular function in humans. Duchenne was one of the first to investigate electricity for therapeutic purposes, and is considered to be the father of electrotherapy.

In 1922, Herbert Gasser (1888–1963) and Joseph Erlanger (1874–1965) used a cathode ray oscilloscope in place of the galvanometer to investigate conduction along the nerve axons that allowed them to observe action potentials. Great impact in the field was made by Edgard Adrian (1889–1977) and Detlev Bronk (1897–1975) in 1929, when they introduced the first needle electrode to record the electrical signal generated during a muscle contraction. It was a groundbreaking innovation with immediate impact on the clinical community. Since then, a series of technological advances have fostered the investigation of muscle function through the analysis of the electrical activity associated with muscle contraction.

So how are muscle contraction and electricity related?

10.1.2 Muscles, Motor Units and EMG

Skeletal muscles are composed of thousands of muscle fibres that contract when they are activated by nerve pulses. These pulses are generated in motoneurons, whose cell bodies are located in the anterior horn of the spinal cord. Nerve pulses travel down the nerve axons of the motoneuron and come in contact with the muscle at the neuromuscular junction. Each motoneuron innervates a number of muscle fibres, which contract almost simultaneously when they are excited by a nerve pulse. Thus, a single motoneuron and all the fibres innervated by this motoneuron constitute the functional unit of muscles, called motor unit (MU).

The muscle fibre membrane has a resting potential of approximately −70 mV. A nerve pulse at the innervation point (i.e. at the neuromuscular junction), temporarily alters this resting potential to +40 mV. The change in membrane potential, referred to as depolarization, propagates in both directions along the muscle fibres from the innervation point. The depolarization voltage travelling down the muscle fibres of a motor unit can be detected with an electrode placed near the muscle fibres or on the skin above the contracting fibres. The characteristic shape of the depolarization voltage is called motor unit action potential (MUAP).

In order to sustain a muscle contraction, motor units must be repeatedly activated. The resulting sequences of MUAPs are called motor unit action potential trains (MUAPTs).

The EMG signal is the electrical signal associated with a muscle contraction that can be detected with a sensor positioned inside a muscle or on the skin on top of a muscle. It represents the summation of the electrical activity from a number of MUAPTs that are simultaneously activated during a muscle contraction. The EMG signal ranges between a few μV to several mV depending on the level of muscle activity (Fig. 10.1).

The activation of a motor unit transforms the nerve signals into force and movement so that the electrical manifestation of a MUAP is accompanied by a contraction of the innervated muscle fibres. The force generated by the contraction of a single motor unit is called force twitch. For sustained contractions, force twitches superimpose if they occur in close enough succession, producing a tetanic, or fused, force. The force produced by a contracting muscle is composed of the individual forces generated by all active motor units within the muscle.

10.1.3 Muscles and Fibre Types

The number of fibres comprising the individual motor units varies widely within a muscle and between muscles. As a general rule, smaller muscles that control fine movements, such as the muscles around the eyes,

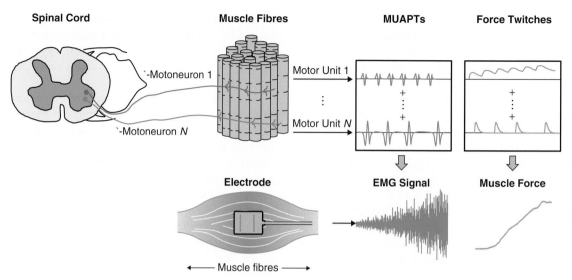

FIGURE 10.1 ■ Schematic of motor units, the generation of the electromyographic (EMG) signal and of muscle force. MUAPT, Motor unit action potential train

are composed of smaller motor units containing a smaller number of muscle fibres. In contrast, muscles that are used to generate strong forces and act on large body masses, such as limb muscles, are composed of larger motor units containing a greater number of fibres per motor unit.

Muscle fibre types can be characterized as slow twitch (type I) or 'slow twitch oxidative' and fast twitch (type II). Type I fibres are characterized by a slower speed of force production than type II fibres and they produce smaller amplitude and longer duration force twitches. Fast twitch fibres can be further categorized into type IIA and type IIB fibres, which are also known as 'fast twitch oxidative' and 'fast twitch glycolytic', respectively. Type IIB fibres are characterized by a higher force and speed production but are fast to fatigue, whereas type IIA fibres are more resistant to fatigue.

Type I fibres have a reddish appearance due to the rich vascularization and blood supply providing oxygen, allowing them to be more resistant to fatigue. Type IIA have larger-diameter muscle fibres and have a paler appearance due to lower vascularization and blood supply. The limited amount of energy stored in their mitochondria renders them fast fatiguing.

Although muscle fibres are classified into distinct groups (type I, IIA and IIB), each human muscle generally contains motor units of varying size and contractile properties that create a continuum among these groups. The proportion of slow to fast fibres in skeletal muscles varies depending on the type of muscle usage.

10.1.4 Motor Unit Recruitment, Firing Rate and Force Modulation

Within each muscle, motor units are hierarchically activated in order of increasing size, a property known as the Size Principle (Henneman, 1957). Smaller motor units, characterized by smaller-amplitude and longer-duration force twitches, are recruited at lower contraction forces, i.e. they have a lower recruitment threshold. Larger motor units, characterized by larger-amplitude and shorter-duration force twitches, are recruited at higher contraction forces, i.e. they have a higher recruitment threshold.

When a motor unit is activated during a sustained contraction, it fires (contracts) repeatedly. The series of firing times can be represented with a train of bars over time, called the firing train, where each bar represents one firing or MUAP of the motor unit (Fig. 10.2). The number of pulses that occur over one second is a measure of the motor unit mean firing rate. The mean firing rate is a time-varying function calculated over a unit-area window that slides through the

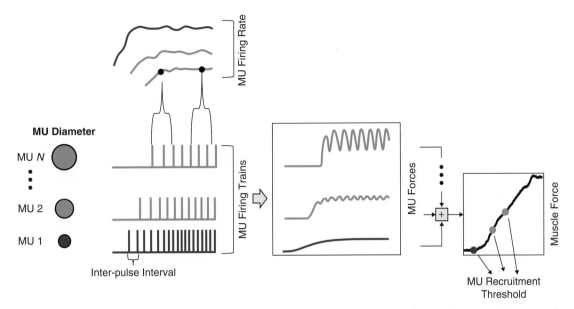

FIGURE 10.2 ▪ Hierarchical activation of motor units and schematic of the motor unit firing trains, mean firing rates, forces and muscle force

motor unit firing train. The time interval between consecutive firings of a motor unit is called the motor unit inter-pulse interval (IPI) (see Fig. 10.2).

The force produced by a muscle is modulated through two main processes: (1) by varying the number of active motor units, and (2) by varying their firing rate. When greater amount of contraction force is needed, a greater number of motor units are activated, which are progressively larger (higher threshold) and generate larger forces. The firing rates of the active motor units also increase with increasing force and, consequently, their average motor unit IPIs decrease.

These two phenomena translate in an increasing amplitude of the recorded EMG signal during increasing force contractions (see Fig. 10.1), due to the increasing number of active motor units and to the increase in their firing rate. In addition, note that the higher-threshold larger motor units are generally characterized by greater amplitude MUAPs in virtue of the greater number of muscle fibres per motor unit.

Over the past five decades, there has been a common acceptance of the notion that higher-threshold motoneurons are characterized by greater firing rates than lower-threshold ones (Eccles et al., 1958; Kernell, 1965, 2003). This notion stems from the observation that,

when the nerves of anaesthetized cats are electrically stimulated, the larger-diameter (higher-threshold) motoneurons exhibit a shorter after-hyperpolarization and greater firing rates than the smaller-diameter (lower-threshold) ones. This firing rate organization would match the motor unit contractile properties: lower-threshold motor units have wider and smaller amplitude force twitches than higher-threshold ones and require lower firing rates to produce tetanic force. This arrangement would conceivably 'optimize' the force-generating capacity of the muscle because motor units would fire at a rate that produces the greatest amount of force.

However, we now know that, in humans performing voluntary contractions, motor unit firing rates are organized in an opposite manner: lower-threshold motor units have greater firing rates than higher-threshold ones. This firing rate arrangement is known as the Onion-Skin scheme (De Luca & Erim, 1994) and it demonstrates that the physiological firing behaviour of motor units when they are *voluntarily* activated in humans is dramatically different than that observed when isolated motoneurons are *electrically stimulated* in animals. One ostensive shortcoming of the Onion-Skin scheme is that the motor unit firing rates do not match the motor unit contractile properties, so that motor units do not always produce fused force and they do not always take full advantage of their force-generating capacity. Why is that? One reason is that the high-threshold motor units are fast-fatiguing and would not be able to sustain force production for a long period if they were to fire fast. Thus, the Onion-Skin scheme is not designed to 'optimize' muscle force. Instead it provides a lower maximal force with the capacity to sustain it over longer time (De Luca & Erim, 1994, Contessa & De Luca, 2013).

10.2 METHODS OF DETECTING THE EMG SIGNAL

There are two main methods of detecting the EMG signal: surface EMG, which involves placing the electrodes on the skin overlying the muscle, and intramuscular EMG recording, which involves inserting an electrode into the muscle itself.

These methods can be used in two main recording configurations: the monopolar and bipolar (or single differential) configurations (Fig. 10.3A, B). The monopolar configuration uses only one detection surface, or electrode, and records the electrical potential in this area with respect to a reference electrode. The drawback of this configuration is that it does not remove the noise or unwanted electrical signals originating from sources other than the muscle being investigated. For this reason, it is rarely used except when specifically required, such as when the electrical potential at a particular point or area is needed. It must be used with great care and constant monitoring to ensure adequate signal quality. The bipolar, or single differential arrangement, overcomes this problem by recording the differential signal detected from two closely placed electrodes with respect to a reference electrode. The differentiation (or subtraction) removes the signal that is common to both electrodes. This configuration has the advantage of removing or minimizing external noise or unwanted signals from sources other than the muscle because these disturbances are detected as a common signal from both electrodes and are eliminated.

Note that electrode refers to the area detecting the electrical signal. The recording object is indicated as sensor. For instance, a sensor with a bipolar configuration is composed of two electrodes.

A reference electrode is needed in both configurations. The purpose of this electrode is to give an electrical reference of the surrounding electrical activity, which can include electrical noise from equipment or internal electrical activity such as the beating heart. Historically, this reference electrode was placed on the body in an area that was either electrically quiet (typically above a bony area) or that contained electrical signals unrelated to those being detected. More recently, the higher-quality electronics and design used in some systems allow the reference electrode to be placed anywhere on the skin.

A third configuration, the double differential arrangement, uses three recording electrodes and the electrical signals detected by the three electrodes go through two levels of differentiation (Fig. 10.3C). This configuration is used for specific applications and will be described in Section 10.3.7: Crosstalk.

10.2.1 Intramuscular EMG Recording Technique

In previous years, intramuscular EMG was the most common EMG detection technique. Recently, non-invasive surface EMG techniques have become the preferred recording method. However, the use of intramuscular EMG is still required to investigate deep muscles whose signal cannot be reliably detected from the surface of the skin. In intramuscular EMG recording, either needle or fine-wire sensors can be used.

Needle sensors (Figs 10.4 and 10.5) are inserted through the skin and fat layers and into the muscle beneath. The needle is thin (typically 0.3–0.5 mm) and can be made from a variety of materials, most commonly Teflon-coated surgical-grade stainless steel. One or two detection areas on the needle pick up the electrical signal in the monopolar and bipolar configurations, respectively. Concentric needle electrodes are commonly used: they differ from standard needles in that they contain a platinum-iridium wire at the centre of the cannula. The tip of the wire acts as the detection surface in the monopolar configuration and picks up the electrical signal between the tip of the wire and the cannula. The bipolar configuration contains a second wire in the cannula that provides a second detection surface.

Fine-wire sensors (Figs 10.6 and 10.7) were popularized in the early 1960s (Basmajian & De Luca, 1985).

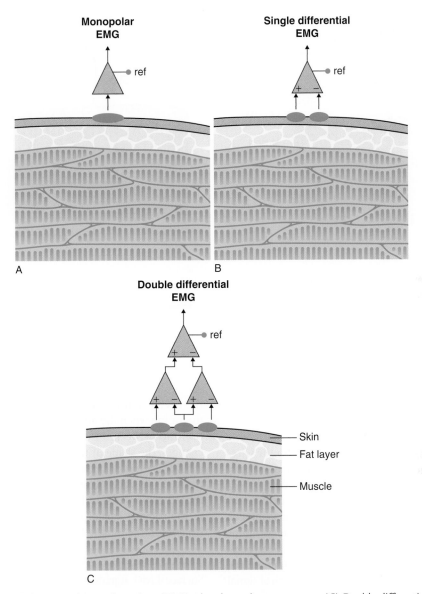

FIGURE 10.3 ■ (A) Monopolar configuration. (B) Bipolar electrode arrangement. (C) Double differential arrangement

They are also inserted through the skin and fat layers into the muscle. However, they are implanted using a hypodermic needle, usually of diameter between 0.4 and 0.5 mm, and the needle is then withdrawn leaving the electrode positioned in the muscle. Fine-wire electrodes can be made from a number of materials including bi-filament Teflon-coated silver wire, nickel chromium alloy, or surgical-grade stainless steel wire.

The electrode wires are often barbed at an angle to ensure that they remain fixed in position after the hypodermic needle is removed. One or two wires are inserted in the muscle in the monopolar and bipolar configuration, respectively.

Fine-wire sensors are less painful than needle sensors because the wires are hardly felt by the subject once the cannula is removed. Thus, they can be more

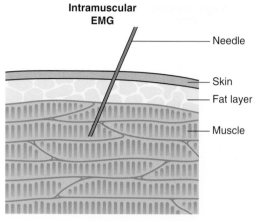

FIGURE 10.4 ■ Schematic of a needle EMG sensor

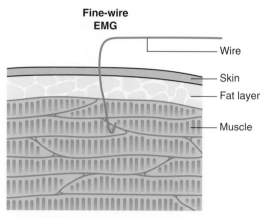

FIGURE 10.6 ■ Schematic of a fine-wire EMG sensor

FIGURE 10.5 ■ Needle Electrodes, Nicolet®

FIGURE 10.7 ■ Fine-wire EMG sensor

easily used during strong contractions or dynamic movements. However, the risk of damage to the wires makes any readjustments in their position after insertion more difficult than for needle electrode.

Intramuscular EMG sensors are generally very selective: the small size of the electrodes' pick-up area detects the EMG signal generated by only a few motor units located in close proximity to the electrodes. Because of their close proximity to the muscle fibres, the distance between the source of the electrical signal (muscle fibres) and the detection area (needle/fine-wire) is relatively short, ranging from less than 0.5 mm to 3 mm. As a result, the frequency spectrum of the intramuscular EMG signal contains more energy at higher frequencies than that of the surface EMG signal, which is low-pass filtered when travelling from the muscle fibres to the surface electrodes.

It is important to remember that the insertion of the needle damages the muscle fibres along its track; hence the observed intramuscular EMG signal may contain consequences of the damage. In addition, both the needle and the wire electrodes may be subject to migration when the muscle fibres contract. As the relationship between the electrodes and the muscle fibres changes, the shape of the MUAPs change, resulting in a modification of the amplitude and frequency spectrum of the signal. This alteration of the signal is not related to physiological factors within the muscle but to the varying spatial relationship between the electrodes and the muscle fibres.

10.2.2 Surface EMG Recording Technique

Surface EMG signals can be detected with electrodes of various shapes and sizes positioned on the skin above the muscle of interest.

The monopolar arrangement (Fig. 10.3A) is the most basic configuration. It requires one electrode to be placed over the muscle that detects the surface EMG signal with respect to a second reference electrode. This arrangement is sometimes used in clinical EMG biofeedback devices and is subject to a variety of contaminations from ambient noise and other sources. The bipolar or single differential arrangement (Fig. 10.3B) uses two electrodes, generally placed 1 to 2 cm apart along the orientation of the muscle fibres, to measure the EMG signal with respect to a third reference electrode.

Surface electrodes can be either passive or active. In the passive configuration, the electrode consists of a detection surface that senses the current on the skin through the skin–electrode interface. The signal is then transmitted to the amplifier for conditioning. In the active configuration, the amplifier stage is at the sensor site, decreasing the path length between detection area and signal conditioning. This characteristic greatly increases the input impedance of the electrodes, reducing noise and artefacts and ensuring high-quality signal recording.

Electrodes are either of solid reusable construction with pure silver bar electrodes (Fig. 10.8A, B) or disposable silver/silver chloride (Ag/AgCl) electrodes (Fig. 10.9A, B).

The shape and size of the electrodes and their inter-electrode distance determine the area of muscle tissue being examined and, therefore, the number of motor units that may be detected within this area. The greater the area and the inter-electrode distance, the greater the electrode pick-up area and the number of detected motor units. However, if the electrodes are too large or if they are placed too far apart, they are more susceptible to detecting signals from surrounding muscles in addition to the target muscle. Interference from electrical signals originated by nearby muscles are referred to as crosstalk signals and can mislead the identification of the activity of the target muscle.

Surface EMG electrodes are exclusively used to pick up electrical signals from superficial muscles located close to the skin. Signals from deeper muscles cannot be recorded reliably. Even for superficial muscles, a thick layer of fatty tissue between the electrode and the muscle acts as an insulator and can affect both the magnitude and frequency response of the EMG signal recorded. In these cases, there are no options other than using intramuscular EMG recording techniques.

Grids of surface electrodes are used when the spatial properties, rather than temporal properties, of a muscle are under investigation. These sensors are commonly referred to as high-density EMG sensors (Fig. 10.10) and contain small electrodes generally spaced between 4 and 10 mm apart. The concept of electrode arrays was first introduced in the late 1970s and 1980s

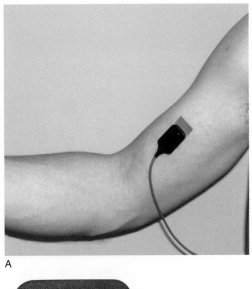

A

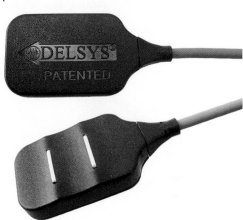

B

FIGURE 10.8 ■ (A and B) Fixed bar reusable single differential electrodes by Delsys Inc.

(Nishizono et al., 1979; Monster & Chan, 1980; Hilfiker & Meyer, 1984; Broman et al., 1985; Blok et al., 1999; and others) and was popularized by Stegemann and Zwarts in the late 1990s. This type of sensor can provide either monopolar or bipolar signals and carry information on the spatial properties of a muscle and the spatial arrangement and direction of MUAPs within the muscle. This spatial information is used, for instance, for conduction velocity studies, or in localization of the innervation zones.

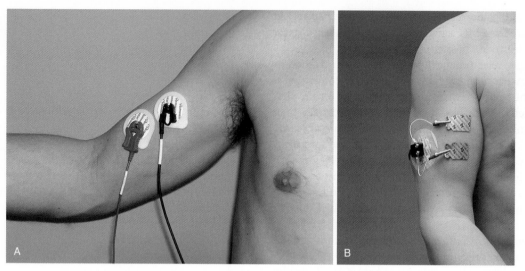

FIGURE 10.9 ■ (A and B) Ag/AgCl disposable electrodes

FIGURE 10.10 ■ High-density EMG sensor by TMSi

10.2.3 EMG Systems

There are a variety of EMG systems available that can be categorized as wireless (Fig. 10.11A, B and C) or tethered (Fig. 10.12). Wireless systems require no wires connecting the subject to the data collection/processing unit. Some of them (Fig. 10.11B) require the subject to wear a small transmitter, which sends signals to the receiver. Others (Fig. 10.11A) are fully wireless and each sensor transmits the EMG signal to a separate receiver. The benefit of a fully wireless device is that the subject is free to move with little or no restriction. The wireless EMG system transmission needs to be reliable to avoid loss of packets of data and to minimize interference from radio frequencies, Bluetooth, microwaves, etc. The wired systems require a physical connection, usually in the form of a cable that connects each EMG sensor to a separate amplifier unit.

The drawbacks of the wired systems are that they restrict the subjects' movements and that the cables need to be well shielded to avoid introduction of electrical noise in the recording.

EMG signals are low voltage and need to be amplified, with values for the amplification typically between 1000 and 10 000 times. Amplifiers are often positioned close to or as part of the electrodes, in the case of active electrodes, reducing the length of the connecting wire that can pick up and introduce electrical interference. Most EMG systems are capable of measuring EMG signals at sampling frequencies of 1000 Hz or higher. This is the minimum sampling frequency required to correctly sample the EMG signal and analyse its frequency content, which is negligible above 450 Hz. For intramuscular EMG recordings, which have higher frequency content, a sampling rate of 2000 Hz is recommended. When individual motor unit action potentials are being investigated, up to 10 000 Hz or 20 000 Hz (10 kHz or 20 kHz) is generally needed.

For biomechanical applications, or any application involving dynamic contractions, the bandwidth provided by a surface EMG system should ideally be between 20 and 450 Hz. Surface EMG signals are known to have no or little energy at frequencies greater than 450 Hz. At frequencies lower than 20 Hz, the disturbances from noise and motion artefacts are prevalent and complicate the interpretation of the EMG

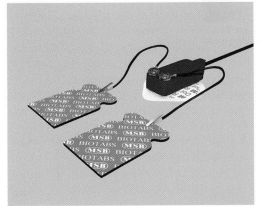

FIGURE 10.11 ▪ (A) Wireless EMG Trigno System by Delsys Inc. (B and C) MIE Telemetry EMG System: Preamplifier, Transmitter and MT8

intramuscular EMG signals, which present considerable energy up to 1000 Hz.

10.3 WHICH FACTORS AFFECT THE QUALITY OF THE EMG SIGNAL?

The signal-to-noise ratio (SNR) is the best generally accepted measure of the quality of the EMG signal. It indicates the ratio of the wanted EMG signal versus the unwanted signal (baseline noise, line noise, movement artefacts) (Fig. 10.13).

The amplitude of the wanted EMG signal depends on the contraction strength, but also on the location of the sensor on the muscle of interest, its orientation with respect to the muscle fibres, and the sensor characteristics, such as the size of the electrodes and the inter-electrode distance. The amplitude of the unwanted signal depends mainly on the electronics and the quality of the skin–electrode contact. To achieve the best SNR, it is also important to minimize disturbances from motion artefacts, physiological sources, external noise sources and unwanted EMG signals from muscles different than the one under investigation (crosstalk signals).

10.3.1 Electrode Position

The location of the sensor on the muscle is the single most important factor for obtaining good SNR. This advice was first provided by Basmajian and De Luca in 1985 in their book *Muscles Alive* and has been confirmed by numerous studies since. The muscle, fat and skin tissues present an internal impedance to the propagation of electric currents and act as a low-pass filter on the detected EMG signal. The filtering characteristics of these tissues are a function of the distance between the active muscle fibres and the detection surfaces of the electrodes. Thus, the location of the sensor on the muscle renders dramatically different surface EMG signal characteristics. For instance, locating the sensor in the proximity of the tendon origin or the innervation zones (Fig. 10.14) yields lower amplitude signals. The fibres in the middle of the muscle belly generally have a greater diameter than those at the edges of the muscle or near the origin of the tendons. Because the amplitude of action potential from the muscle fibres is proportional to the diameter of the fibre, the amplitude of the EMG signal will be

data. This low-frequency cut-off is particularly relevant for biomechanists who work with dynamic contractions where movement artefacts contribute energy components below the 20 Hz range. For more detailed information on this point refer to De Luca and colleagues (2011). The bandwidth should be greater for

FIGURE 10.12 ▪ Delsys Bagnoli Wired EMG System

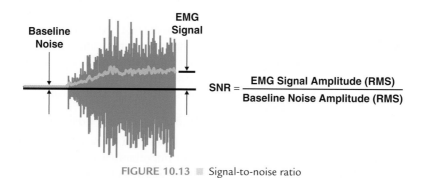

$$SNR = \frac{EMG\ Signal\ Amplitude\ (RMS)}{Baseline\ Noise\ Amplitude\ (RMS)}$$

FIGURE 10.13 ▪ Signal-to-noise ratio

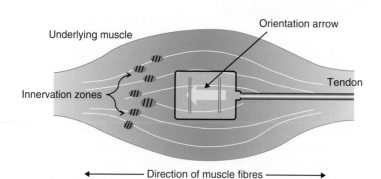

FIGURE 10.14 ▪ Positioning of the surface EMG sensor on the skin above the muscle of interest *(Image modified from Delsys Inc.)*

greater in the middle of the muscle. A sensor located on the innervation zone will detect the cancellation of the action potentials travelling in opposite directions, and will therefore generally have lower amplitude. The preferred location is away from all these areas, towards the middle of the muscle surface. For better signal quality, the bars of the sensors should be aligned perpendicularly to the muscle fibres when possible or along the line of action of the muscle fibres for other electrode types.

10.3.2 Sensor Characteristics

The bipolar configuration is preferred in most cases to remove external noise disturbances. The selectivity of an electrode depends on the area of the detection surface, and, in the case of bipolar electrodes, on the distance between the two detection surfaces. The greater the area and the inter-electrode distance, the greater the pick-up area of the sensor (Fig. 10.15) and, as a consequence, the greater the amplitude of the detected EMG signal. However, the electrode area and inter-electrode distance cannot be too large in order to minimize the detection of unwanted EMG signals from adjacent muscles, including muscles deep to the one of interest. This interference of EMG signals from muscles other than the one under the electrode is referred to as crosstalk. Crosstalk contamination can distort the target muscle signal and mislead the interpretation of the activation timing and magnitude. For single differential recordings, minimum crosstalk contamination can be achieved with an inter-electrode distance of 1 cm (De Luca et al., 2011). More

details on crosstalk are presented in Section 10.3.7: Crosstalk.

It is also preferred to use EMG sensors with a fixed spacing between the electrodes for signal consistency. When movement occurs, the skin over the muscle can move or stretch and the inter-electrode distance between the electrodes may vary if they are separately attached on the skin. However, the amplitude and frequency components of the EMG signal are influenced by the inter-electrode distance in a bipolar configuration, which acts as a band-pass filter (a larger distance will result in lower bandwidth). Thus, a fixed inter-electrode distance during EMG recording ensures consistency of the amplitude and frequency information in the signals.

10.3.3 Baseline Noise and Skin–Electrode Interface

The baseline noise originates in the electronics of the amplification system (thermal noise) and at the skin–electrode interface (electrochemical noise). It can be observed when a sensor is attached to the skin and the muscle is completely relaxed.

The thermal noise is generated by the first stage of the amplifiers and is due to a physical property of the semiconductors. This noise source cannot be eliminated or reduced other than by using advanced technology. The electrochemical noise is generated by the ionic exchange between the metal in the electrode and the electrolytes in the salts of the skin (also known as the electrolyte–electrode interface). The magnitude of this noise is proportional to the square root of the

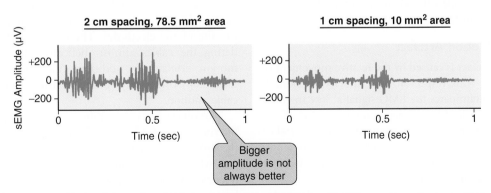

FIGURE 10.15 ■ Amplitude of the surface EMG signal recorded with sensors with different electrode area and inter-electrode spacing

resistance of the electrode surface. It cannot be completely eliminated but can be reduced by increasing the electrode area and, more importantly, by cleaning the electrode surface and the skin.

With some commercially available passive electrodes, electrical contact can be greatly improved by using a saline gel or paste. Active surface electrodes generally do not require the use of conductive medium and are often referred to as 'dry' electrodes.

Effective preparation of the skin is critical to ensure that the EMG sensor is tightly attached and makes good contact with the skin. Adequate skin preparation includes the removal of the surface layer of dead skin cells along with its protective oils. This is best done by cleaning the skin with alcohol at the site chosen for electrode application and by light abrasion of the skin, if needed. Excessive hair must also be removed to lower the electrical impedance and to ensure a secure contact between sensor and skin.

Another key aspect for minimizing baseline noise is the use of appropriate filtering. The frequency spectrum of the baseline noise ranges from 0 Hz to a frequency range much greater than the surface EMG signal (several thousand Hz), with an amplitude that is greater at the low-frequency end and decreases to an approximately constant level by 10 to 20 Hz. A study by De Luca and colleagues (2011) showed that a band-pass filter between 20 and 450 Hz is the optimal filter to truncate the contribution from baseline noise while minimizing the attenuation of the low-frequency components of the surface EMG signal. See the reduction in the amplitude of the baseline noise on surface EMG signals recorded from the quadriceps muscle with a band-pass filter between 20 and 450 Hz in Fig. 10.16.

10.3.4 Motion Artefact

Motion artefact is caused by the relative movement of the sensor with respect to the underlying skin. It usually contaminates the EMG signal at low frequencies and may lead to erroneous interpretation of the EMG data. Signal contamination from motion artefact is a special concern for biomechanics investigations that include recording EMG signals during dynamic contractions or vigorous activities.

One source of motion artefact is at the electrode–skin interface, when the chemical balance of the

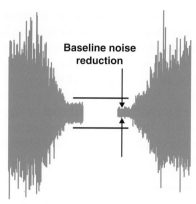

FIGURE 10.16 ■ Reduction in baseline noise after band-pass filtering the surface EMG signal between 20 and 450 Hz

skin–electrode interface is altered due to volumetric changes during muscle contraction (shortening and stretching of the muscle). Another source is the force impulses that are transmitted to the electrodes through the muscle and the skin, as in the case of a jerk movement or a heel-strike while walking. Any gel or electrolyte material that is used between the electrode and the skin can greatly affect this noise issue.

Tethered EMG sensors are commonly affected by an additional source of motion artefact, when the cables connecting the EMG sensors to the amplifier move through electromagnetic fields in the environment and generate a potential that is amplified by the recording system. The cables themselves can also produce displacement of the EMG sensors if not properly positioned. Modern EMG technology now uses sensors that have the first stage of amplification located on-board or within centimetres of the site of the electrodes. The low impedance, rendering the cables almost ineffective in generating motion artefact. In addition, fully wireless EMG technology completely eliminates artefact noise from cable motion.

Motion artefact can be minimized with appropriate filtering (Fig. 10.17). Specifically, high-pass filtering the surface EMG signal at 20 Hz reduces motion artefact, whose energy is mostly concentrated at low frequencies, while minimizing the attenuation of the low-frequency components of the surface EMG signal (De Luca et al., 2011).

Motion artefact

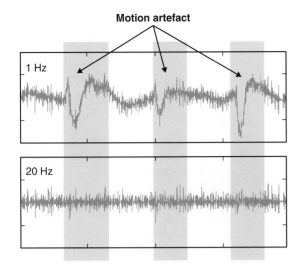

FIGURE 10.17 ■ Reduction in motion artefacts after high-pass filtering the surface EMG signal at 1 Hz and at 20 Hz *(Modified from De Luca et al., 2010)*

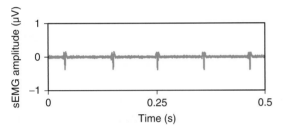

FIGURE 10.18 ■ Surface EMG baseline contaminated by heart rate

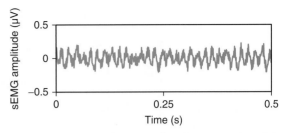

FIGURE 10.19 ■ Surface EMG baseline contaminated by 60 Hz noise

10.3.5 Physiological Noise

Physiological noise originates from tissues other than muscles that generate electrical signals, such as electrocardiogram (EKG) (Fig. 10.18) and electro-oculogram (EOG). This noise can be reduced by properly locating the EMG sensor further away from the source of the noise, if possible. Alternatively, rotating the sensor so that the electrodes align on equipotential planes (i.e. both electrodes are equidistant from the source) generally reduces physiological noise in the EMG signal.

10.3.6 Power Line and Electrical Noise

Noise from power lines (50 or 60 Hz), fluorescent lights and electrical devices originate from the electromagnetic radiation that is pervasive in all environments. It is generally not a concern with modern differential amplification technology and with the application of a reference electrode on the subject that can virtually eliminate this noise source. However, in environments where the electrical power lines do not have a proper ground return path, as can be the case in old wiring layout, the line noise may be a concern (Fig. 10.19). In these cases, a local electrician should be consulted. Line interference can also contaminate the EMG recording if the reference electrode is not properly attached or does not make good contact with the skin.

The bipolar configuration and the characteristics of most amplifiers in EMG system technology nowadays are designed to reduce disturbances from external noise that may contaminate the recording. The ability to reject common signals (disturbances) is usually reported as common mode rejection ratio (CMRR). A commonly accepted value for CMRR is above 90 dB.

10.3.7 Crosstalk

Crosstalk refers to the unwanted electrical activity generated by muscles close to the one under investigation that is picked up and recorded by the EMG sensor positioned on the target muscle. Crosstalk EMG signals have characteristics and frequency bandwidth similar to that of the target muscle and are therefore very difficult to identify or completely eliminate. Crosstalk contamination confuses the interpretation of the magnitude and activation timing of the target muscle, and should be a special concern when recording from small muscles located in close proximity to bigger muscles that might be active at the same time.

To minimize crosstalk contamination, the EMG sensor should be positioned on the belly of the target muscle, as far as possible from nearby muscles. The

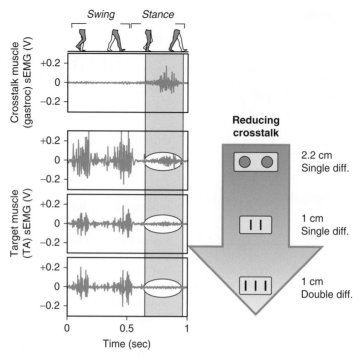

FIGURE 10.20 ■ Decreasing amount of crosstalk from the gastrocnemius (gastroc) muscle detected from a sensor positioned on the target tibialis anterior (TA) muscle when using a larger (2.2 cm) and a smaller (1 cm) inter-electrode distance single differential sensor and a double differential sensor *(Adapted from De Luca et al., 2011)*

characteristics and inter-electrode distance of the sensor technology are also critical because they determine the electrode pick-up area on the muscle. The influence of inter-electrode spacing on the degree of crosstalk contamination using bar and disk electrodes was investigated in a study by De Luca and colleagues (2011). Surface EMG signals were recorded from the tibialis anterior muscle (target muscle) during walking and during constant-force isometric contractions. The amount of crosstalk contamination from the triceps surae muscle (crosstalk muscle) was assessed by recording single-differential surface EMG signals with inter-electrode spacing ranging from 5 to 40 mm. Results showed that, for single differential recordings, minimum crosstalk contamination can be achieved with an inter-electrode distance of 1 cm (De Luca et al., 2011). This distance reduces crosstalk contamination and ensures adequate signal reliability when selectively recording from individual muscles.

When a small electrode and inter-electrode spacing are not enough to guarantee minimal crosstalk

disturbances, a double differential configuration of the electrodes may be used. The double differential arrangement (Fig. 10.20) requires the use of three electrodes, usually in one single solid unit. These electrodes work in much the same way as the single differential electrodes, only now the signals from three electrodes are differentiated in a two-stage process, causing an even greater reduction in noise. The additional differentiation of the double differential configuration has the added benefit of reducing the pick-up volume and, therefore, reducing the possible interference of surrounding muscles and crosstalk.

10.3.8 Recommendations for Good-Quality Surface EMG Signals

Sensor technology: use active differential sensors of high-quality technology. Ensure that there is no instrumentation or electrical connection introducing line interference in the recording.

EMG sensor: use sensors with small (1 cm or less), and possibly fixed, inter-electrode spacing.

Reference electrode: ensure the reference electrode, if separate from the EMG sensor, is connected to the instrumentation and to the skin of the subject.

Sampling rate: ensure that the surface EMG signal is sampled at least at 1000 Hz.

Amplifier gain: a gain of 1000 is usually appropriate for surface EMG signals. Ensure that no clipping is present in the signal, i.e. that the signal amplitude does not exceed the maximum value allowed by the recording system (saturation). If this occurs, check that the EMG sensor and reference electrode are adequately attached and connected. If possible, reduce the gain or reposition the EMG sensor to reduce the signal level.

Skin preparation: remove excessive hair; clean the skin with alcohol to improve electrical contact with the sensor; if necessary peel off the dead skin (i.e. with a hypo-allergic adhesive tape).

Sensor placement: place the EMG sensor in the in the middle of the muscle belly, away from known innervation zones and tendon origins, with the electrodes aligned parallel to the muscles fibres. Position the sensor as far as possible from sources of physiological noise, ensuring that both electrodes are equidistant from the source in single differential sensors.

Sensor attachment: ensure good contact between the electrodes and the skin. When using solid metal electrodes, avoid using gel electrolyte.

Filtering: a Butterworth filter with corner frequencies of 20–450 Hz and a slope of 12 dB/oct is recommended for general use in surface EMG recording.

10.4 PROCESSING THE EMG SIGNAL

10.4.1 Raw EMG Signal

The raw EMG signal consists of the superposition of the electrical voltage associated with the propagation of action potentials along the contracting muscle fibres that can be recorded with a sensor on the skin above a muscle or inside a muscle. It is measured in volts (V).

The waveform of the observed action potential for each motor unit depends on the orientation of the sensor contacts with respect to the contracting fibres. If the electrodes are aligned parallel to the muscle fibres, motor unit action potentials generally have a biphasic shape, and the sign of the phases will depend on the direction from which the fibre depolarization approaches the detection site (Fig. 10.21A).

In human muscle tissue, the amplitude of the action potentials is dependent on the diameter of the muscle fibres, the distance between the muscle fibres and the detection site, and the filtering properties of

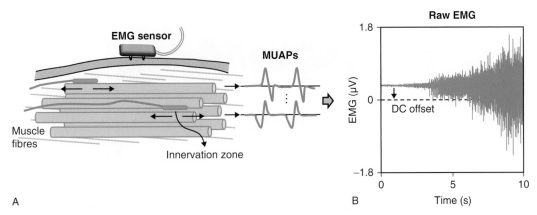

FIGURE 10.21 ■ (A) Graphics representing how the shape of the MUAP changes with the position of the sensor on the contracting muscles. (B) Raw EMG signal

the electrode. The duration of the action potential is inversely related to the conduction velocity of the muscle fibres, which ranges between 3 and 6 m/s. Muscle fibres of larger, higher-threshold motor units generally have greater conduction velocities, greater amplitude and shorter time duration action potentials.

The raw EMG signal has a time-varying amplitude that fluctuates around zero and represents the amount of muscle activation. Higher amplitude indicates greater muscle activation, i.e. a greater number of contracting motor units at higher firing rates (Figs 10.21B and 10.22A). Although the amount of activity relates to the magnitude of the force produced by a muscle, the amplitude of the EMG signal is generally **not** linearly proportional to the force generated by the muscle.

Note also that the EMG signal often oscillates at either side of a DC level different than zero, referred to DC offset, which is due to the recording

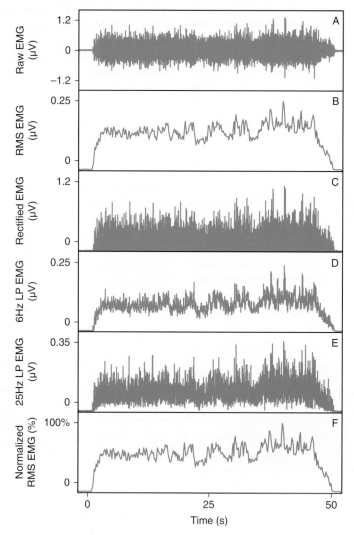

FIGURE 10.22 ■ (A) Raw EMG signal. (B) Time-varying root mean square (RMS) calculated with a window of length 0.25 ms. (C) Rectified EMG signal. (D) Envelope calculated with a 6 Hz low-pass (LP) filter. (E) Envelope calculated with a 25 Hz LP filter. (F) Time-varying RMS of the EMG signal normalized and displayed in %

instrumentation. Removing the DC offset simply consists in high-pass filtering the EMG signal above DC.

10.4.2 Analysis of EMG Amplitude

The amplitude of the EMG signal varies over time during a contraction and with the contraction force. After the DC offset is removed, the average value of the EMG signal is around zero and, thus, averaging the EMG signal does not provide useful information. Other measures are usually employed for the amplitude analysis of the EMG signal in the time domain.

Root Mean Square

The calculation of the root-mean-square (RMS) value of the EMG signal is the preferred method of estimating the signal amplitude. It consists of a three-step calculation: each data point in the signal is squared, the average value over a specified window length is determined, and the square root of this value is then calculated for the moving window (Fig. 10.22B). The RMS value provides a measure of a physical property of the EMG signal that is the energy of the signal. Thus, it is a more meaningful measure of EMG amplitude than other commonly used mathematical functions in the past, such as the mean rectified value or the integrated value.

Rectification

Rectification takes the absolute value of the EMG signal, so that all negative values are transformed into positive voltage values. It can be achieved by squaring the signal and computing the square root. This is a common operation before determining the threshold for the onset or offset of the muscle (Fig. 10.22C).

Envelope

The amplitude of the raw EMG signal is very variable. Low-pass filtering is a common procedure applied to smoothen the signal and to provide a more indicative representation of the time-varying EMG amplitude. The envelope calculation maintains the lower frequencies in the EMG signal while removing the higher frequencies and rendering the EMG fluctuations less variable.

The amount of smoothing depends on the cut-off frequency of the low-pass filter, with lower cut-off resulting in a smoother signal. Fig. 10.22D, E shows the effect of low-pass filtering the rectified EMG shown in Fig. 10.22C at cut-off frequencies of 6 Hz and 25 Hz, respectively. The lower cut-off for low-pass filtering produces a smoother pattern and facilitates visual identification of the mean value, though it removes most of the frequency component of the EMG signal and reduces the signal amplitude.

Amplitude Normalization

Comparing the amplitude of EMG signals recorded from a specific muscle among individuals, or even in different test days for the same individual, is very difficult. This is due to the number of confounding variables that can affect the amplitude of the EMG signal, such as simply the location of the sensor on the muscle. However, amplitude comparison is frequently needed, for instance when assessing the effect of different conditions and treatments, or when evaluating the contribution of muscles to different tasks. Normalizing the amplitude of the EMG signal with respect to a reference value (Fig. 10.22F) allows comparison of the EMG amplitude for signals detected from the same muscle and subject or among different subjects under different conditions or interventions.

One of the most common normalization methods uses maximum voluntary contraction (MVC) to identify the reference value for normalization. It involves recording the EMG signal during a maximal isometric contraction and normalizing the subsequent EMG data as a percentage of the maximal value obtained during the MVC. There are a number of drawbacks to this method. For instance, it is very difficult to establish whether a subject has produced a truly maximal contraction, especially in the case of neurological involvement or joint pathology. In addition, the amplitude of the EMG signal is unstable during maximal efforts so that normalization at a submaximal contraction level, such as at 80% MVC, is generally preferred. Another issue relates to the normalization of the EMG signal recorded during anisometric contraction (i.e. concentric and eccentric) to the maximal value obtained during an isometric MVC, because isometric and anisometric contractions produce different EMG signals.

A preferred method for amplitude normalization consists in using the maximum observed EMG signal during an activity as reference value. This is best done by identifying the maximum value among a set of trials for a specified activity: for instance three, or better, five trials.

Note that the issue of normalization can sometimes be avoided by analysing variations in EMG amplitude as a percentage change rather than as difference in absolute amplitude.

EMG Amplitude, Isometric Muscle Force and Joint Moment

The relation between the amplitude of the EMG signal and the force exerted by the muscle has caused many debates in the literature. It is known that the amplitude of the EMG signal increases as the contraction strength increases, but how?

This is best studied during isometric contractions, that is, during contractions where the joint angle is maintained fixed so that the length of the muscle does not change. Isometric contractions are better suited to relate EMG amplitude and muscle force because some confounding factors are minimized, such as the changes in the EMG amplitude as the relative position between the sensors and the muscle fibres changes during anisometric contractions.

A study by Lawrence and De Luca (1983) investigated the relation between EMG and force for different muscles (biceps, deltoid and first dorsal interosseous), rate of force production, and training level of the subjects (pianists, long-distance swimmers, power lifters and normal subjects). Only the muscle type influenced the relation between EMG signal and muscle force, with the smaller muscle (first dorsal interosseous) showing an almost linear relation with the force, that was instead non-linear for bigger muscles (Fig. 10.23).

The difference among muscles is due to several factors that affect the relation between EMG and force: the motor unit recruitment and firing rate properties; the relative amount and location of slow-twitch and fast-twitch fibres with respect to the recording electrodes; the crosstalk contamination from other muscles; the muscle viscoelastic properties; and the force contributions from agonist and antagonist muscles. This last factor is particularly important because a single muscle may not be solely responsible for producing a moment around a joint. Additionally, there is often a high degree of variation between subjects, not only between different muscles in the

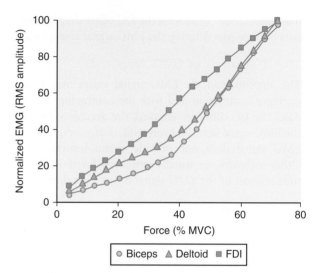

FIGURE 10.23 ■ EMG signal amplitude (RMS) vs isometric muscle force (in % maximum voluntary contraction or MVC) for the biceps, deltoid and first dorsal interosseous (FDI) muscles. *Figure modified from Lawrence and De Luca (1983)*

same subject but also in how the surface EMG signal translates to muscle force.

For anisometric (dynamic) contractions, the relation between EMG amplitude and muscle force is not linear. When the muscle contracts, the length of the muscle shortens, the joint angle decreases, and the moment arm of the muscle (the distance from the muscle to the centre of rotation of the joint) increases. If we apply a constant force, as the moment arm increases, the force produced by the lengthening muscle must decrease, resulting in a non-linear EMG–force relation. In addition, two other factors contribute to the non-linearity between the force output and the EMG signal amplitude. The first factor is the change in the relative position of the source of the EMG signal and the sensor which remains attached to the skin as the muscle moves below the skin. The second factor is the non-linear relationship between the force produced by the muscle and the varying length of the contracting muscle fibres.

EMG Amplitude During Concentric and Eccentric (Anisometric) Muscle Contractions

Muscle length remains more or less constant during isometric contractions, whereas it shortens during concentric contractions, and it lengthens during eccentric contractions. Concentric contractions are generally weaker and require greater muscle activation than isometric and eccentric contractions for a particular load.

If a surface EMG signal is recorded from the biceps brachii muscle when flexing (concentric) and extending (eccentric) the arm at the elbow while holding a fixed load, the raw and RMS EMG patterns in Fig. 10.24 are generated. The highest muscle activity is observed during the concentric phase and the lowest during the eccentric movement.

It should be noted that the difference in EMG amplitude during dynamic eccentric and concentric contractions may be affected by several factors, including changes in the length of the muscle fibres. The relative movement between the surface sensors on the skin and the shifting muscle fibres underneath also cause changes in the amplitude of the EMG. Thus, the use of EMG amplitude as an indication of muscle

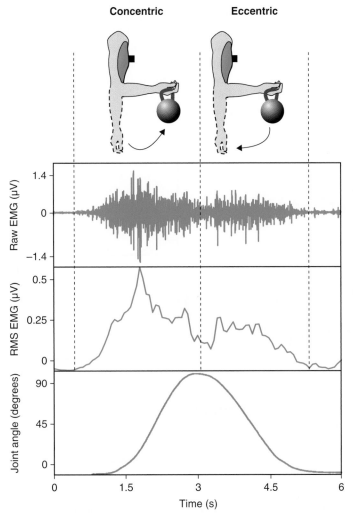

FIGURE 10.24 ■ EMG signal (raw and RMS amplitude) during concentric and eccentric movement

activity during anisometric contractions should be considered with caution.

Analysis of Muscle Contribution

The amplitude of the surface EMG signals, most commonly estimated as the RMS of the EMG signals, is analysed to obtain an indication of the amount and timing of the relative contribution of different muscles during a specific activity (i.e. during standing, walking, running, and so on).

As there may be considerable variability among different trials even for the same subject, the EMG amplitude can be averaged over multiple trails before analysis and interpretation of the data. For repetitive activities or for contractions that can be repeated over time, the average and standard deviation of the EMG amplitude are calculated and commonly presented as a function of the percentage of the cycle time (activity duration) rather than as a function of time in seconds. See Fig. 10.25 for an example of the contribution of different muscles during walking as a function of the gait cycle.

10.4.3 Analysis of the EMG Activation Timing

The amplitude of the surface EMG signal is commonly used to determine when a muscle contraction begins and ends. This information establishes in which phases of an activity specific muscles are engaged and contribute to force production, or when different muscles act as agonists or antagonists.

The analysis of the activation timing is generally performed by identifying the time instants when the EMG amplitude increases above (start) or decreases below (end) a predetermined threshold that indicates the baseline level. The rectified EMG or the RMS of the EMG is generally used rather than the raw EMG signal (Fig. 10.26A, B). In this method, the lowest value of the EMG amplitude (or RMS) is identified as the baseline level. Anything with amplitude greater than this baseline level is identified as signal. Another method consists in identifying the time instant when a regression line, calculated over the EMG RMS on a short window of time (less than 50 ms), increases

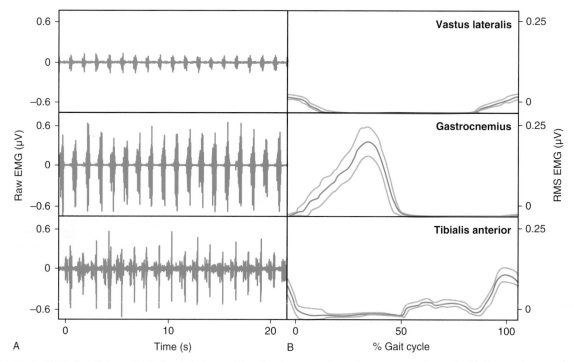

FIGURE 10.25 ■ (A) Raw EMG signal during walking for the vastus lateralis, gastrocnemius and tibialis anterior muscles. (B) Mean and standard deviation of the EMG RMS in percentage of the gait cycle for the three muscles

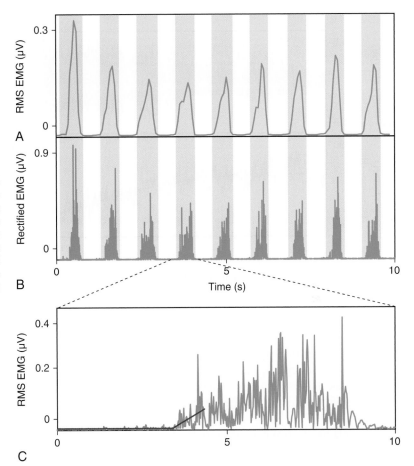

FIGURE 10.26 ■ (A and B) EMG RMS signal and rectified EMG signal from the gastrocnemius muscle during walking and superimposed activation timings identified as the signal increased above 3 standard deviations of the baseline noise. (C) EMG RMS and superimposed regression lines. The time instant where the two line meets indicates the ON time

above the regression line during baseline. The calculation includes the following steps: find a segment of the signal with a constant lowest-level value, calculate the RMS value with a window of 50 ms or less, calculate the regression line for this segment, follow the signal until it begins to increase and calculate the regression for the segment with the increased signal. The point where the regression lines meet is the ON time (Fig. 10.26C). The reverse calculation will find the OFF time. The advantage of this method is that it is independent of the amplitude of the noise level, as long as it is lower than the signal level (De Luca, 1997).

10.4.4 Analysis of the EMG Frequency Content

Most of the EMG signal content is in the range of frequencies between 20 and 450 Hz (Fig. 10.27A). The frequency components of the EMG signal can be investigated by computing the fast Fourier transform (FFT) of the signal, which identifies the content of the signal at each frequency. Another commonly used measure is the power spectrum density, which represents the distribution of power (square value of the magnitude of the FFT) as a function of frequency.

The frequency spectrum of the EMG signal depends only minimally on the firing rates of the MUAPTs detected. It depends primarily on the shape of their action potentials, which is related to the conduction velocity of the muscle fibres (Basmajian & De Luca, 1985). If the conduction velocity decreases, such as when the muscle fatigues, the depolarization current requires more time to traverse the fibres and the detected MUAPs will have longer time durations. Hence, the frequency spectrum of the MUAPs and the

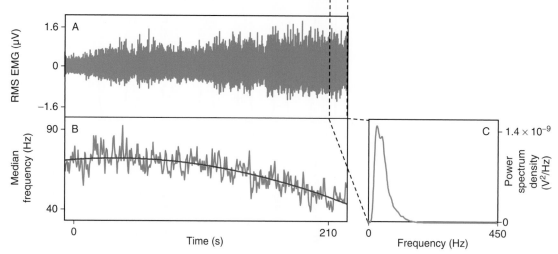

FIGURE 10.27 ■ (A) Raw EMG signal from an isometric contraction sustained with the biceps brachii muscle while holding a heavy load for over 3 minutes. (B) Power spectrum density of the raw EMG in first and last 10 minutes of the contraction. (C) Time-varying decrease in the median frequency

EMG signal, which they comprise, will have a relative increase in the lower-frequency components and a decrease in the higher-frequency components. In other words, a shift toward the low-frequency end would occur (Fig. 10.27B).

This explains why two parameters of the frequency spectrum of the EMG signal are often reported as measures of muscle fatigue: the central (median) frequency and the mean (average) frequency, as described later.

Frequency Changes During Muscle Fatigue

During sustained muscle contractions, muscle fatigue develops and can lead to muscle failure, that is, to a point when the muscle cannot produce a required force.

Let's consider a muscle contraction that is sustained at a given force level for a prolonged period, as in Fig. 10.27A. The EMG signal was recorded from the biceps brachii muscle of a subject while holding a fixed weight for over 3 minutes. Muscle fatigue built up throughout the contraction, so that a progressively stronger effort was required over time to perform the same time (i.e. hold the same weight). The increasing effort is apparent from the progressive increase in the amplitude of the EMG signal (Fig. 10.27A), indicating a greater

amount of muscle activation. The increase in EMG amplitude is mostly due to the recruitment of new motor units and to the increase in their firing rate as the muscle attempts to maintain the required tasks while its force-generating capacity decreases with fatigue (Contessa et al., 2013).

The frequency analysis of the surface EMG signal in this circumstance gives important information about muscle fatigue. The frequency shift that occurs with muscle fatigue is mostly due to the modifications in the action potential shapes of the active motor units that have a longer duration as the conduction velocity decreases with fatigue (the sodium and potassium movement slow down through the channels in the muscle) (Stulen & De Luca, 1981).

The amount of fatigue can be quantified by considering either the mean or the median frequency. Specifically, a decrease in the median or mean frequency of the EMG spectrum during the task is evidence of muscle fatigue.

The median frequency is the middle frequency of the magnitude versus frequency graph, or the power versus frequency graph. This may be calculated by finding the sum of all the magnitudes at each frequency and dividing by the total sum. It expresses the contributions up to a given frequency as a percentage of the total magnitude; for example, the percentage of the power spectrum up to a frequency of 80 Hz is the sum of all the power magnitudes from 0 to 80 Hz, divided by the total power present in all frequencies. The median frequency is the frequency at which 50% of the sum of the magnitudes is below and above: that is, at 50% of the percentage of the total magnitude. The median frequency is usually preferred over the mean frequency due to the skewed nature of the frequency spectrum.

10.5 EMG DECOMPOSITION

Decomposition of the EMG signal is the procedure by which the raw EMG signal is separated into the action potential trains of the constituent motor units. This concept is depicted in Fig. 10.28.

Decomposition of the EMG signal is a tool for exploring the workings of the nervous system in normal and dysfunctional conditions. It involves the tasks of identifying the MUAPs which are present in the EMG signal and uniquely assigning them to individual motor units active during the same contraction.

EMG decomposition provides information on the morphology, recruitment and firing behaviour of motor units during muscle contractions. These can be used to explore motor control strategies that the central and peripheral nervous system use to govern motor unit activation and, as a result, how muscle force is controlled.

The decomposition process should identify the greatest number of motor units that constitute the EMG signal. The number of identified motor units is referred to as motor unit yield. In addition, the decomposition should identify the greatest number of action potentials for each motor unit. The number of action potentials, or firings, that are correctly identified is a measure of the accuracy of the process.

10.5.1 Challenges of EMG Decomposition

The decomposition task is relatively simple when only one or two MUAPTs with distinct and clearly identifiable action potential shapes are present in the EMG signal. However, it becomes a more complex task as the number of active motor units present in the EMG signal increases. The greatest challenge of EMG decomposition is motor unit superpositions: during any muscle contractions, a number of motor units are simultaneously active, and their MUAPs generally superimpose in time. The shape and number of action potentials present at any given time is thus confounded

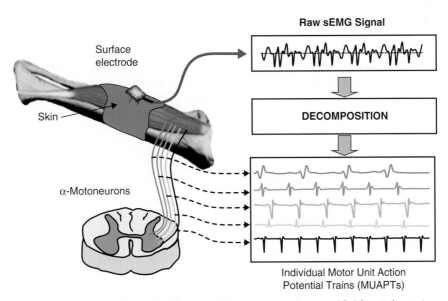

FIGURE 10.28 ■ Surface EMG decomposition process. *(Picture modified from Delsys Inc.)*

by this superposition. The number of simultaneously active, and possibly superimposing, motor units could vary from a just few to several hundred, depending on the specific muscle, the force level of the contraction, and the proportion of the muscle under investigation. A second challenge is the presence of low-amplitude MUAPs, which can often be almost indistinguishable from the EMG baseline noise. A third challenge is changes in the shape of the action potential of a motor unit, which could occur as a result of physiological processes (such as muscle fatigue) or as a result of movement of the detecting electrodes with respect to the contracting muscle fibres. Finally, another challenge is represented by the possible similarity among the shapes of action potentials belonging to different motor units. These factors are shown in Fig. 10.29.

10.5.2 Intramuscular EMG and Surface EMG Decomposition

Early attempts at decomposing EMG signals have been limited for many years to decomposition of intramuscular EMG signals, recorded with either needle or fine-wire electrodes. Adrian and Bronk (1929) developed the first concentric needle electrode to identify the shape and firing rate of MUAPs. Identification of single MUAPs from intramuscular EMG was performed either visually (e.g. Claman, 1970; De Luca & Forrest, 1972, among others) or with automated software for motor unit identification (e.g. Andreassen, 1977; Dill et al., 1972, among others) that aimed at improving the accuracy and objectivity of the identification, as well as decreasing the processing time.

In 1978, LeFever and De Luca introduced the first quadrifilar electrode for intramuscular EMG decomposition. It consisted of a needle with four wires encased in the tip and located approximately 200 μm apart. This technology simultaneously recorded three single differential EMG signals from three different pairs of intramuscular electrodes located in the same muscle area and detected during the same contraction. This technique increased motor unit discrimination power (as will be explained later), and allowed decomposition of up to 9 concurrently active motor units during relatively low-level muscle contractions, with an accuracy for automated decomposition up to 65%. The accuracy value could be improved with manual editing. The technology performance was later improved, providing up to 11 concurrently active MUAPTs identified with accuracy greater than 85% in contractions ranging from minimal to maximal forces.

However, decomposition of intramuscular EMG signals is very difficult to perform and provides limited information. It is an invasive technique that requires trained operators and is uncomfortable for the subject. It can only be used for contractions performed at low forces levels, as strong efforts can be painful. In addition, the needle or fine-wire electrodes inserted into the muscle are very selective and record the EMG activity of only a few motor units. Whereas this selectivity provides a less complex EMG signal that may be easier to decompose accurately, it limits the number of detected motor units. These invasive electrodes are also very susceptible to noise and displacement inside the muscle during a contraction. The result of this displacement is that detection of the same set of motor units during a contraction is rarely possible.

For all these reasons, recent efforts have focused on the decomposition of surface EMG signals. The use of surface EMG signals offers some obvious advantages: it is not invasive, it provides no discomfort or risk for

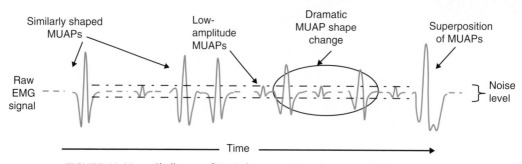

FIGURE 10.29 ■ Challenges of EMG decomposition. *(Picture modified from Delsys Inc.)*

the subject, it can be used for contractions ranging from minimal to maximal force levels, and it uses electrodes that are less selective and thus provides a greater amount of information. The potential of detecting a greater number of motor units from a larger portion of the muscle provides a better understanding of the behaviour of the entire muscle. The downside of surface EMG decomposition is the added difficulty in discriminating the MUAP shapes from surface EMG recording.

The greater number of motor units detected with surface EMG sensors increases the chances of superpositions even at low force levels. Additionally, surface EMG sensors are located further from the source of the signal (i.e. the contracting muscle fibres) and the detected EMG signals undergo a greater amount of spatial filtering. As a result, the shapes of the detected

MUAPs are wider, more similar to each other and more difficult to discriminate. A comparison of the differing complexity of a surface EMG signal and an intramuscular EMG signal is shown in Fig. 10.30.

De Luca and colleagues (2006) and Nawab and colleagues (2010) recently developed a novel surface EMG decomposition technique that uses a special surface sensor composed of five small pin electrodes located at the edges and in the middle of a 5 × 5 mm square (Fig. 10.31A). Four simultaneous bipolar surface EMG signals are recorded from different pairs of electrodes during the same contraction (Fig. 10.31B), and they are decomposed into the constituent motor unit trains by using sophisticated artificial intelligence algorithms and template matching algorithms that are able to identify action potentials belonging to different motor units based on their shape similarity.

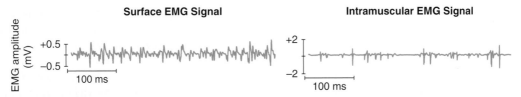

FIGURE 10.30 ■ Decomposition of surface and intramuscular EMG signals. *(Picture modified from Delsys Inc.)*

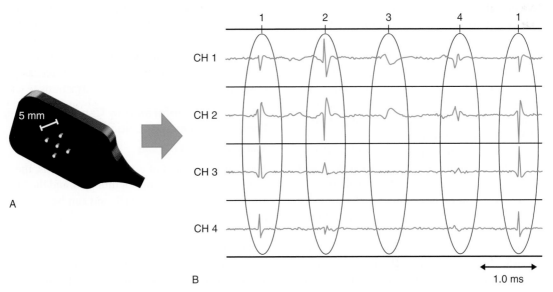

FIGURE 10.31 ■ (A) A 5-pin surface sensor for surface EMG decomposition. (B) Four channels of surface differential EMG signals. *(Picture modified from Delsys Inc.)*

Four different surface EMG channels are needed to improve the motor unit detection power: the electrical activity (action potential) generated by the same motor unit will be detected in all four channels, though motor units will have a different shape in each channel by virtue of the different orientation of the pin electrode with respect to the contracting muscle fibres. This distinction is used by the decomposition algorithm to identify the individual MUAP firings and assign them to a particular motor unit.

Fig. 10.31B shows the assignment of four distinct motor units based on their action potential shape similarity in the four channels of EMG recording. Note that multiple channels are required even for this simple case, as, for instance, the action potential shapes of motor unit #1 and #2 are similar in channel 2. Thus, their discrimination based uniquely on this channel would be very difficult. However, in the other channels their shapes are distinguishable and this additional information can be used to resolve the ambiguity. Note also that the individual MUAPs are visually identifiable in this EMG segment. However, surface EMG signals are generally more complex and the numerous superpositions make the identification process not trivial.

This recent surface decomposition technology is able to decompose up to 50 concurrently active motor units with recruitment threshold spanning the entire range of contraction forces, providing a numerous and heterogeneous pool of motor units for investigation. Contractions ranging from minimal to maximal force levels can be investigated with very high accuracy, on average 95% (Nawab et al., 2010). This value indicates that on average 95% of the firings are correctly identified, including false identification (a MUAP is mistakenly identified in the signal) or missed firing (a MUAP in the EMG signal is not detected).

The major limitation of this surface EMG decomposition technique at this time is that it can be used to accurately decompose surface EMG signals recorded during isometric contractions only. During isometric contractions, the muscle length remains approximately constant, as does the relative position of the electrode with respect to the contracting muscle fibres. This characteristic ensures that the MUAP shapes do not change dramatically during a contraction, unless the electrodes are displaced. The decomposition algorithm can thus more accurately track the motor unit shapes

over time. In addition, force contractions with trapezoidal shape are typically used to provide accurate results: the subject increases the contraction force gradually up to a desired level, and maintains the force at this level for a few seconds before decreasing the force gradually back to no effort (see Fig. 10.32 for an example). These types of contractions are commonly studied as they provide information on the activation and de-activation behaviour of motor units during the ramp-up and ramp-down regions of the contraction force.

In contrast, the decomposition task during dynamic contractions is even more complex due to the changing shapes of the MUAPs as a consequence of the relative movement of the electrodes with the lengthening and shortening of muscle fibres during the contraction. Studies are currently being performed to expand the decomposition capabilities to dynamic, more functional tasks. Successful approaches at decomposing surface EMG signals collected from dynamic contractions have recently been reported for cyclic movements, such as flexion and extension of the elbow or gait tasks (De Luca et al., 2014). The decomposition of these repeated dynamic tasks had an accuracy of approximately 90%, with motor unit yield as high as 25.

These first attempts at decomposing surface EMG signals collected during anisometric contractions are promising. Future development and improvements will open new areas of motor control studies. It will soon be possible to investigate the control of motor units during dynamic activities that address issues related to movement and not only to force, as previously limited during isometric contractions.

10.5.3 What Information Can Be Obtained From EMG Decomposition?

Motor Unit Firing Trains

The firing instances of a motor unit, that is, the time instances at which the motor unit fires, and the MUAPs are found in the EMG signal, and can be represented by vertical bars (representing the MUAPs) on a horizontal time axis.

Fig. 10.32A shows the results obtained from the vastus lateralis muscle during an isometric contraction at 50% MVC. The black line represents the leg extension force produced by the subject and recorded with a force sensor. The force increased linearly from 0% to

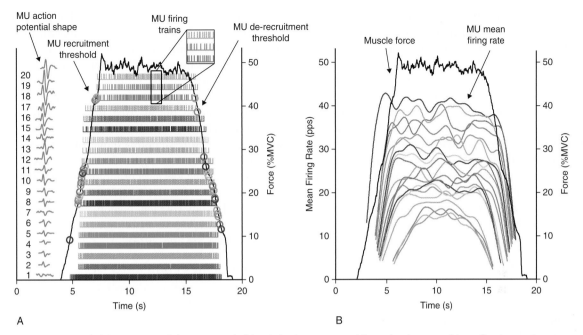

FIGURE 10.32 ◼ (A) Parameters of the motor unit firing behaviour extracted from the decomposition of an isometric contraction: action potential shapes, recruitment and de-recruitment threshold, firing trains. (B) Motor unit mean firing rates. The contraction force is shown in *black*

50% of the subject's MVC, was maintained at the 50% MVC force level for approximately 20 s, and then decreased back to zero. 20 MUAPTs were identified in this contraction. The firing trains for each identified motor unit are shown with vertical bars of different colours.

The time interval between consecutive firings indicates the IPI of the motor unit. Note that, even during the middle phase of the contraction, where the force is maintained constant, the motor unit IPIs vary slightly between consecutive firing: that is, the bars in the firing trains are not uniformly spaced. The variation in the IPIs is an indication of the synaptic noise that influences the firing times of motor units.

Recruitment and De-recruitment Threshold

The recruitment and de-recruitment threshold of a motor unit indicate the force levels at which the motor unit starts and stops firing during a contraction: that is, the activation and de-activation thresholds. They are identified as the force corresponding to the first and last firing (first and last bar in Fig. 10.32) of the motor unit firing train.

In Fig. 10.32A, the detected motor units are arranged in order of recruitment threshold, with motor unit #1 indicating the motor unit recruited at the lowest force level and motor unit #20 indicating the motor unit recruited at the highest force level during the ramp-up phase of the contraction. The recruitment and de-recruitment thresholds are indicated with coloured circles superimposed on the black force trace.

Note that, during linearly increasing and decreasing isometric contractions, motor units are recruited and de-recruited in the opposite manner: the first motor unit to be recruited during the force ramp-up phase is the last motor unit to be de-recruited during the force ramp-down phase (i.e. it has the lowest activation and de-activation threshold).

MUAP Shape

The shape of the action potential generated by a motor unit in the EMG signal provides indication on the morphology of the motor unit and the state of the muscle fibres. The amplitude, time duration and number of phases in the action potential are the parameters that neurologists are accustomed to evaluating during a standard clinical EMG examination.

In Fig. 10.32A, the action potential shapes for each identified motor unit are shown on the left side of each firing train. Note that, as commonly observed during voluntary isometric contractions in human subjects, the amplitude of the MUAP shape increases for progressively later-recruited motor units. This is an evidence of the Size Principle property introduced in Section 10.1.4: Motor Unit Recruitment (Henneman, 1957), which states that smaller motor units—the ones composed of fewer muscle fibres and producing the lower-amplitude MUAP shapes and force twitches—are recruited at lower force levels than bigger motor units.

Motor Unit Mean Firing Rate

From the firing train of a motor unit, it is possible to calculate the number of pulses per unit time: that is, the motor unit firing rate. This parameter is expressed in pulses per second (pps). This can be calculated as a time-varying mean firing rate trajectory by computing the mean firing rate of the motor unit over a short (typically 0.4–2 seconds) time window that is moved along the duration of the contraction (Fig. 10.32B). In Fig. 10.32B, the time-varying mean firing rates of the identified motor units are calculated by filtering the firing trains with a Hanning window of 2 seconds and are presented with different colours.

Note that the earliest-recruited motor unit (motor unit #1) has the highest firing rate, whereas the latest recruited motor unit (motor unit #20) has the lowest firing rate at any time and force during the contraction. This property of the motor unit firing behaviour is called the Onion-Skin property and will be described in more detail later.

Other Parameters of the Motor Unit Firing Behaviour

Other measures that can be obtained from the decomposition of EMG signals are the cross-correlation between the mean firing rates of concurrently active motor units and the level of synchrony among their firings.

Cross-correlation is a mathematical operation that measures the amount of common behaviour among the time-varying firing rates of two motor units (see Section 10.5.4: Findings from EMG Decomposition

for more details). The existence of a high-degree of cross-correlation implies a similarity in the firing variations of motor units: that is, motor units modulate their firing rates in unison. It does not imply that the individual firings are synchronized. This property is named synchronization and it refers to the tendency of motor units to fire at fixed time intervals with respect to each other. Synchronization has often been associated with force smoothness, with a higher degree of synchronization causing the force to be more variable. However, the low level of motor unit synchronization indicates that synchronization may be best understood as an epiphenomenon with a currently unknown physiological significance (De Luca et al., 1993; De Luca & Kline, 2014).

10.5.4 Findings From EMG Decomposition

The following section describes recent findings obtained from EMG decomposition that advanced our knowledge of muscle force generation strategy and motor control.

The Onion-Skin Scheme of Motor Unit Firing

For over six decades, starting with the work of Eccles (1958) and Kernell (1963) on the axons of anaesthetized cats, it has been believed that later-recruited motor units fire faster than earlier-recruited ones, so that each motor unit fires at a rate that produces optimal force. This concept refers to the fact that later-recruited motor units are fast-twitch fibres and require greater firing rates to produce tetanized force than earlier-recruited slower-twitch motor units do.

De Luca and colleagues (1982a) were among the first, together with Seyffarth (1940), Person and Kudina (1972) and Tanji and Kato (1973), to document that the opposite firing rate arrangement occurred in humans performing voluntary contractions. Lower-threshold earlier-recruited motor units display greater firing rates than higher-threshold later-recruited ones at any time and force during a voluntary isometric contraction. This property is known as the Onion-Skin scheme because when the firing rates of motor units are plotted as a function of time, they form overlapping layers resembling the structure of the skin of an onion (see Fig. 10.32B). However, until only recently, the data supporting this Onion-Skin firing rate scheme were sparse and based on only a few motor units obtained from intramuscular techniques. More recent data obtained with surface decomposition technology on a much greater number of motor units (De Luca & Hostage, 2010; De Luca & Contessa, 2012) have undoubtedly proven that the Onion-Skin arrangement describes the physiological firing behaviour of motor units during voluntary isometric contractions.

This property can be quantified by calculating the average firing rate of motor units at a given contraction force and plotting this value as a function of the motor unit recruitment threshold. An inverse relationship between motor unit firing rate and recruitment threshold is observed and can be quantified with the slope and intercept of the regression line (Fig. 10.33).

The fact that later-recruited motor units maintain lower firing rates than those needed to produce tetanized force, as shown in the Onion-Skin scheme, suggest that the muscle control strategy does not maximize force production. It would seem reasonable to expect motor units to fire at a rate that matched their force-generating capacity, so as to maximize force. However, it should be noted that the later-recruited motor units are fast-fatigable and they would likely become exhausted in a very short time if they were to fire fast. By maintaining low firing rates for these fast-fatigable motor units, the Onion-Skin scheme ensures that force contractions can be sustained even for

prolonged periods of time. In addition, a reserve capacity of force-generation is maintained within the muscle for the extreme circumstances when it might be needed.

From these evidences, De Luca and Erim (1994) suggested that the control of motor units has developed not to maximize force alone, but instead to maximize a combination of contraction force and contraction time, and to maintain a reserve capacity of force generation.

The Common Drive

The common drive is another property of motor unit firing that was observed thanks to the decomposition technique (De Luca et al., 1982a, b; Miles, 1987; Stashuk & deBruin, 1988; among others). This property refers to the tendency of motor unit firing rates to fluctuate in unison with almost no time delay (Fig. 10.34).

The common drive property indicates that the central nervous system has evolved a relatively simple strategy for controlling motor units: it does not individually control the firing behaviour of each motor unit, but instead, it modulates the behaviour of the entire pool of motor units in a similar manner.

The strength of the common drive received by concurrently active motor units can be quantified by computing the cross-correlation function between the

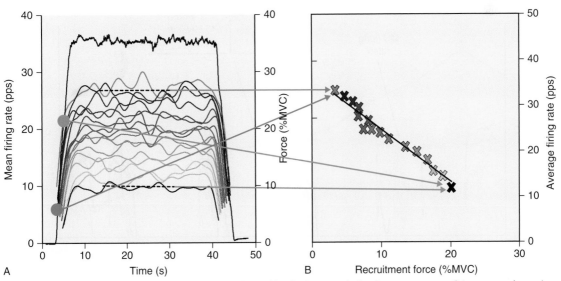

FIGURE 10.33 ■ Motor unit Onion-Skin property evidenced by the inverse relation between average firing rate and recruitment threshold

firing rates of pairs of concurrently active motor units (Fig. 10.34). For each motor unit pair, the amount of common drive is generally quantified as the peak of the cross-correlation function computed between the mean firing rates of the two motor units for time lags between ± 100 ms. Note that not all motor unit pairs necessarily show a cross-correlation peak within the specified window (i.e. common drive) and that

the strength of common drive varies among motor unit pairs.

Motor Unit Firing Behaviour During Muscle Fatigue

Motor unit firing behaviour was studied in the vastus lateralis muscle of the quadriceps of three young subjects during a series of isometric knee extensions performed to exhaustion (Adam & De Luca, 2003, 2005; Contessa et al., 2009).

Results showed that the recruitment threshold of all motor units decreased and that additional higher-threshold motor units were progressively recruited as the contractions were repeated over time until the endurance limit. The recruitment order was maintained throughout the contraction protocol. Additionally, the firing rates of the active motor units decreased in the first minute of the contraction and then increased progressively with time until the endurance limit. Changes in the firing rates complemented the changes in the amplitude of the muscle force twitch, which first increased and then decreased (Fig. 10.35). The inverse relationship between motor unit firing rate and recruitment threshold (Onion-Skin scheme) was maintained throughout the fatigue series.

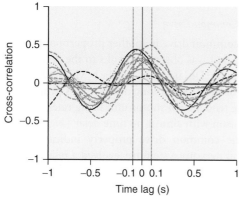

FIGURE 10.34 ■ Cross-correlation function between pairs of motor unit mean firing rates

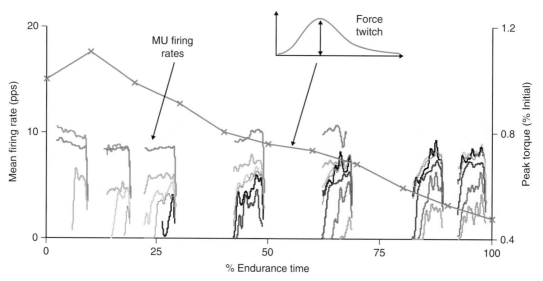

FIGURE 10.35 ■ Motor unit firing rate and force twitch behaviour with fatiguing repeated contractions performed to exhaustion

The observed firing rate and recruitment adaptations complemented the mechanical changes in the force generation capacity of the muscle (force twitch), indicating that the firing rates of motor units adapts to counteract the change in the force produced by the muscle fibres during the contraction series. With fatigue, an increased activation of the motor unit pool (i.e. an increase in the number of active motor units and in their firing rate) is necessary to compensate the loss in force generation capacity.

SUMMARY: ELECTROMYOGRAPHY AND MEASUREMENT OF MUSCLE FUNCTION

■ The EMG signal is an electrical signal associated with the contraction of a muscle and the signal is produced by the depolarizing of motor units, often referred to as motor unit action potential (MUAP).

■ EMG can be collected using a variety of techniques. The most common is surface EMG as this does not require electrodes to be pushed through the skin, as with intramuscular EMG.

■ EMG signals may be processed in a number of ways. These can give information about the amount of motor unit activity or fatigue within a muscle.

Section 3

CLINICAL ASSESSMENT

THE BIOMECHANICS OF CLINICAL ASSESSMENT

JAMES SELFE ▪ JIM RICHARDS ▪ DAVE LEVINE ▪ DOMINIC THEWLIS

This chapter covers the biomechanics of common movement tasks used in clinical assessment of the lower limb. This includes step and stair ascent and descent, sit to stand, timed up and go, gait initiation and squats and dips.

AIM

To consider the biomechanics of common movement tasks used in clinical assessment of the lower limb.

OBJECTIVES

- To consider the concept of kinetic chains in the clinical assessment of lower-limb tasks
- To explain the theoretical aspects of the different movement tasks used in clinical assessment
- To describe the movement patterns during the different movement tasks
- To critically evaluate the different movement tasks in relation to joint function and control.

11.1 KINETIC CHAINS

Steindler (1955) is credited with introducing the concept of open and closed kinetic chains to the study of human movement (Ellenbecker & Davies, 2001). Steindler's *Kiniesiology of the human body* (1955) introduced useful terminology adapted from mechanical engineering. In mechanical engineering, the link concept considers rigid overlapping bars that are connected in series by pin joints. The system is considered closed if both ends are connected to an immovable framework, thus preventing translation of either the distal or proximal joint centre. This creates a system where movement at one joint produces movement at all other joints in a predictable manner (Palmitier et al., 1991).

When applied to human movement it is apparent that there can never be a situation where there is a truly closed kinetic chain (CKC), and certainly the movement of the knee joint, for example, is more complex than that occurring around a pin joint. In the lower limb an open kinetic chain (OKC) is said to occur when the foot is free to move in space with little or no resistance. A CKC occurs in the lower limb when the foot meets considerable resistance, for example, the ground (Palmitier et al., 1991). However, Ellenbecker and Davies (2001) point out that there is some controversy over the definition of considerable resistance, particularly when it comes to classifying activities such as cycling as either open or closed chain. CKC exercises, such as standing squats and step ascent and descent, are assumed to be more functional than OKC exercises, such as seated leg extensions or straight leg raises (SLR) (Doucette & Child 1996).

The combined simultaneous segmental motion and movement in multiple planes around multiple axes increases the demand for dynamic stabilization and joint control, which increases muscle co-contraction activity (Ellenbecker & Davies, 2001). This is particularly important at the knee when the hamstrings provide an active dynamic restraint to prevent excessive anterior shearing at the tibiofemoral joint during

241

strong quadriceps contractions. Without the hamstrings co-contraction, the anterior cruciate ligament is potentially exposed to dangerously high forces. Another important difference in motor control between OKC and CKC is the contraction of the bi-articular muscles. During CKC exercise, simultaneous hip and knee extension occur when rising from the flexed position, causing rectus femoris to simultaneously eccentrically lengthen across the hip, but concentrically shorten across the knee. Conversely the hamstrings shorten across the hip but lengthen at the knee. This form of pseudoisometric muscle contraction has been referred to as 'concurrent shift' (Palmitier et al., 1991).

11.2 SITTING TO STANDING

11.2.1 Introduction

Rising from a seated position is a significant mechanical challenge and, in particular, requires considerable quadriceps activity. Quadriceps demand is increased in two ways: firstly, the quadriceps are required to generate enough concentric moment to extend the knee against the combined effects of gravity and body weight; secondly, they have to resist the antagonistic action of the hamstrings. In order to rise from a chair, trunk flexion with associated hip flexion occurs. Excessive hip flexion is resisted by contraction in the hamstrings, which simultaneously induces knee flexion; this 'unwanted' knee flexion then has to be overcome by additional quadriceps activity (Ellis et al., 1980).

11.2.2 Biomechanics of Sit-to-Stand

Schenkman and colleagues (1996) described three phases of sit-to-stand activity:

- Initial phase – used to generate upper-body momentum. Centre of mass (CoM) predominantly translates horizontally forwards.
- Transitional phase – momentum from upper body is transferred to the whole body as the CoM changes from horizontal to vertical translation.
- Extension phase – vertical ascent of body takes place.

In this study, they investigated the effects of four different chair heights on a group of healthy young adults (25–36 years) and a group of healthy older adults (61–79 years). All subjects increased trunk flexion velocity by nearly 50% to overcome the mechanical difficulty associated with lower chair height. The authors explain that due to conservation of momentum, this allows the upper-body momentum to be harnessed to assist the hip and knee extension required to initiate lift off from the seat, which they refer to as momentum transfer. If chair height is lower, the starting position of the centre of gravity is lower, making lift off from the seat more demanding. Older subjects tend to move more slowly; therefore, they generate less upper-body momentum during the first phase, which leads to difficulties actually getting out of the chair.

The motion of the ankle, knee and hip joints may be studied during sit-to-stand activities by considering the angle against time and the angular velocity of each joint. All three movement patterns are distinctly different and the interaction of these joints gives a very detailed means of assessment of the sit-to-stand task.

Ankle Joint Motion During Sit-to-Stand Task

The ankle is initially in slight dorsiflexion, although this will vary slightly with different initial foot positions. The ankle then moves smoothly into an increasing dorsiflexed position. As the person leaves the chair the ankle moves back towards the ankle neutral position. The angular velocity graph shows an initial dorsiflexion velocity (positive) followed by a plantarflexion velocity (negative) (Fig. 11.1A, B).

Knee Joint Motion During Sit-to-Stand Task

The knee joint is initially flexed at 90°; then after a short delay smoothly extends to near full extension when the person is upright. The angular velocity graph shows a smooth increase and decrease in the extension velocity (negative), demonstrating a controlled movement into extension (Fig. 11.2A, B).

Hip Joint Motion During Sit-to-Stand Task

The hip joint is initially flexed at 90° at the onset. There is an immediate movement into further flexion as the trunk is moved forward over the feet. Then, at approximately the same time as the onset of knee extension, the hip starts to extend until the upright

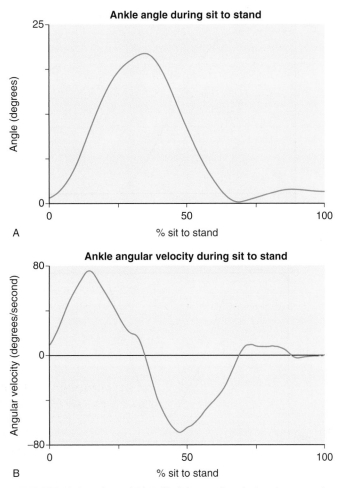

FIGURE 11.1 ◼ (A and B) Ankle joint motion during sit to stand

position is attained. The velocity graph shows an initial flexion angular velocity (positive) as the trunk is inclined forwards, followed by an extension velocity (negative) until the upright position is attained (Fig. 11.3A, B).

11.3 THE TIMED UP-AND-GO TEST

The timed up-and-go test (TUG) is commonly used to measure functional mobility in older adults. As the name implies, the time taken to rise from a chair, walk 3 m, turn, and return to the chair and sit down is measured. Subjects who are able to complete the test in less than 20 s have been shown to be independent in activities of daily living and walk at speeds that are sufficient for community mobility. Those subjects requiring greater than 30 s to complete the test tend to be more dependent in activities of daily living and often require gait aids (Podsiadlo & Richardson, 1991). The test has also been shown to be a sensitive and specific measure for identifying community-dwelling adults who are at risk for falls (Shumway-Cook et al., 2000).

The advantages of using a functional test of this type are many. For example, it has excellent face validity, as subjects can relate to the activity and easily see the relevance of the test to their own mobility. It is easy and simple for clinicians to extract the relevant data (i.e. time) from the test. In addition from a clinical point of view it is an excellent test, as it provides a

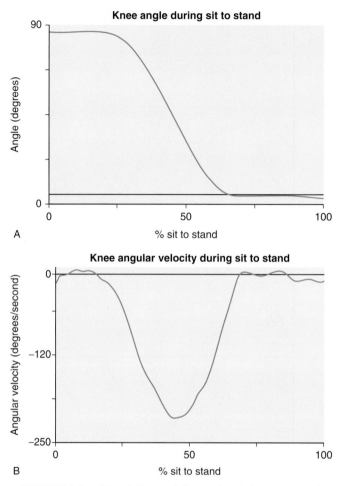

FIGURE 11.2 ■ (A and B) Knee joint motion during sit to stand

number of highly relevant challenges (see the following list), biomechanical analysis of the whole task is a challenge. However, the different activities may be studied in isolation:

- Rising from the seat
- Attaining balance
- Gait initiation and acceleration away from the seat
- Steady-state gait
- Deceleration and preparation for turning
- Turning through 180°
- Acceleration
- Gait cessation
- Descent to the seat.

11.4 STEPS AND STAIRS

Steps and stairs are an important aspect of daily life whether we are at home or out in the community. Access to many buildings and internally within buildings is often determined by our ability to negotiate steps and stairs. In many parts of the world legislation is in place to ensure alternative access arrangements are in place. However, there are also a number of recommendations for and regulations governing stair design. For example, the international building code states that the maximum riser height on a stair should be no more than 21 cm (International Building Code, 2003), whereas in the UK the guidelines for inclusive mobility recommend a maximum riser

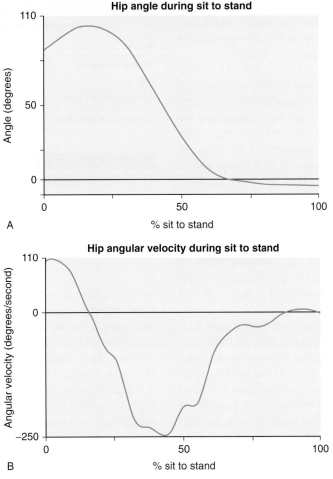

FIGURE 11.3 ■ (A and B) Hip joint motion during sit to stand

height of 17 cm (Department for Transport, 2004). The reason for this focus on stairs is that going up and down stairs from a mechanical view point is quite different to level walking.

That stairs are much more challenging than level walking is evidenced by the inclusion of step and stair activities in the assessment of lower-limb functional status (Cowan et al., 2000; Salsich et al., 2002; Selfe et al., 2001b) and their inclusion in rehabilitation regimes (Cook et al., 1992; McConnell, 2002; McGinty et al., 2000). Patients with conditions where muscle function is impaired may be particularly prone to difficulties with stair climbing and, in fact, in some populations negotiating stairs can be quite hazardous. In the UK in 1992 one-quarter of non-fatal falls in older

people, living in ordinary housing, were from stairs (Wright, 1994). Fatal falls on or from steps or stairs demonstrate an increased incidence with increased age. Older adults aged 65 or over accounted for 68% of the total of deaths in this category of fall in England and Wales in 1994–95 (Dowswell et al., 1999). Whilst it is acknowledged that the cause of falls from steps or stairs is likely to be multifactorial, this section will focus on reviewing the biomechanics of step and stair climbing. In the context of this chapter, step or stair climbing refers to both ascent and descent.

11.4.1 Step and Stair Ascent

When ascending, subjects are required to raise their centre of gravity during the pull up and then actively carry it forward to the next step. This is achieved through concentric muscular contraction, which displaces the centre of gravity vertically, a by-product of which is the generation of potential energy.

Clinically, it is important to note that McFadyen and Winter (1988) identified toe off during the pull-up phase as the greatest point of instability in the ascent of stairs (Table 11.1). There are two explanations for this: firstly, the effect of the external flexion moments, and secondly, the articular geometry of the joints, in particular the knee joint. At toe off, all the body weight is transferred onto the stance limb, where the hip, knee and ankle joints are all in a flexed position. In this position, all the external moments applied to the major lower-limb joints are flexion moments; subjects, therefore, require the generation of considerable concentric muscle activity to overcome the collapsing effect of these external moments. As pull up continues, stability increases due to the decreased effect of these external flexion moments.

When ascending, the stance knee starts from a relatively unstable position of semi-flexion, with the ligaments lax and little congruence between the articular surfaces of the femur and tibia. As the pull up takes place, the stance knee moves towards a more extended position, approximating to the close pack position of the joint (the most stable joint posture), with the ligaments tightening and the articular surfaces becoming more congruent (Shinno, 1971).

11.4.2 Step and Stair Descent

Stair descent is more challenging than stair ascent. From a clinical perspective Shinno (1971) suggests that during descent stability is more dependent on quadriceps function; therefore, any weakness in the quadriceps may show up as an impaired ability to descend stairs compared with ascending stairs. McFadyen and Winter (1988) support this view and state that the hip musculature contributes little to the work of lowering the body; this is accomplished predominantly by eccentric contraction of the quadriceps (Tables 11.1 and 11.2).

When descending, subjects must actively carry their centre of gravity forwards and then resist gravity during the controlled lowering phase. This is achieved through eccentric muscular contraction, which controls the rate of lowering of the centre of gravity by

TABLE 11.1	
Phases of Stair Climbing	
Ascent	**Descent**
Weight acceptance	Weight acceptance
Pull up	Forward continuance
Forward continuance	Controlled lowering

McFadyen & Winter, 1988

TABLE 11.2	
Key Muscles Involved in Stair Climbing	
Ascent	**Descent**
Vastus lateralis	Vastus lateralis and medialis
Gluteus medius and soleus	Gastrocnemius and soleus

McFadyen & Winter, 1988

absorbing kinetic energy. If strong eccentric contractions were not employed, the centre of gravity would accelerate under the influence of the gravitational pull of the Earth (Jevsevar et al., 1993).

From biomechanical and anatomical points of view, stair descent is the reverse of ascent; interestingly, this is partly what makes stair descent more challenging than ascent. During the controlled lowering phase, the hip and knee joints start from a relatively extended position and then flex, which causes a progressive increase in the external flexion moments. In order to prevent collapse, these external flexion moments have to be matched by the generation of progressively higher levels of eccentric muscle contraction. In a study of 10 healthy males, analysing the motions, forces and moments of the major joints of the lower limb, it was found that the mean flexion–extension moment at the knee for stair ascent was 57.1 Nm. For stair descent, the mean flexion–extension moment was nearly three times greater at 146.6 Nm (Andriacchi et al., 1980). Anatomically, the stance knee starts in a relatively stable extended position and progressively moves into a more unstable position of flexion as controlled lowering takes place. This also causes a progressive demand for increased muscular control.

In addition to the reasons outlined earlier, there is another factor that contributes to the explanation as to why step descent is more challenging compared to step ascent, which is related to the quadriceps extensor mechanism, in particular to the role of the patellofem-oral joint (PFJ). It is important to consider the role of the patella in the extensor mechanism and the contact zones on the patella as the knee moves into flexion. When moving from full extension to full flexion, dis-crete parts of the patella articulate with the femur; these are referred to as contact zones. Patellar contact zones have a horizontal orientation which are spread over approximately one-third of the articular surface of the patella. Fig. 11.4 shows that the contact area moves proximally towards the superior pole or base of the patella as the knee moves from extension into flexion. The resulting force and pressure (force/contact area) are referred to as patellofemoral contact force and patellofemoral contact pressure, which may be estimated using equations and models (Ward & Powers, 2004).

The link between knee flexion angle, knee flexion moment, patellofemoral contact force and patellofem-oral contact pressure has been used to explain the pres-ence of patellofemoral pain in active and athletic populations (Bonacci et al., 2014). As the knee flexion angle increases, the knee flexion moment also increases, resulting in a greater patellofemoral contact force or PFJ reaction force (PFJRF). Fig. 11.5 shows an indi-vidual at two different stages of a step descent; firstly at the point where all the weight has just been trans-ferred to the stance limb, and secondly at the point just before the swing limb makes initial contact with the step below. Initially the flexion angle and moment are

small, resulting in a low quadriceps, patellar tendon force and PFJRF; as the person moves into greater knee flexion the moment, quadriceps force, patellar tendon force and PFJRF all increase.

Such loads are not just restricted to tasks such as stair descent. During more high-risk sporting manoeu-vres, such as running, cutting, jumping and landing, the knee flexion moments are significantly higher, leading to higher patellofemoral contact forces and pressures. These, coupled with cumulative loading, are a significant risk factor for tissue overload and onset of patellofemoral pain in sporting populations. This is highlighted by the fact that patellofemoral pain is the most common chronic injury in recreational runners, which is characterized by pain and linked to the contact of the posterior surface of the patella with the femur (Besier, 2005).

When comparing the effects of loading between the hip and the knee in the same subjects, dramatic changes occur depending on activity. The knee is exposed to 33% higher compressive force than the hip during level walking, and 116% higher compressive force than the hip during stair climbing (Taylor et al., 2004a,b). Having an insight into the differential mag-nitude of the loads applied to the joints of the lower limb is important for a number of reasons. Coupling anatomical knowledge about the positions in which joints are inherently either stable or unstable, along with understanding the magnitude of loading, helps to present a clearer picture as to why joints get injured. Having this understanding should in turn help clinicians to plan logical progression through rehabilitation programmes, incrementally loading joint structures during different functional activities.

Using stair climbing as an example gives us a useful biomechanical insight into the importance of weight control and the changes in loading in the lower-limb joints that can occur with fairly small, but clinically significant, changes in bodyweight. Reilly and Martens (1972) report that the PFJ compression force is in the region of three times the bodyweight when involved in stair climbing. If we then apply these data to a very simple example of a patient who is 2 kg over their optimum bodyweight going up a flight of stairs that has 10 steps, we can see what a disproportionate and potentially detrimental effect the extra weight has on the PFJ (see Table 11.2).

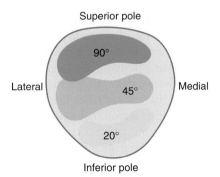

FIGURE 11.4 ■ Posterior aspect of the patella showing size and orientation of patellar contact zones at three different angles of knee joint flexion *(Adapted from Fulkerson & Hungerford, 1990.)*

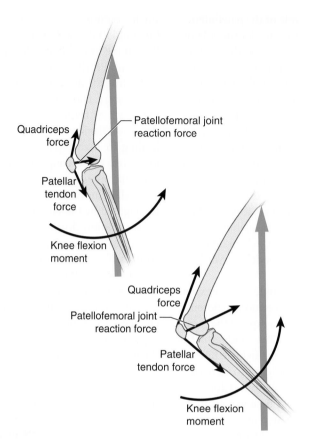

FIGURE 11.5 ■ Knee flexion moment, quadriceps force, patellar tendon force, patellofemoral joint reaction force during a step-down task

There are two groups of patients that this knowledge is particularly pertinent to: the obese and the intensive sports player. When considering the obese patient, the magnitude of these forces is going to be greater due to body mass. It is therefore important to consider weight loss strategies as part of the rehabilitation. There often emerges a 'catch 22' situation as one of the keys to weight loss is exercise; however, this may aggravate joint problems. Clinicians need to be sensitive to this issue and plan rehabilitation activities carefully in order not to provoke the very problem that the patient is seeking help for.

Intensive sports players present a slightly different rehabilitation challenge. The problem is often that they have a very high calorie intake, which is fine while they are playing a lot of sport. When they have an injury, they are unable to use as many calories, but often their appetite is undiminished and they maintain a very high calorie intake, which means that their weight increases. In terms of rehabilitation this can be problematic. Another reason that weight may increase is due to 'comfort eating', which occurs due to depressed mood because of being unable to play sport and because of boredom as 'there is nothing else to do!' The subject of weight control is a very sensitive one and problems associated with weight control can be associated with other underlying emotional problems, so clinicians have to proceed carefully in this area.

Finally, in this section it is also worth considering the work of Nissel and Ekholm (1985), who reported significant gender differences in the loading across the PFJ. They found that women compared to men had shorter patellar tendon moment arms, which lead to a 20% increase in patellar force. They argue that females are, therefore, exposed to higher PFJ stress than men of the same weight. This may account for the higher reported prevalence of patellofemoral disorders in females.

11.4.3 Motion of the Lower Limbs During Stair Descent

We will now consider the functional movement of the ankle and knee joints during stair decent. These will be described by angle against time graphs and angular velocity graphs to gain an understanding of not only the range of the motion of the different phases of joint motion but also control of the joint motion.

Ankle Joint Motion

The ankle joint starts off in slight dorsiflexion during initial standing. The ankle then moves into dorsiflexion slowly as the body moves over the foot to take the first step. The toe then comes off the ground, and a second smaller movement into dorsiflexion ensures foot clearance of the step. The ankle joint then plantarflexes to prepare for the contact with the next step down. At contact with the next step, the ankle rapidly dorsiflexes as the foot takes the load and the body moves the tibia over the ankle joint. The foot then comes off the ground and the ankle joint plantarflexes to prepare for the next step. The angular velocity graph shows the rate at which the ankle joint is plantarflexing (negative) and dorsiflexing (positive) (Fig. 11.6A, B).

Knee Joint Motion

The knee joint starts off flexed as the subject prepares to take a step. The knee then flexes to clear the step, then extends to move the tibia forwards and down to the next step. At contact with the next step, the knee flexes to take the load and to control the descent of the CoM to the lower level. The knee then moves towards extension once again to move the tibia forwards and down to the next step. The angular velocity graph shows the rate at which the knee joint is flexing (positive) and extending (negative) (Fig. 11.7A, B).

11.5 SQUATS AND DIPS

Squatting is a fundamental human movement and resting postural activity. It is a basic component of many sporting activities and is, therefore, found in many training and rehabilitation regimes in various altered formats. Although not so common in Western societies, it is a very common resting posture adopted regularly by millions of people worldwide. During squatting, a number of significant biomechanical events take place.

Many variations of squat exist (Tables 11.3 and 11.4). Earl and colleagues (2001) performed mini squats with simultaneous hip adduction, which increased quadriceps activity by 25%. Stuart and colleagues (1996) compared power squat, front squat and lunge. They found that during the lunge, quadriceps activity was significantly increased and hamstring activity was significantly decreased.

11.5.1 Quadriceps Wrap

Quadriceps wrap (tendofemoral wrap or wrap around effect) is the term coined to describe the point at which the quadriceps tendon comes into contact with the femur and starts to bear some of the load. Quadriceps wrap is considered to be a protective mechanism for the patella, by causing an unloading of the PFJ. The greatest rate of loading of the PFJ occurs just before quadriceps wrap. After quadriceps wrap has occurred, the rate of PFJ loading plateaus (Gill & O'Connor, 1996). There is no consensus as to the precise knee angle at which quadriceps wrap occurs (Table 11.5).

To date, no references to quadriceps wrap have appeared in the rehabilitation literature, which may be because the functional range that is of primary interest in rehabilitation is the first 30° of flexion. However, many athletic and occupational activities require the knee to be flexed beyond the point at which quadriceps wrap occurs, and it may be relevant to consider this phenomenon when rehabilitating these patients.

11.5.2 Quadriceps Neutral

The quadriceps neutral angle is defined as 'the knee flexion angle for which quadriceps contraction results in no anteroposterior shear on the tibia' (Singerman et al., 1999). At low flexion angles, due to the anterior orientation of the patellar tendon, the effect of the

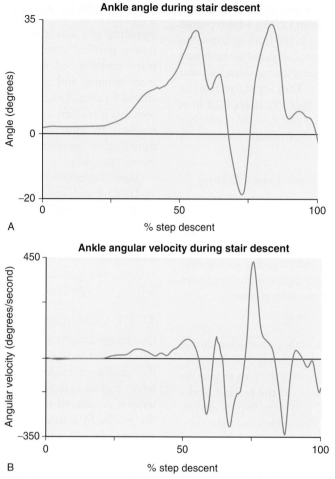

FIGURE 11.6 ■ (A and B) Ankle joint motion during step descent

TABLE 11.3				
Knee Forces During Squatting				
	Mean Body Weight (N)	Mean Load Lifted (N)	Normalized Peak Tibiofemoral Shear % (BW+load)	Normalized Mean Peak Tibiofemoral Compression % (BW+load)	Normalized Mean Peak Patellofemoral Compression % (BW+load)
Stuart et al., 1996	798	223	29	54	
Ariel, 1974	888	1982	56	276	
Escamilla et al., 1998	912	1437	80	133	194
Escamilla et al., 2001	917	1309	99	154	210
Wilk., et al 1996	912	1442	76	261	
Toutoungi et al., 2000	765	0	353		
Nissell and Ekholm, 1985	932	2453		198	191
Hattin et al., 1989	790	339		367	
Wretenberg et al., 1996		650			324
Reilly and Martens, 1972	834	0			765

N, Newtons; BW, bodyweight.
Adapted from Escamilla, 2001.

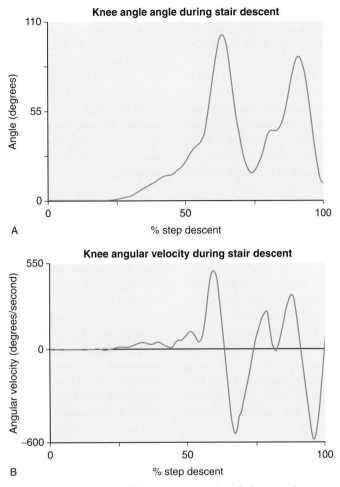

FIGURE 11.7 ▪ (A and B) Knee joint motion during step descent

TABLE 11.4		
Muscle Activity During Squatting		
	Muscle	**Peak Activity Knee Flexion Angle**
Escamilla et al., 1998	Quadriceps	80°–90°
	Hamstrings	10°–60° (ascent)
	Gastrocnemius	60°–90°
Stuart et al., 1996	Hamstrings	30°

Adapted from Escamilla, 2001.

TABLE 11.5	
Angle at Which Quadriceps Wrap Occurs	
Ellis et al., 1980	80°–105°
Nissel and Ekholm, 1985	60°

TABLE 11.6	
Angle at Which Quadriceps Neutral Occurs	
Escamilla, 2001	50°–60°
Singerman et al., 1999	50°–55°

quadriceps is to shear the tibia in an anterior direction. At high angles of flexion, due to the posterior orientation of the patellar tendon, the effect of the quadriceps is to shear the tibia in a posterior direction. The changing direction of pull of the patellar tendon is due to articular geometry of the femoral condyles, which are cam shaped. It is interesting to consider that, although quadriceps neutral appears to be quite a significant biomechanical event of the knee, no references to the effect it has on the PFJ have been found (Table 11.6). Although quadriceps wrap and quadriceps neutral appear to be quite significant biomechanical events of the PFJ, neither event is referred to in mainstream rehabilitation literature. Future researchers should consider whether quadriceps wrap and/or quadriceps neutral are important phenomena when investigating patient populations.

Dahlkvist and colleagues (1982) analysed the joint and muscle forces of deep squatting, and differences between descent and ascent were found. During a slow descent, both the PFJRF (7.41 × bodyweight) and the quadriceps force (5.27 × body weight) were greater than when slowly ascending, with the PFJRF being 4.73 × bodyweight and the quadriceps force being 4.94 × bodyweight (Dahlkvist et al., 1982). The authors attributed this phenomenon to the larger momentum that occurs when descending.

11.5.3 Joint Moments and EMG Activity During a Single Limb Squat

The use of eccentric activities for rehabilitation associated with tendinopathy has been well documented (Cook & Khan, 2001; Panni et al., 2000; Roos et al., 2004). A number of authors (Alfredson et al., 1998; Jonsson & Alfredson, 2005; Khan et al., 1998; Purdam et al., 2004) have suggested that a 25° decline squat in comparison to a flat squat produces a significant improvement in the ability of the individual to participate in sports and a reduction in pain. There is, however, no scientific justification given as to why 25° was chosen for the decline. The suggested depth of squat based on the angle of knee flexion varies between 50° and 90° (Alfredson et al., 1998; Onishi et al., 2000; Young et al., 2005). Initially, Purdam and colleagues (2004) proposed 50°, with the basis for this being that the force in the patellar tendon is equal to that of the

quadriceps tendon when in this particular orientation. However, subsequently, Purdam and colleagues (2004) proposed 90° of flexion, with Jonsson and Alfredson (2005) using 70° of flexion. Based on this variation in the range of flexion within the literature, there is no consensus within contemporary research. However, clinically, there can be considerable differences in the amount of knee flexion different individuals are able to achieve during eccentric squat activities, so the relevance of controlling the amount of knee flexion is debatable.

To investigate the optimum angle of decline Richards and colleagues (2008) considered four angles: 0°, 8°, 16° and 24°. The subjects were instructed to perform the squat as slowly as possible to approximately 90°. Whilst performing the squat movement, force and analysis data electromyography (EMG) were collected from rectus femoris and gastrocnemius. External joint moments were calculated using inverse dynamics methods. The enveloped EMG magnitudes at maximum knee flexion (Figs 11.8 and 11.9), the iEMG (integrated EMG) during the squat, and the ankle and knee moments at maximum knee flexion were recorded (Figs 11.10 and 11.11). EMG and iEMG were normalized to the maximal dynamic contraction (Kellis & Baltzopoulos, 1996).

The ankle moments during a squat showed a significant decrease with the introduction of a declined angle. It is interesting to note that the ankle moment is not significantly different between 0° and 8°, but for all subsequent increases, a significant difference was seen in the pairwise comparisons. This can be explained by examining the nature of the change in ankle angle in relation to the base of support from the foot. As the decline increases, so the base of support decreases; however, the further away from the horizontal, the larger the effect of the angular changes on the base of support. The reduction of the base of support has the effect of reducing the moment arm between the ground reaction force (GRF) and the joint and thus the moment.

The knee moments during a squat also showed significant increase with the introduction of a declined angle. However, knee moments showed significant changes in the comparisons for all angles apart from between 16° and 24°. Although the moment about the ankle is significantly reduced, this does not correspond

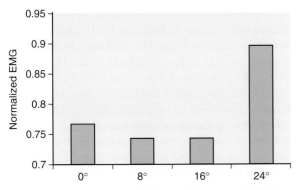

FIGURE 11.8 ■ Electromyography (EMG) from gastrocnemius

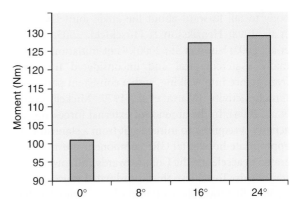

FIGURE 11.11 ■ Knee joint moment at different decline angles

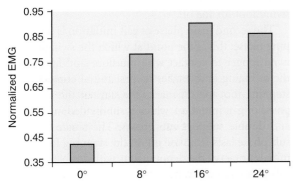

FIGURE 11.9 ■ Electromyography (EMG) from rectus femoris

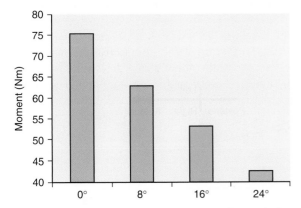

FIGURE 11.10 ■ Ankle joint moment at different decline angles

to the level of muscle activity seen in the gastrocnemius. This, in effect, requires an increase in stability leading to an increase in the activity of the gastrocnemius to stabilize the ankle in this position; an alternative explanation is the shortening of the muscle, which would also lead to a greater EMG signal in relation to the joint moment. This would suggest that increasing the angle of decline to 24° increases the activity of the gastrocnemius, and does not decrease it as previously thought (Jonsson & Alfredson, 2005; Purdam et al., 2004).

The effectiveness of different angles of decline has been established for targeting the knee extensors, concluding that a 16° decline produces an exercise that specifically targets the knee extensors with minimum effect about the ankle. At 24° of decline an increase in gastrocnemius activity is evident, which implies that the increased angle challenges the stability of the ankle. However, further studies are required to establish the effectiveness in specific exercises used in clinical practice.

11.6 GAIT INITIATION

11.6.1 Normal Phases of Gait Initiation

Gait initiation is the mechanical and neurological process by which the body's CoM decouples or separates from the centre of pressure (CoP), causing the body to fall forward about the ankle joint (Halliday et al., 1998; Henriksson & Hirschfeld, 2005; Martin et al., 2002; Viton et al., 2000). Gait initiation is normally a stereotypical and unconsidered transition from stance into walking, with a consistent pattern of muscle activity (Mann et al., 1979; Mickelborough et al., 2004). In the absence of external forces, muscle activity is required to initiate gait from a standstill. An appropriate horizontal GRF component must be generated to accelerate the CoM forwards and towards the stance side. To achieve this, coordinated muscle activity initially moves the CoP towards the swing leg. Jian and colleagues (1993) extended gait initiation beyond support limb toe off to the time when the swinging limb becomes the stance limb and toe off occurs. Therefore, gait initiation was broken down into preparatory phase, take off phase and stabilizing period. Elble and colleagues (1994) described gait initiation as movement from a steady state to the point up to and including when the swinging toe left the ground. Brunt and colleagues (1999) defined gait initiation as the transition from quiet stance to steady-state gait. These events have generally been identified based on either force platform data or EMG data. Force analysis has identified events based on the projected CoP during contact phases of gait initiation. The process of gait initiation has therefore been generally accepted to consist of two main phases: the preparatory (postural) phase and stepping (monopodal) phase (Fiolkowski et al., 2002; Mickelborough et al., 2004; Viton et al., 2000), with the preparatory and stepping phases being of similar duration.

The preparatory phase is when the body begins the decoupling process, shifting the CoP initially in the direction of the swinging limb and then in the direction of the stance limb (Halliday et al., 1998). The preparatory phase lasts from onset until the toe off of the stepping foot and is divided into two sub-phases: a release phase and an unloading phase (Archer et al., 1994). During the release sub-phase, the CoP is moved towards the swing foot, which has the effect of increasing the horizontal GRF components that accelerate the CoM in the opposite direction (Polcyn et al., 1998). This release sub-phase lasts until the furthest point of posterolateral CoP movement, when the CoP abruptly changes direction, marking the start of the unloading sub-phase. During the unloading sub-phase, the CoP moves rapidly across to the stance foot, unloading the swing foot for toe off.

The second main phase of gait initiation is the stepping phase; this is the point at which the swinging leg is no longer in contact with the floor and lasts until the swinging limb makes its first initial contact. The stepping-foot toe off marks the start of the stepping phase of gait initiation, which is subdivided into single and double support sub-phases. The single support sub-phase lasts from toe off on the stepping foot until initial contact of the stepping foot, with double support sub-phase lasting from initial contact of the stepping foot until the toe off of the original supporting foot.

Fig. 11.12 shows the movement of the CoP *(black line)* and CoM *(blue line)* during gait initiation. This shows the direction of walking from left to right with the right foot being the initial swing foot. Onset is

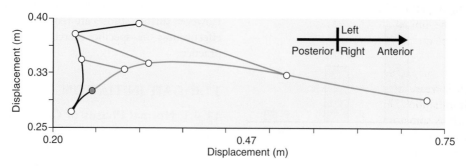

FIGURE 11.12 ■ The characteristic pattern of horizontal centre of pressure (black line) and centre of mass (blue line) displacement during gait initiation

marked with a blue circle and the major foot contact events are shown as open circles on each curve. During the release phase the CoP moves from onset to its furthest point in the posterolateral direction with little movement of the CoM. During the unloading phase, the CoP moves across to the stance limb while the CoM moves towards the midline between the two feet. The stepping phase sees a movement of the CoP forwards from heel to toe on the stance foot; at the same time the CoM moves forward as the swing limb is no longer in contact with the floor and as the body moves forward. This decoupling of the CoM and CoP generates an acceleration vector which can be shown as a line between the CoP and CoM.

11.6.2 Gait Initiation, Freezing of Gait and Parkinson's Disease

Freezing of gait (FOG) remains one of the most common debilitating aspects of Parkinson's disease. It has been linked to injuries and falls and is a main contributory factor in reducing quality of life (Giladi & Nieuwboer, 2008; Moore et al., 2007). FOG causes temporary cessation of effective stepping and a sensation of 'feet being glued to the floor' (Giladi et al., 1992) and occurs when people turn (63%), initiate walking (23%), walk through narrow spaces (12%) and reach destinations (9%) (Schaafsma et al., 2003).

There are multiple factors that can induce and overcome components of FOG (Lebold & Almeida, 2010), with pharmacological and surgical intervention often unable to ameliorate symptoms (Griffin et al., 2011). The European guidelines for Parkinson's disease strongly recommend using cues for the improvement of walking speed; however, they weakly recommend against cueing of gait for improvement of FOG. This can be due to the limited literature that is available on this topic and the variety of cues used to improve FOG. Transverse lines (TLs) on the floor have been shown to improve gait in people with Parkinson's disease, including an increase in stride length and improvement in gait initiation (Lim et al., 2006). Other external cues, such as somatosensory, visual and auditory stimuli, have also been used with mixed results; however, these studies focused mainly on steady-state gait and not on overcoming gait initiation failure (Frazzitta et al., 2009).

Gait initiation failure or 'start hesitation' is a component of FOG which is described as a difficulty in initiating gait in the Unified Parkinson's Disease Rating Scale (UPDRS). Gait initiation is normally a stereotypical and unconsidered transition from stance into walking. Giladi and colleagues (1992) explored the presence of motor blocks in a sample of 990 people with Parkinson's disease; 318 were found to have FOG, 86% of these had blocks in initiation of gait.

Jiang and Norman (2006) investigated the effects of visual and auditory cues on gait initiation in people with Parkinson's disease. They found differences in maximum horizontal force between people with Parkinson's disease who freeze and do not freeze, and between the different cues. The auditory cues used were rhythmic sounds matched to the participant's average step time and the visual cues were high-contrast TLs on the floor adjusted for the participant's height and first step length, which, although beneficial, have a limited practical value outside of the laboratory setting (Donovan et al., 2011). Moreover, the auditory cues in this study did not produce a significant difference when compared to the no cue condition. Unfortunately, the authors grouped the individuals with and without gait initiation difficulty together when studying the effect of the different cues, therefore diluting the effect and the potential clinical relevance of the findings.

Van Wegen and colleagues (2006) investigated the use of a rhythmic somatosensory cueing device attached to the wrist on gait initiation in people with Parkinson's disease. This showed that participants were able to modify their stepping pattern. The authors suggested that such cues draw attention to the act of walking. Dibble and colleagues (2004) considered the effects of different sensory cueing methods on gait initiation in people with Parkinson's disease. The cueing methods were a single and repetitive auditory signal from an electronic metronome and an electrical stimulus from a neuromuscular stimulator. They found that both these sensory cueing modalities had a negative effect on displacement of the body and swing limb. Cubo and colleagues (2004) examined the effects of a metronome in 12 patients with freezing when in their 'on' state and reached similar conclusions: walking time increased when using the metronome.

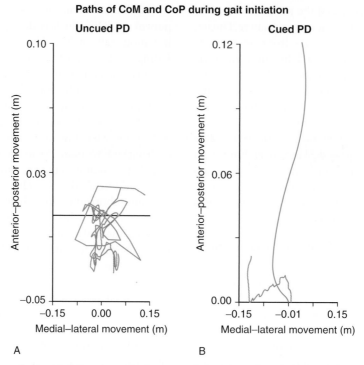

Paths of CoM and CoP during gait initiation

A

B

FIGURE 11.13 ■ Centre of mass (CoM) and centre of pressure CoP pathway in a Parkinson's disease patient during a single stride (A) during a freezing of gait episode, (B) using a visual cue (lasercane)

McCandless and colleagues (2016) assessed 20 people with Parkinson's disease; 12 of the participants had freezing episodes whilst been tested in a biomechanics laboratory, and from these participants, 100 freezing and 91 non-freezing trials were recorded. This study identified clear measurable differences in the mechanisms and control between freezing episodes and non-freezing episodes and the effect of commercially available portable cueing devices, which showed improvements in step length and CoM velocity (Fig. 11.13). This provided important information about the immediate effect of cues on gait initiation for people with Parkinson's disease who are affected by freezing, and showed a significantly lower number of freezing episodes when using cueing devices (Fig. 11.14). This could be used to inform clinical practice about the effectiveness of such cues, but future research is needed to determine possible effects of cue training and

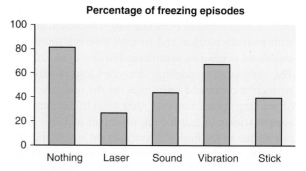

FIGURE 11.14 ■ Percentage of freezing episodes using the different cueing methods

the longer-term effect of using cueing devices. Although clinicians may be reluctant to provide Parkinson's patients with sticks, as it is perceived as detrimental due to possible effects on an already flexed posture, they need to balance the possible adverse effects against the possible beneficial effect as a cue.

11.7 MUSCLE STRENGTH AND POWER ASSESSMENT

11.7.1 What Affects Strength and Power Assessment?

One way in which we often talk about muscle and joint performance is *strength*. But what exactly *is* strength? The dictionary tells us that strength is the capacity for exertion or endurance or the power to resist force. However, a better way of thinking about *muscle* strength is the amount of force a particular muscle or muscle group can produce. However, when evaluating muscle strength, the measures taken are not directly measuring the actual strength of the muscle or muscle group. What is usually recorded is the effective moment being produced by the muscle. This is because muscle forces are hard to measure, requiring information about the position of the muscle, position of the body segment, muscle insertion points, and the line of pull of the muscle, all of which will be constantly changing during dynamic activities.

Most measures taken in the clinical setting do not go as far as to estimate actual muscle forces. However, there are a number of methods of indirect evaluation. Indirect evaluation of the force produced by a muscle can be influenced, however, by a number of factors. The following section will consider the different permutations of these factors when considering upper-limb muscle strength.

These factors include:

- body segment inclination
- load position and size
- muscle insertion
- angle of pull of the muscle
- type of contraction
- speed of contraction.

Body Segment Inclination

The inclination of body segments can have a very large effect on joint moments. The effect of the weight of the forearm in the three positions shown in Fig. 11.15A–C is very different. The maximum moment about the elbow is when the forearm is level; when the forearm is inclined either up or down the moment reduces, and when the forearm is vertical there will be no moment about the elbow at all, as the entire weight of the segment will be acting through the joint. It is also important to note the direction of the 'stabilizing'

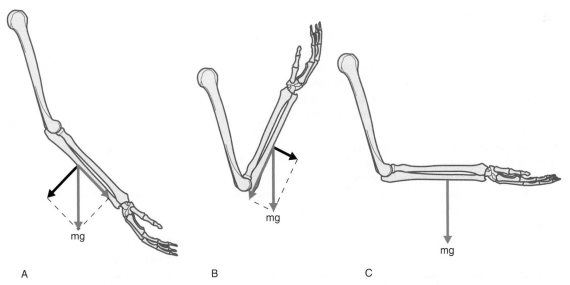

FIGURE 11.15 ■ Effective moment caused by the weight of the limb: (A) forearm angled down, (B) forearm angled up and (C) forearm horizontal

component that acts along the forearm. When the forearm is angled down, the component acting along the forearm will try to pull the forearm away from the upper arm, whereas with the forearm angled up the forearm is pushed into the upper arm, which will have the effect of reducing and increasing the joint force at the elbow, respectively (Fig. 11.15A, B, C).

The Position and Size of the Applied Load

The position and size of the load applied has an important effect on the moment about the elbow and will, in turn, have a significant effect on the muscle and joint forces. When assessing muscle strength both these factors should be measured and taken into account. If you position a load at the end of a subject's arm to see if he/she can support it, the moment will depend on the size of the load and the subject's limb length (Fig. 11.16A, B).

Both the size of the load and the subject's limb length need to be considered when assessing an individual's muscle performance or strength. For example, if two individuals of different heights and, therefore, different tibial (shank) lengths conducted the same leg raise activity with the same loads, the shorter of the two would, in fact, use less muscle force (strength) to lift the same load, assuming the muscle insertion points were not significantly different.

Muscle Insertion Points

Different muscles will have different insertion points. The position of these insertion points will have a large effect on the muscle force required to support a given turning moment. Fig. 11.17A, B shows two examples: (A) the muscle insertion point is close to the elbow joint and (B) the muscle insertion point is much further away.

For a particular load, there will be a larger force in the muscle if its insertion point is close to the joint. Conversely, if there is a maximum force that a muscle group can cope with, then larger loads will be able to be carried with the insertion point further away from the joint.

This leads us to an interesting point when we consider weight lifters. Is a weight lifter able to lift a larger load because he or she can support larger muscle forces, or is this due to a difference in the muscle insertion points? If the latter is the case, are we actually

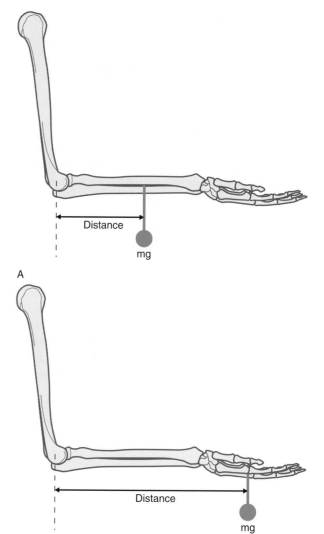

FIGURE 11.16 ▪ (A and B) Position and size of the applied load

assessing something different from *strength* (the force in the muscle) with the task?

The Effect of the Angle of Muscle Pull

As the body segment moves relative to the ground, so the angle of the muscle moves relative to the body segment. The maximum moment that the muscle can produce is when the elbow is at 90°, as this makes an

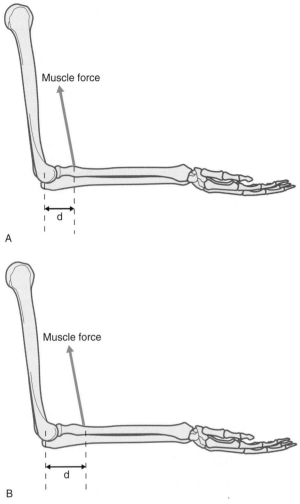

Muscle force

d

A

Muscle force

d

B

FIGURE 11.17 ■ (A and B) Position of muscle insertion point

smallest. It is also interesting to note the direction of the 'stabilizing' component of the muscle force when the elbow is flexed. This appears to be pulling the forearm away from the joint and will not provide a compressive stabilizing force into the joint; however, in this position the rotary component will be providing a compressive force into the joint.

Type of Muscle Contraction

The type of muscle contraction affects the resistance that can be controlled, held or overcome. The three types of muscle contraction are isometric, concentric and eccentric. Isometric contractions are stabilizing contractions, where the muscle length remains virtually constant. Concentric contractions are where the muscle shortens during the activity. These are generally the weakest muscle contractions, requiring more motor unit recruitment than isometric and eccentric for a particular load. Eccentric contractions are where the muscle lengthens during the activity. These are generally the strongest muscle contractions, requiring less recruitment than isometric and concentric for a particular load.

The Effect of the Speed of Contraction

There are three ways of classifying speed during exercises: isotonic, isokinetic and isometric. Isotonic is when a constant load is applied but the angular velocity of the movement may change, which allows an infinite variation in the rate of contraction of a muscle. Although this is closest to real-life muscle and joint function, the changing speed continually affects the amount of force that a muscle can produce and makes the exact muscle function quite hard to assess. Isokinetic is when the velocity or angular velocity of the movement is kept constant, but the load may be varied. This setting of the speed of working helps improve our assessment of muscle performance, but the speed or velocity of the joints are being restricted to only one set speed at any one time. Isometric relates to the force varying, but the joint is held in a static position; therefore, muscle length remains the same as no movement occurs. This tells us what static moment may be supported; however, this does not necessarily relate to the moments that can be produced or supported dynamically.

approximately 90° angle between the muscle and the body segment and, therefore, produces the greatest rotary component from the muscle force. As the elbow joint is moved away from this position, either flexed or extended, the moment that the muscle can produce is reduced as the rotary component of the muscle force acting at 90° to the forearm is reduced. When the forearm is vertical with the elbow fully extended, the muscle would find it much harder to produce a moment as the rotary component will be at its

11.8 CLINICAL ASSESSMENT OF MUSCLE STRENGTH

11.8.1 The Oxford Scale

The Oxford scale is a common clinical assessment method for muscle strength. The Oxford scale classifies muscle strength, which is categorized using the following criteria:

0 = No contraction
1 = Flicker of a contraction
2 = Active movement with gravity eliminated
3 = Active movement against gravity
4 = Active movement against light resistance
5 = Full functional strength (full range of motion against strong resistance).

Scores 0–3 give a very useful functional progression by assessing if the individual can support against gravity; however, care must be taken that the body segments are constantly placed to ensure a consistent and 'correct' effect due to the weight of the body segment and associated muscle action. Scores 4 and 5, however, are open to considerable variability as the exact definition of light resistance and strong resistance will vary from clinician to clinician. In addition, the position of the applied load on the body segments, or the length of the body segment, will vary the moment arm, which will have a significant effect on the muscle force needed to resist. The velocity with which the joint moves will also have a direct effect on the joint power. This leads to difficulties in the comparison of assessments using Oxford scale between clinicians doing the testing and the individuals being tested.

Although the Oxford scale has come under criticism, it does offer a rough guide for the assessment of muscle strength. One way of improving the concept of the scale is to introduce a degree of measurement to the assessment. This could be as simple as measuring the distance from the resistance to the joint and being consistent in this distance for the testing of a particular joint for all individuals, or bringing in some form of force measurement to assess what force can be produced. This could give a more objective means of assessing change in muscle function through a rehabilitation programme.

11.8.2 Hand-Held Dynamometers

One method of improving the measurements taken in clinical assessment is using a hand-held dynamometer, which measures force and can be attached to different body segments using a sling. A wide variety of these relatively inexpensive and portable instruments are available. They consist of a variety of force-detecting systems, including hydraulics, springs, load cells and strain gauges. Similar to testing when using the Oxford scale, care is required in the positioning of the force sensors on the body segments to ensure repeatable and useful measurements. Bohannon (1990) stated that when using hand-held dynamometers it is particularly important that the dynamometer should be placed perpendicular to the tested limb. It is also very important that the tester stabilizes the tested limb appropriately during the test in order to prevent unwanted substitution movements. Clear explanation of test procedures is also required and an opportunity to practice is also crucial and should be standardized. Clinicians should, however, be aware of the potential for greater patient discomfort during break tests compared to make tests, and should use appropriate clinical reasoning to inform their decision as to which type of test is appropriate to conduct for individual patients. Some examples are shown of the use of a hand-held dynamometer in the assessment of knee extensor and hip abductor strength, which has been used to identify clinical subgroups in patellofemoral pain (Selfe et al., 2016) (Fig. 11.18A, B).

11.8.3 Free Weights and Springs

DeLorme is credited with introducing the concept of repetition maximum (RM) in 1945. RM is defined as the weight that can be moved a given number of times and no more; therefore, a 1 RM is the maximum amount of weight that can be lifted once, whereas a 10 RM is the maximum weight that can be lifted 10 times.

The use of spring balances to measure muscle strength is also clinically popular and is incorporated into a number of standardized functional assessment protocols. One of the best-known examples of these is the Constant score for the shoulder (Constant & Murley, 1987).

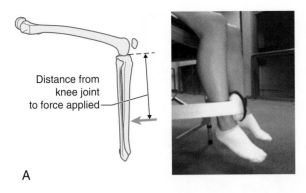

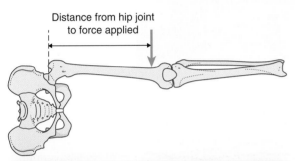

FIGURE 11.18 ■ (A) Measurement of knee extensor moment. (B) Measurement of hip abductor moment

11.9 ISOKINETIC AND ISOMETRIC TESTING

The use of isokinetics allows for a standardized assessment by controlling, or pre-setting, the angular velocity and measuring the resistance that can be produced by an individual. In controlling the angular velocity and measuring the resistance, the muscle power produced becomes very easy to find. Isokinetics machines also allow concentric, eccentric and isometric moments (commonly referred to as torque in isokinetics) and concentric and eccentric power to be found separately. Many isokinetic machines are also capable of isotonic assessment—isotonic referring to constant load, or torque—throughout the range of motion. Isotonic

testing is a simulating of free weights, although isokinetic machines also have the ability to allow for the weight of the segment through the range of motion being tested, therefore giving a closer representation of the torque and power provided by the muscles (Fig. 11.19).

11.9.1 Measurements Taken in Isometric Testing

The term isometric refers to exercise where a force is applied, but no movement occurs. Therefore, the type of contraction being tested is always isometric. Variables that have been used in research include:

- Maximum torque during a contraction
- Maximum torque at different joint angles
- Ratio of maximum torque of antagonistic muscle
- Maximum torque-to-bodyweight ratio
- Impulse torque.

Torque Measures During a Contraction

A single value of peak torque may be recorded during a contraction. This does not tell us how the torque is produced over time, but just the maximum value observed. The peak torque at various time points may also be considered, which gives a measure of the endurance or fatigue: that is, the ability to not only produce but also maintain a particular force. This is often carried out at a percentage of the maximum force, or torque, which can be generated.

The peak torque at different joint angles can also be assessed. This is a series of measures of peak torque that are taken at different joint angles. This shows the effect of the angle of pull of the muscle relative to the body segment and the muscle length. Both these factors affect the maximum amount of torque that can be produced. Fig. 11.20 shows how the peak torque during a knee extension exercise changes as the knee is moved from 90° of flexion to 0° of flexion.

Fig. 11.21 shows the relationship between the peak torque produced and the joint angles. From this second figure we can see that, near to full extension, much less force can be produced, and as the knee flexion angle gets over 60° the extensor torque begins to plateau, although increases in isometric torque are seen up to a knee flexion angle of 90°.

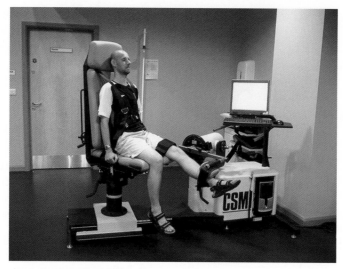

FIGURE 11.19 ■ Isokinetic machine

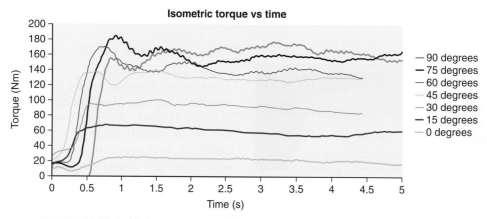

FIGURE 11.20 ■ Maximum torque at different joint angle for a knee extension exercise

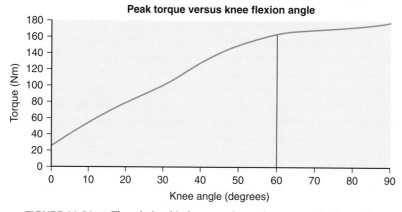

FIGURE 11.21 ■ The relationship between the peak torque and joint angles

Ratio of Maximum Torque of Antagonistic Muscle

The maximum torque of antagonistic pairs of muscles and the ratio between them may be found. However, care is advised as this does not relate to work done or power production as no movement is occurring during isometrics. However, the study of different antagonistic pairs of muscles can produce interesting data of the maximum torque values at different joint angles.

Maximum Torque-to-Bodyweight Ratio

Sometimes it is necessary to compare results from different individuals. Therefore, the maximum torque values may be divided by bodyweight. This is an attempt to allow for larger subjects having larger muscle bulk and, therefore, should be able to produce larger torque values. By normalizing in this way, we are measuring the maximum torque that can be produced for a given muscle bulk. This allows different individuals to be compared, although it can be susceptible to errors due to anthropometric variations.

Impulse Torque

The impulse torque is the area under the torque-versus-time graph. This relates to the maximum torque produced and how long it may be sustained. This has been erroneously referred to as work done by the muscle. Isometric testing does not tell us about the balance between the work done by the muscles or the power generation and absorption, as no movement takes place. However, it can be useful to look at the sustainability of a particular torque during isometric fatigue testing (Fig. 11.22).

11.9.2 Typical Measurements Taken in Isokinetic Testing

Isokinetic refers to movement at a constant angular or linear velocity. For both concentric and eccentric contractions, variables that have been used in research include:

■ Peak torque and angle at peak torque
■ Peak torque-to-bodyweight ratio
■ Angle-specific torque
■ Work done by muscles
■ Peak power
■ Average power.

Peak Torque

This is simply the peak moment or torque recorded throughout the range of motion. With isokinetics the angular velocity remains constant, so the magnitude of torque for any particular angular velocity is proportional to the maximum power that muscle group can produce throughout the range of motion. It is interesting to note that the peak torque occurs at approximately 60–70°. Above this critical angle the torque decreases, which can be related to a decrease in the control of the joint in close chain exercises, such as knee flexion into a squatting position, where at angles greater than 60°, the subject can no longer hold the position and the knee collapses into flexion. The importance of the 60° knee flexion angle can be related to the quadriceps lever arm, which works with a mechanical advantage up to 60°; however, at angles greater than 60° knee flexion, the quadriceps work at

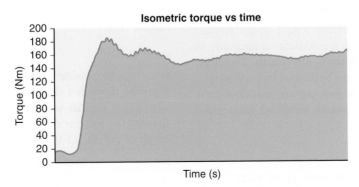

FIGURE 11.22 ■ Impulse torque: area under torque vs time

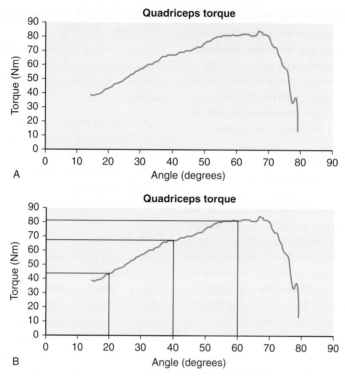

FIGURE 11.23 ■ (A) Torque versus angular displacement for quadriceps (peak torque = 84 Nm) and (B) angle-specific torque measurements for quadriceps

a mechanical disadvantage (Gill & O'Connor, 1996; Nissel & Ekholm, 1985). The nature of this mechanical disadvantage can be seen by the rapid decrease in the torque that may be produced by the quadriceps beyond 60°. For this reason, the angle or time-to-peak torque is sometimes recorded, which would indicate where the joint is acting at its maximum torque development or mechanical/physiological advantage (Fig. 11.23A).

Peak Torque-to-Bodyweight Ratio

As with isometric maximum torque-to-bodyweight ratio, in isometric testing it is sometimes necessary to compare results from different individuals. Therefore, the peak torque values may be divided by body weight. By normalizing in this way, we are measuring the maximum torque that can be produced for a given muscle bulk.

Angle-Specific Torque

Another method of assessment is to measure the torque at specific angles, sometimes referred to as 'angle-specific torque' (AST). This is where the clinician picks two or more angles within the range of motion. The torque produced at these angles is then recorded. This allows key points within the range of motion to be studied and not just the time the subject reaches peak torque production. This is a useful technique if the profile of the torque-versus-angle graph is of interest, and is particularly useful if the subject being tested has joint instability due to an injury that may only affect a small part of the total range of motion (Fig. 11.23B).

Work Done

Work may be considered to be occurring when a force is moved through a distance. However, when we are considering angular work, the distance moved is the distance the force is moved through an arc, and the length of the arc is $\mathbf{r}\,\theta$. Therefore:

$$\text{Work} = \text{Force} \times \text{r} \times \theta$$

or

$$\text{Work} = \text{Torque} \times \theta$$

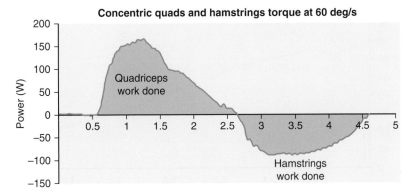

FIGURE 11.24 ■ Concentric quads and hamstrings torque at 60°/s

Therefore, work done may be found by calculating the area under the torque-versus-angle graph, using the technique shown in Chapter 3. It should be noted that the angle should, in fact, be in radians in order to calculate the work done; however, to aid communication of the results, this is more often than not hidden in the isokinetics software (Fig. 11.24).

As with peak torque-to-bodyweight ratio, work may also be expressed with respect to body weight. This is an attempt to allow for larger subjects having larger muscle bulk. They, therefore, should be able to produce larger torque and do more work for a given range of motion. The value of lean bodyweight is sometimes used to normalize the data rather than actual bodyweight to take into account the amount of subcutaneous fat.

Peak Power and Average Power

Peak power is found in much the same way as peak torque. If the torque throughout a range of motion is known, and the angular velocity, then the power can be calculated (see Section 5.3.2: Angular power). Fig. 11.25A, B shows isokinetic testing of the quadriceps muscle group concentrically at 30°/s. The two graphs show the same data but displayed in different ways: Fig. 11.25A shows the power versus time and Fig. 11.25B shows power versus angle. Both provide useful data, although the power versus angle is generally considered more useful as it provides an anatomical reference of the power production in relation to the joint position. Average power may also be found over the entire range of motion or time, although it is questionable how useful this is.

11.9.3 Muscle Testing Using Isokinetics

Testing of Antagonistic Pairs of Muscles

The testing of antagonistic pairs of muscles is one of the most common uses of isokinetics. Fig. 11.26 shows a data set from testing the concentric/concentric plantarflexor/dorsiflexor torque over the range of motion of the ankle joint. The positive torque values represent the plantarflexors, and the negative torque value the dorsiflexors. The positive angle values indicate when the ankle is in plantarflexion and the negative when the ankle is in dorsiflexion (Fig. 11.26).

The nature of this loop of data demonstrates the 'interaction' of this antagonistic pair of muscles, although antagonistic isokinetic data are not often presented in this loop form and it is more usual for the data of each of the muscles to be graphed separately, and for specific values to be recorded as discussed in Section 11.9.2: Typical measurements taken in isokinetic testing.

It is often important to assess the balance between antagonist muscle groups. To assess this balance, the peak torque or peak power of the two muscle groups can be found and a ratio calculated. A common ratio to see quoted is the concentric/concentric hamstring/quadriceps ratio or HQ ratio. It is well documented that the HQ ratio is approximately 0.67 : 1, the hamstrings being two-thirds of the strength of the quadriceps in normal subjects. However, in many sporting activities one muscle group may be disproportionately

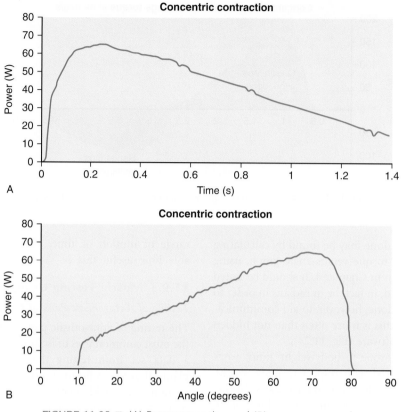

FIGURE 11.25 ■ (A) Power versus time and (B) power versus angle

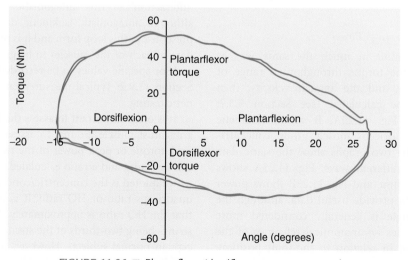

FIGURE 11.26 ■ Plantarflexor/dorsiflexor torque versus angle

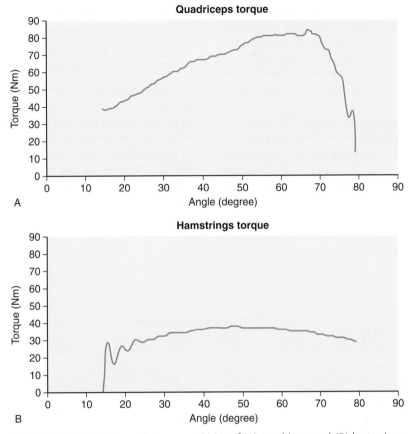

FIGURE 11.27 ▪ Concentric torque at 30°/s of (A) quadriceps and (B) hamstrings

trained and this may lead to an imbalance between the antagonistic muscles. The HQ ratio gives a measure of this muscle imbalance. This may be found for any antagonistic pair of muscles (Fig. 11.27A, B).

$$\text{Peak torque quads} = 84 \text{ Nm}$$

$$\text{Peak torque hams} = 38 \text{ Nm}$$

$$\text{HQ ratio} = 0.45 : 1$$

Fig. 11.27A and b shows a marked muscle imbalance between the quadriceps and hamstrings, the quadriceps having an increased peak torque compared to the hamstrings relative to that of normal subjects. This reduced capacity of the hamstrings relative to the quadriceps would mean there is more likelihood of anterior shear of the tibia forwards as the quadriceps overpowers hamstrings. Such an anterior draw would lead to the pre-stressing of the anterior cruciate ligament. It is interesting to note that professional soccer players have very well-developed quadriceps, but also often have relatively poor defined hamstrings, leading to a very low HQ ratio. Therefore, this imbalance between quadriceps and hamstrings should be considered as one of the possible mechanisms for both anterior cruciate ligament injury and hamstring tears.

However, it is questionable how functional the comparison of concentric/concentric torque of power is for antagonistic pairs. If we take, for instance, the activity of kicking a ball, the quadriceps will be acting concentrically to accelerate the tibia forwards towards extension and, therefore, generating the power to kick the ball, whereas the hamstrings will be required to act eccentrically to decelerate the tibia to stop the knee going into hyperextension, which again could cause ligament and joint damage. This will not necessarily be due to the kicking of the ball but more due to the deceleration of the tibia by the hamstrings after the ball has left the foot. A more functional assessment, therefore, would be to examine the antagonistic pair of muscles by testing the extensors concentrically and the flexors eccentrically, or vice-versa, depending on the antagonistic pair being assessed or the activity being replicated.

So, should the concentric power and torque be more balanced with the eccentric? Previously we considered that concentric muscle action was the weakest, then isometric, with eccentric muscle action being the strongest. The data in Fig. 11.28 show that greater power may be obtained by working a particular muscle eccentrically when compared to concentrically at the same angular velocity.

$$\frac{\text{Concentric}}{\text{Eccentric}} = 65:78$$

$$\frac{\text{Concentric}}{\text{Eccentric}} = 0.83:1$$

This increase in strength and power of eccentric muscle action when compared to concentric is also supported by the EMG data we saw in Chapter 10, Fig. 10.24. This demonstrates that eccentric contractions are stronger than concentric contractions. This would, therefore, lead us to the conclusion that eccentric flexors/concentric extensors should indeed be more balanced; however, this relationship will also depend on the angular velocity being tested (i.e. as the speed increases, do we get a greater discrepancy between

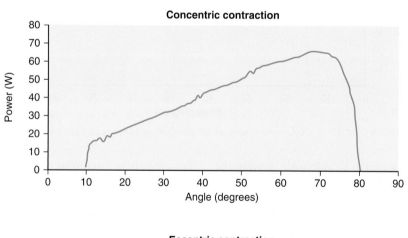

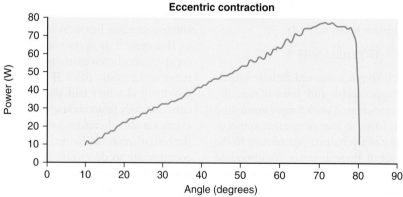

FIGURE 11.28 ■ Concentric and eccentric power (peak concentric power = 65 W, peak eccentric power = 78 W)

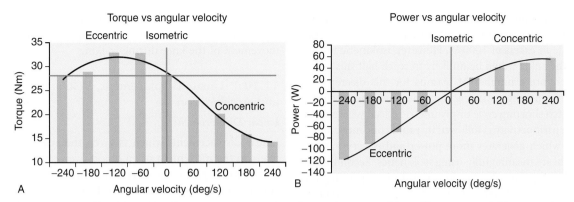

FIGURE 11.29 ■ Effect of angular velocity on (A) concentric and eccentric torque of the ankle evertors and (B) concentric and eccentric power of the ankle evertors

concentric, eccentric and isometric muscle torque and power?).

Aagaard and colleagues (1995) found that the eccentric/concentric hamstrings/quadriceps ratio approached 1:1 at 240°/s, whereas other authors (Osternig et al., 1996) found that the eccentric/concentric hamstrings/quadriceps ratio approached 1:1 at 60°/s post anterior cruciate ligament reconstruction, both of which support the differences in the balance between concentric/concentric ratio and the eccentric/concentric ratio of the antagonistic pair of muscles, with the latter considered as being more functionally relevant.

The Effect of Angular Velocity on Concentric and Eccentric Torque and Power

When considering the effect of angular velocity on concentric and eccentric torque and power we have to be very careful what we mean by torque (moment) and power, as power is a product of both moment and angular velocity. If we are testing at a set angular velocity, then this becomes less important; however, the big question is whether flexion and extension velocities will be the same during functional tasks and whether we need the torque or the power to be balanced at the different stages of acceleration and deceleration during the functional tasks. This is particularly important during ballistic tasks where angular velocities may exceed 1000°/s. At this point, isokinetics may not be the best tool, as currently most devices struggle to assess above 400°/s and the vast majority of research has been carried out at angular velocities less than

300°/s. However, isokinetics does allow us to control and isolate different angular velocities to gain a very useful insight into the relationship between joint angles, angular velocity, torque and power, albeit at comparably low angular velocities. So, what happens to the concentric and eccentric torque and power when angular velocity increases?

Figs 11.29A and B show the torque and the power versus the angular velocity for the ankle evertors in degrees/second. The negative angular velocities indicate eccentric and the positive concentric. These data show a number of relationships, perhaps the most noticeable being the difference in the torque produced concentrically and eccentrically. At 60°/s this shows that the concentric torque is 70% of the eccentric value, with the eccentric torque increasing from the isometric state up to 120°/s, which is in agreement with Hortobagyi and Katch (1990), and the concentric torque decreasing from the isometric state.

As the concentric angular velocity increases, the torque significantly reduces; however, at the same time the power continues to rise, which is due to power being a product of both the torque and the angular velocity: P = Mω (see Section 5.3: Angular work, energy and power). At 240°/s the power plateaus, which indicates the maximum concentric power available occurs at around 240°/s. However, a different pattern is seen during eccentric power, with the torque initially increasing before decreasing more slowly, leading to higher values for eccentric power. At this point we have to be very careful about stating 'the maximum amount of power a muscle can produce about a joint'. For instance, in most amateur sprinters a concentric power can often be produced about the ankle in excess of 1000 W. However, isokinetic testing of the ankle would be unlikely to reach half this value. There are a number of reasons for this discrepancy. Firstly, during sprinting the power is generated in a stretch shorting cycle or plyometric effect, where rapid eccentric activity is followed by rapid concentric activity, which generates more power and is very hard to replicate meaningfully using isokinetics. Secondly, the ankle angular velocities involved during sprinting are in the order of 500°/s during stance phase and 1000°/s during swing phase. The power absorption and generation which occurs during stance phase is at similar speeds attainable by isokinetic testing. However, this potential for a discrepancy in the angular velocities between isokinetic and functional sporting tasks can cause problems in interpretation and has led to subsequent criticism.

There has been much debate on the nature of the results obtained from isokinetics versus functional tasks, much of which has focused on the nature of power production. However, the ability to measure in a controlled environment, such as isokinetics, is invaluable in the assessment of improvement through both sports training and rehabilitation. More functional tasks may be susceptible to variations but give a better understanding of how useful the power is. After all, power is nothing without control!

11.10 ASSESSMENT OF JOINT CONTROL AND QUALITY OF MOVEMENT

Joint control may be considered in terms of linear and angular movement. Linear movement relates to the movement of the body or a body segment in the vertical, anterior–posterior or medial–lateral directions or a combination of all three. Angular movement relates to flexion–extension, abduction–adduction and transverse plane motion of a joint. For both linear and angular movements, joint displacement, velocity and acceleration may be measured and assessed.

The methods described in this section include the calculation and interpretation of linear and angular joint displacement, velocity and acceleration. These can be applied to any joints of the body; however, to demonstrate this we will consider the movement of the hand during an upper-limb reaching task and the movement of the knee during walking.

11.10.1 Linear Displacement, Velocity and Acceleration

Linear Displacement

Linear displacement, which is given the symbol (s), refers to the movement of an object over a particular distance in a particular direction. Displacement may also be calculated by the average velocity multiplied by time. The average velocity may be calculated by adding the initial and final velocities and dividing by two:

$$\text{Average velocity} = \tfrac{1}{2}(u+v)$$

$$\text{Displacement} = \tfrac{1}{2}(u+v)t$$

Linear Velocity

Linear velocity is the rate of change of displacement; that is, the distance covered in a particular time. This is the speed of movement in any particular direction or anatomical plane.

$$\text{Velocity} = \frac{\text{change in displacement}}{\text{time}}$$

This is sometimes written as $\dfrac{ds}{dt}$

Linear Acceleration

Acceleration is the rate of change of velocity; that is, the change in velocity over a given time. Acceleration tells us about the rate of change of velocity. This is an important aspect of all movement which relates to the muscles overcoming inertial forces to either start or stop movement. This relates directly to Newton's second law of motion, F = ma, which states that the rate of change of velocity (acceleration) is directly proportional to the forces applied on the body, which, in the case of human movement, come from the action of muscles. Therefore, muscle forces can either cause an acceleration or deceleration of a body segment.

$$\text{Acceleration} = \frac{\text{change in velocity}}{\text{time}}$$

This is sometimes written as $\dfrac{dv}{dt}$

11.10.2 Kinematics of a Reaching Task

The previous equations can be used to examine the quality of movement of different tasks. We will consider linear control by evaluating what information may be gained from the study of the movement of the hand forwards during a reaching task, such as reaching to pick up a cup, in a subject who is pain and pathology free and a patient who has a painful unstable shoulder. The motion of the upper limb during reaching can be examined by studying the displacement, velocity and acceleration graphs. All these are derived from the same displacement data; however, they all yield significantly different information which may be used to help us to describe functional aspects of the task.

Linear Displacement of the Hand During Reaching with and without Shoulder Dysfunction

The graphs in Figs 11.30A, B show the hand starting at a position zero and moving forwards in a reaching motion. The gradient of the curve indicates the velocity at which the hand is moving throughout the task. Fig. 11.30A shows an individual who is pain and

pathology free, and Fig. 11.30B shows an individual with a painful unstable shoulder. Both graphs show a similar pattern indicating a similar amount of hand movement. The individual with the painful unstable shoulder appears to have a less smooth pattern of movement; however, we cannot measure this directly from the linear displacement graph.

Linear Velocity of the Hand During Reaching with and without Shoulder Dysfunction

The velocity graph is found by measuring the change in the linear displacement over successive time intervals. The linear velocity graph for the hand of the individual who is pain and pathology free shows a bell-shaped curve (Fig. 11.31A). Initially the velocity of the hand is zero; the hand then accelerates to its maximum velocity at approximately the mid-point of the reaching movement. The hand then decelerates; this takes slightly longer than the acceleration phase to ensure accuracy of hand positioning. The individual with a painful unstable shoulder (Fig. 11.31B) shows a marked difference, with a continuously varying velocity which is followed by a decrease then an

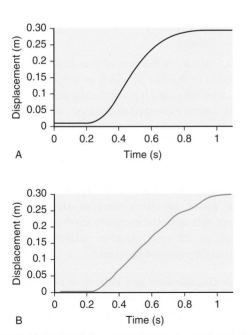

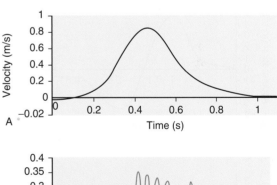

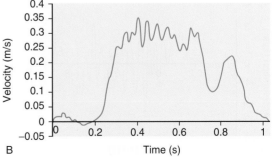

FIGURE 11.30 ■ Displacement versus time of a reaching task: (A) pain and pathology free; (B) with pain and shoulder dysfunction

FIGURE 11.31 ■ Velocity versus time of a reaching task: (A) pain and pathology free; (B) with pain and shoulder dysfunction

increase in velocity indicating either an unstable or painful part of the movement. The peak velocity may be measured from this graph indicating the level of performance of the task. The unsmooth nature of the pattern gives us a further insight to the control of the task; however, we cannot measure this directly from the linear velocity graph.

Linear Acceleration of the Hand During Reaching with and without Shoulder Dysfunction

Acceleration is calculated by measuring the change in the linear velocity over successive time intervals. The pain- and pathology-free individual (Fig. 11.32A) shows an initial acceleration peak early in the movement. The acceleration then decreases to zero as the hand reaches its maximum velocity. The hand then goes into a deceleration phase as it reaches its target. The peak deceleration is lower than the acceleration phase, but it lasts longer as shown with the velocity curve; this is to ensure accuracy of positioning the hand at the target. The individual with a painful

unstable shoulder (Fig. 11.32B) shows a marked difference when considering the linear acceleration graph; this shows a rapidly changing graph, indicating a lack of smooth controlled movement with no clear acceleration and deceleration period. This lack of smoothness gives us important information about a lack of control which could arise from poor control at the shoulder. One way of considering this lack of stability is to measure the frequency of oscillation or the number of zero crossings of the acceleration graph. If we consider the number of zero crossings, we can see that the pain- and pathology-free individual only has one zero crossing, dividing the graph into an acceleration and a deceleration phase. However, the individual with a painful unstable shoulder has more than 10 zero crossings, giving a clear objective measurement of the poor of control.

11.10.3 Angular Displacement, Velocity and Acceleration

Angular Displacement

Angular displacement is given the symbol (Θ), and refers to the movement of an object through an angle. Angular displacement can be measured in two ways, either in degrees or in radians.

Angular Velocity

Angular velocity is the rate of change of angular displacement, or the rate at which an angle is covered in a particular time, and is given the symbol (ω). Angular velocity can be expressed in degrees/s or radians/s.

Angular Acceleration

Angular acceleration is the rate of change of angular velocity and is given the symbol (α). As with linear acceleration, this relates to the muscles overcoming inertial forces to either start or stop movement, although this is not commonly used as an outcome measure on its own. Angular acceleration can be written in degree/s^2 or radians/s^2.

Angular Displacement and Velocity in the Assessment of Quality of Movement in Stroke Survivors

Richards and colleagues (2003) compared different knee kinematic characteristics of stroke patients and age-matched healthy volunteers. The stroke patients were all able to score at least 8/13 on the Rivermead

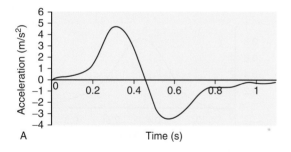

A

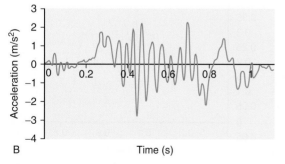

B

FIGURE 11.32 ■ Acceleration versus time of a reaching task: (A) pain and pathology free; (B) with pain and shoulder dysfunction

motor assessment gross function section and had therefore made good recovery with very mild residual disability. The purpose of this was to determine which out of the different knee kinematic characteristics was sensitive enough to pick out the differences between the paretic and non-paretic sides, and between the stroke patients and age-matched healthy volunteers. Three tasks from the Rivermead motor assessment were used: walking, sit-to-stand and stepping on a block. Significant differences were found between patients and volunteers for only some of the timing and joint angle characteristics but were seen in all angular velocity characteristics; however, no timing or angle measures were sensitive enough to pick up differences between the paretic and non-paretic sides. In all tasks, significant differences were seen between the paretic and non-paretic sides for angular velocity (Fig. 11.33A, B, C), indicating that this is a far more sensitive measure of movement control than angle and timing alone. The development of inertial measurement units (IMUs) covered in Chapter 8 offers a potentially clinically useful, quick method of measuring segment angular velocity, which needs to be further explored as a measure of movement control in clinical populations.

11.11 BIOFEEDBACK

The term biofeedback was first coined in 1969 by a group of workers who went on to found the Biofeedback Society of America. They defined biofeedback as 'the use of appropriate instrumentation to bring covert physiological processes to the conscious awareness of one or more individuals' (Wolff, 1978).

Biofeedback provides instant feedback as to whether an exercise is performed correctly. This is particularly useful when the exercises are not particularly easy to perform. With the use of biofeedback, the patient can be taught much more quickly what a particular muscle contraction feels like. This is particularly important as it enables patients to take that sensation away with them out of the clinic, so when they perform the exercise at home they know exactly what it feels like. This instant feedback is especially useful in the early stages of rehabilitation.

During neurological or stress rehabilitation programmes, we often require the patient to relax more.

In this case, the threshold of the EMG signal is reduced, therefore increasing the sensitivity, which is called negative shaping. More commonly used in musculoskeletal work is positive shaping, whereby the threshold of the EMG signal is increased as the patient progresses. This elicits a stronger contraction in order to reach the required level of feedback, thereby making the patient work harder using this process.

11.11.1 Types of Feedback

Feedback which varies continuously as a function of changes in muscle activity is termed analogue feedback and is the type that has been used most extensively in the clinical use of EMG feedback. Regarding the merits of auditory versus visual feedback, auditory feedback has the advantage that the patient does not need to concentrate on a visual display, and can use the feedback even whilst the eyes are closed, which may be found helpful when the patient's attention is focused on internal, proprioceptive cues. In addition, visual feedback is not particularly suitable in gait training or throwing activities. However, very small changes in activity levels may be more sensitively indicated using visual feedback. Musculoskeletal uses of biofeedback tend to be dominated by three areas that target superficial muscles:

- Spinal scanning
- Shoulder instabilities
- Knee problems, especially concerning the extensor mechanism.

Instant Feedback

Biofeedback provides instant feedback as to whether the exercise is performed correctly. This is particularly useful when the exercises are not particularly easy to perform. With the use of biofeedback, the patient can be taught much more quickly what a particular muscle contraction feels like. This is particularly important as it enables patients to take that sensation away with them out of the clinic, so when they perform the exercise at home they know exactly what it feels like. This instant feedback is especially useful in the early stages of rehabilitation.

Shaping

Shaping is achieved by adjusting the threshold setting of the machine. If in neurological or stress

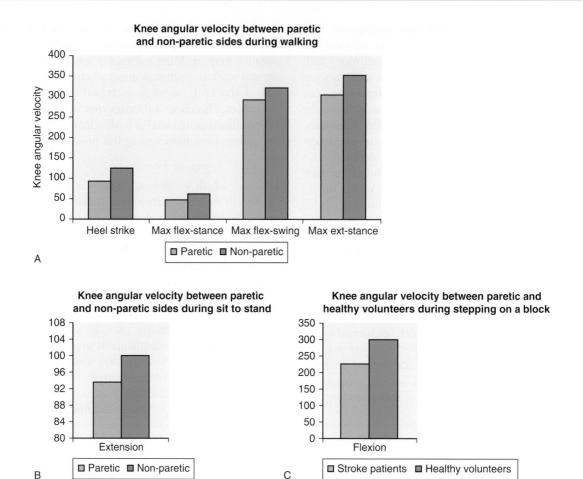

FIGURE 11.33 ■ Knee angular velocity between paretic and non-paretic sides in stroke survivors with very mild residual disability during (A) walking, (B) sit to stand and (C) stepping on a block

rehabilitation programmes you require the patient to relax more, you turn the threshold of the machine down, which makes it more sensitive. In this way, it becomes harder for the patient; this is called negative shaping. More commonly used in musculoskeletal work is positive shaping, whereby the threshold of the machine is turned up as the patient progresses. This will help to elicit a stronger contraction in order to reach the required level of feedback. Patients can be made to work extremely hard using this process.

Identification of Poor Phases of Contraction

Biofeedback enables the therapist to determine whether the contraction is being sustained over the chosen period during isometric exercises and whether the contraction is sustained during both concentric and eccentric phases of an isotonic exercise.

Objective Measurement

Quantitative scoring of EMG activity is desirable for two main reasons. Firstly, it allows the therapist to evaluate the patient's progress over a number of training sessions and to decide whether or not further training is likely to produce worthwhile gains in function. Secondly, it is helpful to the patient to know what progress he/she has made relative to other training sessions, as well as receiving feedback during the session. This information is itself a form of feedback and knowledge of results has been shown to be an effective way of maintaining a high level of motivation.

EMG Biofeedback in Muscle Training

EMG biofeedback is often used in the training of the level of muscle activation in specific muscles. One example of the use of EMG biofeedback training is on vastus medialis and vastus lateralis. This has been used to selectively strengthen one or other or both these muscles with the view to improve medial–lateral tracking of the patella in people with patellofemoral pain. The use of such biofeedback has shown that over an 8-week training programme, significant changes can be seen in the ratio between vastus medialis and vastus lateralis, which indicates that the use of EMG biofeedback into physiotherapy exercise programmes can facilitate the targeted activation of vastus medialis during daily activities (Ng et al., 2008) (Fig. 11.34).

Biofeedback in Using Three-Dimensional Motion Analysis

Using three-dimensional motion analysis and biomechanical models it is now possible to measure and assess the movements of joints in the sagittal, coronal and transverse planes. These can be displayed alongside an avatar in real time which can be used as

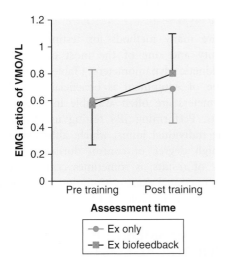

FIGURE 11.34 ■ Vastus medialis/vastus lateralis ratio during a physiotherapy exercise programme with and without biofeedback *(Adapted from Ng et al., 2008.)*

biofeedback; however, due to cost and complexity, this is beyond almost all clinical settings.

In the near future, with the advent of IMUs, which are now in every smart phone, it may be possible to have accurate biofeedback using movement data. IMUs contain triaxial accelerometers, gyroscopes and magnetometers from which it is possible to get a representation of joint angles, and offer a direct measure of angular velocity and acceleration in the different three planes. These can be used to show, for example, the coronal and transverse plane stability of the shank (tibial segment) or thigh (femoral segment). As well as being a potentially useful biofeedback tool during proprioceptive training, this would also help the assessment of movement control and determine if a deficit exists.

11.12 PROPRIOCEPTION

11.12.1 What Is Proprioception?

Proprioception is a broad term describing a range of complex sensorimotor or neuromuscular control parameters. It is not a description of any individual sensory modality (Table 11.7).

Lephart and Fu (2000) define proprioception as 'the acquisition of stimuli from conscious and unconscious processes in the sensorimotor system.' Proprioception can be appreciated and measured consciously by a complex system involving a variety of neural receptors known as mechanoreceptors; these are thought to mediate the sensations of kinaesthesia and joint position sense (JPS) (Table 11.8).

It is interesting to consider Table 11.8 as, traditionally, cutaneous receptors have not been thought to play a significant role in joint stability. However, the table shows that five out of the eight receptors thought to

TABLE 11.7
Components of Proprioception
Detection of movement from joints (joint position sense)
Sensation of force and contraction
Sensation of body segment orientation
Sensation of whole body orientation
Kinaesthesia is a constituent of proprioception; it is a combined sense of movement from a variety of anatomical structures

Based on Williams & Krishnan, 2007; Lephart et al., 1992.

TABLE 11.8		
Mechanoreceptors Involved in Proprioception		
Receptor Type	**Location**	**Stimulus**
Ruffini ending	Capsule, ligament, menisci, skin	Stretch, strain
Pacinian corpuscle	Capsule, ligament, menisci, fat pads, skin	Compression
Golgi tendon organ-like	Capsule, ligament, menisci	Strain
Free nerve ending	Capsule, ligament, menisci, skin	Nociceptive
Muscle spindle	Throughout whole muscle	Stretch
Golgi tendon organ	Musculotendinous junction	Strain
Meissner's corpuscles	Skin	Deformation caused by light touch
Merkel's discs	Skin	Continuous pressure

Based on Williams & Krishnan, 2007

TABLE 11.9
Factors Leading to Decreased Proprioceptive Acuity
Inherited predisposition
Joint effusion
Joint pain
Muscle atrophy
Decreased cutaneous sensitivity
Older age

Based on Williams & Krishnan, 2007; Callaghan et al., 2002.

TABLE 11.10
Common Proprioceptive Tests
Threshold to detect passive motion (TTDPM)
Passive angle reproduction (PAR)
Active angle reproduction (AAR)

be involved in proprioception are found distributed in the skin. Sensory feedback through the skin may therefore have a greater importance than previously assumed. A number of studies, particularly on the knee, have provided evidence which supports this case. Prymka and colleagues (1998), found significant differences between 43 patellofemoral pain (PFP) patients and 30 healthy control subjects in isolated JPS testing of the knee. The poor proprioceptive performance associated with PFPS was improved after applying a simple elastic bandage. A proposed mechanism for this finding was that the bandage stimulated rapidly adapting superficial receptors in the skin during joint motion and increased pressure on the underlying muscles and joint capsule. Proprioceptive deficits have been found in:

- Anterior cruciate ligament deficient knees (Beynnon et al., 1999).
- Osteoarthritic knees (Sharma et al., 1997; Hewitt et al., 2002).
- Knees with chronic effusion (Guido et al., 1997).
- Patellofemoral dislocation (Jerosch & Prymka, 1996).
- Patellofemoral pain (Baker et al., 2002; Callaghan et al., 2007).

11.12.2 Functional Relevance of Proprioception

Proprioception is thought to play a more significant role than pain in preventing injury in the aetiology of chronic injury and in degenerative joint disease (Lephart & Henry, 1995).

11.12.3 Assessment of Deficit in Proprioception

There are many methods for testing proprioceptive acuity and one of the most common is to use isokinetic dynamometers (Table 11.9). Using this type of equipment is beneficial as isokinetic dynamometers are often available in clinical environments. Performing JPS testing usually involves isolating individual joints, which, although providing a high degree of control during testing and accuracy of results, is sometimes criticized due to its non-functional nature. Common proprioceptive tests are listed in Table 11.10.

11.13 ASSESSMENT OF PHYSIOLOGICAL COST

Inman (1967) stated that nature does not care how individuals walk, but they should walk as efficiently as possible. This is an important point in rehabilitation: subjects undergoing rehabilitation may not walk as efficiently if they try to imitate a normal gait pattern. However, this will vary from condition to condition.

Relative efficiency may be studied by measuring physiological cost to determine if efficiency increases or decreases. Many parameters may affect this, such as walking speed, level of disability and the type of surface. There are several ways to measure this, including oxygen consumption, heart rate and mechanical energies.

11.13.1 Oxygen Consumption and Energy Expenditure

Passmore and Durnin (1955) reviewed results obtained from previous studies of human energy expenditure determined by oxygen uptake. This included the energy expenditure of activities such as sleeping, walking, climbing and running. An equation was found which stated that energy expenditure was linearly proportional to walking speed, between 3 and 6.5 km/h, for level walking. Passmore and colleagues stated that age, sex and race had no statistically significant effect on the metabolic cost of the work done or energy. They also stated that the type of surface may have an effect on the energy cost of walking, although, unless the surface is markedly rough, the effect will probably not exceed 10%.

Measures of energy expenditure have been used to compare normal and pathological gait patterns for many years. Simonson and Keys (1947) studied the energetics and motor coordination of two poliomyelitis patients and two normal subjects. The study required the subjects to walk on a driven treadmill at various speeds and grades while a measure of oxygen consumption was taken. Motor co-ordination was investigated using high-speed motion pictures to record the subject's locomotion. This study included a comparison between the use of braces and unaided locomotion. Oxygen consumption was found to be a good index for the level of energy expenditure and that relief was obtained from the use of braces.

Passmore and Draper (1965) recommended the use of the following equation to calculate the energy expenditure from the oxygen percentage concentration of expired air:

Oxygen uptake and energy expenditure

$$E = \frac{4.92 \, V \, (20.93 \, O_e)}{100}$$

where E = energy expenditure in calorie/minute, V = volume of expired air per minute and O_e = percentage oxygen concentration of expired air.

Or the equation below if considered in joules/min:

$$E = \frac{20.59 \, V \, (20.93 \, O_e)}{100}$$

11.13.2 Energy Expenditure During Walking

The relationship between energy expenditure and speed in level walking was studied by Ralston (1958). Since that time other investigators have found similar results (Bobbert, 1960; Corcoran & Brengelmann, 1970; Cotes & Meade, 1960). A general equation for energy expenditure versus walking speed based on these works was reported by Rose and Gamble (1994). This equation indicates that the energy consumed by an individual increases with the square of the walking speed.

The equation for energy expenditure during walking reports energy expenditure in calorie/kg per min rather than calorie/min as reported by Passmore and Draper (1965). Controlling for body mass allows comparison of the energy expenditure per minute per kilogram between individuals of different mass, rather than the total energy consumed, which would be particular to that individual:

$$E_w = 32 + 0.005v^2$$

where E_w = energy expenditure (calorie/kg per min) and v = velocity (m/min).

Or the equation below if expressed in joules/kg per min (Fig. 11.35):

$$E_w = 133.95 + 0.0209v^2$$

Imms and colleagues (1976) studied oxygen consumption of normal subjects and of patients recovering from fractures of the leg with plaster of Paris casts. It was found that walking with a cast almost doubled the energy expenditure at 1.5 m/s. The subjects who walked with crutches without plaster of Paris casts also showed an increase in energy expenditure at all speeds of walking; the gap widened with increased walking speed, but the levels were not as high as when they

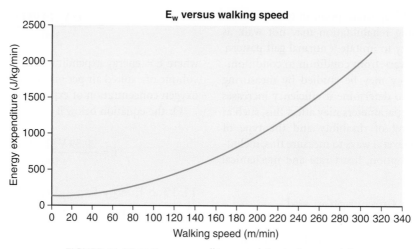

FIGURE 11.35 ■ E_w versus walking speed (joules/kg per min)

walked with the plaster of Paris casts on. Imms and colleagues showed that walking with the aid of sticks required more energy than walking with crutches. It was also reported that unequal stride lengths (i.e. a change in symmetry of spatial parameters of gait) induced by pain, stiffness of joints, or muscular weakness may contribute to the elevation of the energy expenditure in walking. This was due to interference with the normal patterns of kinetic and potential energy changes during vertical and horizontal oscillations of the body. Patients who retained slight asymmetries of gait at the end of rehabilitation also had the persistence of increased energy expenditure when all walking aids had been discarded. This supports the link between mechanical energies and physiological energy expenditure, and implies that a change in the kinematics of a movement will affect the physiological cost of that movement.

Crouse and colleagues (1990) used measures of oxygen consumption and cardiac response of ambulation with short leg and long leg prostheses in a patient with bilateral above-knee amputation, and compared the performance with three 'able-bodied men'. Oxygen uptake (VO_2), minute ventilation (VE) and heart rate (HR) were measured for the amputee and the able-bodied controls during progressive treadmill exercise to maximum capacity. Olree and colleagues (1996) studied the effort required in 11 children with cerebral palsy during treadmill walking. Oxygen

uptake was measured directly at varying walking speeds. The authors concluded that oxygen uptake was the best method of assessing effort and stated that the assessment of effort in children with cerebral palsy is a vital determinant of the efficacy of any given treatment.

It is clear from these studies that the use of oxygen uptake as a measure of physiological cost is well established and in use in research and clinical practice. However, in all these studies the subjects were required to walk on a treadmill and breathe through a mouthpiece.

11.13.3 Energy Expenditure with Respect to Distance Walked

The measurement of energy per unit distance walked provides a quantitative measure of energy economy. This has been viewed in the same way as fuel economy in a motor car:

$$E_m = \frac{E_w}{v} = \frac{32}{v} + 0.005v$$

Or the equation below if expressed in joules/kg/m:

$$E_m = \frac{E_w}{v} = \frac{133.95}{v} + 0.0209v$$

where E_m = energy expenditure per metre (calorie/kg per m or joule/kg per m), E_w = energy expenditure

E_m versus walking speed

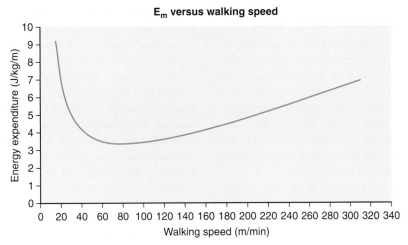

FIGURE 11.36 ■ E_m versus walking speed

(calorie/kg per min or joule/kg per min) and v = walking speed (m/min).

Fig. 11.36 shows how the energy expenditure per metre walked varies with walking speed. When this value is at a minimum, the individual is walking at his most efficient speed for that condition. Therefore, from this relationship we can determine that the most efficient walking speed is 80 m/min or 4.8 km/h (3 miles/h) for normal able-bodied walking. Any increase or decrease in walking speed will cause an increase in the energy expenditure per metre walked and a reduction in efficiency (Corcoran & Brengelmann, 1970; Ralston, 1958). In this way, the most efficient walking speed can be found for a particular individual. Such work has been carried out on many pathological gait patterns, including amputee gait (Crouse et al., 1990).

11.13.4 Heart Rate and Physiological Cost

Astrand and Ryhming (1954) investigated the relationship between oxygen uptake and heart rate, and subsequently formed a nomogram for the calculation of aerobic capacity from pulse rate during submaximal work. This accounted for pulse rate, maximal oxygen uptake, bodyweight, oxygen intake, work level and sex. Using the nomogram, they claimed that maximal attainable oxygen intake (aerobic capacity) can be calculated from the heart rate and oxygen intake (or work level) reached during a test of the submaximal rate of work in a treadmill test, cycle test or step test. It was

suggested that the individual's aerobic capacity/kg bodyweight per min would give a good measure of physical fitness.

Rowell and colleagues (1964) used the nomogram developed by Astrand and Ryhming to predict the maximum VO_2 from the pulse rate and VO_2 at a single submaximal workload. They studied the limitations to the prediction of maximal oxygen intake when using heart rate. The problems with using heart rate were identified. These included: pulse rate varying independently of the O_2 uptake but directly with the emotional state or degree of excitement of the subject; the degree of physical conditioning; the elapsed time after the previous meal; total circulating haemoglobin; the degree of hydration of the subject; alterations in ambient temperature; and hydrostatically induced changes resulting from prolonged erect posture. It was found that the ambient temperature can alter the relationship of the submaximal pulse rate to the VO_2 in such a fashion that estimates of the maximal VO_2 in high environmental temperature are seriously in error. It was found that an ambient temperature of 62°F (16.6°C) provided the most favourable condition for prediction. With the factors that affect heart rate as a predictive tool for oxygen uptake identified, it is possible to use heart rate as a measure of physiological cost. However, the factors reported by Rowell and colleagues (1964) should be controlled as strictly as possible.

11.13.5 Heart Rate and Walking Speed

MacGregor (1979) used the heart rate as an index of physiological cost. This suffers from two principal disadvantages: the relationship is non-linear, and the variations in physical fitness. Reciprocating heart beat interval (RHI) at rest shows considerable inter-individual variation such that the RHIs under specified work rates are not directly comparable. MacGregor stated that if this function is divided by walking speed it gives a physiological cost index (PCI).

It was found after a large number of tests that the PCI tends to be lowest at the walking speed self-selected by the patient to give optimum performance or minimal effort. This is true of normal subjects as well as a wide variety of patient groups. Confidence limits were found for PCI at preferred walking speed and a number of patient groups were compared with the results for the normal subjects.

MacGregor (1979) described a method of long-term ambulatory physiological surveillance using a modified tape recorder. The equipment recorded patterns of postural changes using accelerometers, as well as heart rate, throughout a 24-hour period. This made it possible to determine the time spent in each posture, and enabled subjects' physiological cost to be studied in relation to activities throughout a 24-hour period. It was found that the long-term ambulatory physiological surveillance equipment (or LAPSE) provided a non-invasive, relatively non-obtrusive system for physiological and biomechanical surveillance. This allowed the possibility of repetition of test protocols virtually without ethical restriction, and gave objective measures of physical handicap in a non-laboratory environment.

MacGregor (1981) evaluated patient performance using LAPSE. A patient with polio was used and compared with a normal subject in a laboratory test. This gave values for the normal limits at the 99% level of the PCI and walking speed for normal male and female subjects. Data were also obtained from 10 patients with rheumatoid arthritis and the effect of placebo tablets and a non-steroidal anti-rheumatic agent was studied with respect to PCI. This investigation demonstrated the potential of studying the relationship of physiological cost versus walking performance in subjects with gait pathologies.

Nene and Patrick (1989) used PCI to evaluate locomotion with the ORLAU Parawalker. Prior to this paper few studies had been published on energy expenditure with the use of these devices. ORLAU had, in the past, used the PCI as an indicator of energy expenditure of handicapped gait. However, in cases of traumatic paraplegia with a high thoracic level injury, absence or incomplete function of the sympathetic nervous system can result in an unpredictable heart rate response; consequently, for greater accuracy, evaluation of the energy cost of the Parawalker gait was performed by direct measurement of the oxygen consumption. This demonstrates that PCI is not always an adequate measure of the energy expenditure, and that care should be taken when using PCI to make sure the conditions are correct to use this method of physiological measurement.

Nene (1993) studied the physiological cost index of walking in adolescents and adults. The subjects walked in a figure-of-eight path during which walking speed and heart rate were measured. It was found that both heart rate and walking speed were higher in adolescent subjects than in adults. It should be noted that Nene, when reviewing previous work carried out on treadmills, stated that the artificial contrasts of treadmill walking do not represent the energy expenditure of normal walking adequately, although no reference was provided to back up this statement.

Gussoni and colleagues (1990) used PCI as an indicator of the energy cost of walking in subjects with total hip joint replacement. This study reported confidence limits for healthy controls, which allowed the results from the patient testing to be plotted and compared with normal. Haskell and colleagues (1993) described the conceptual basis and preliminary evaluation of a procedure using simultaneous recording of heart rate and two motion sensors to provide an accurate profile of physical activity. When heart rate is used as a measure of physical activity, it has to be noted that the slope of the relationship between heart rate and oxygen uptake varies between subjects, depending on their endurance capacity or fitness. The relationship between heart rate and oxygen uptake for a particular subject will vary depending on the level of activity. Haskell stated that heart rate can be influenced by the emotional status of the subject and environmental conditions, such as temperature and humidity. As a result, it has been generally accepted that heart rate alone is not an accurate method of assessing physical activity, although it can be used as an indicator.

SUMMARY: THE BIOMECHANICS OF CLINICAL ASSESSMENT

■ A closed kinetic chain is said to occur in the lower limb when the foot meets considerable resistance, e.g. the ground. An open kinetic chain occurs when the foot is free to move in space with little or no resistance.

■ The majority of research literature has been on walking. The study of the movement patterns during different functional tasks is equally important to individuals' participation and quality of life.

■ There are many clinical tests that may be investigated using biomechanics. This allows the critical evaluation of the different tasks used in clinical assessment to be analysed in detail in relation to joint function and control.

■ Investigating these will help to explain the theoretical aspects of the different movement tasks used in clinical assessment.

12

BIOMECHANICS OF ORTHOTIC MANAGEMENT

JIM RICHARDS ■ AOIFE HEALY ■ NACHIAPPAN CHOCKALINGAM

This chapter covers the biomechanics of orthotic management of the lower limb. This includes the theoretical mechanics of indirect and direct orthotic management and clinical case study data of the use of the devices covered.

AIM

To compare the different theoretical methods of controlling and supporting the lower limb with orthotic management and their effect on signal case studies.

OBJECTIVES

■ To describe the theoretical function and the clinical effect of different configurations of foot orthoses

■ To describe the theoretical function and the clinical effect of different configurations of ankle foot orthoses

■ To describe the theoretical function and the clinical effect of different configurations of knee orthoses

■ To explain how moments can be altered about a joint

■ To explain how shear and axial forces can be altered about a joint

■ To explain how the line of action of the ground reaction forces (GRFs) can be altered about a joint.

12.1 FOOT ORTHOSES

So, what are foot orthoses? These are shaped or moulded inserts for the shoe which aim to hold the foot in position, change the foot position, or change the range of motion of either the whole foot or between the different segments of the foot. Foot orthoses can have a direct effect on the segments of the foot, but

they can also have significant clinical effects indirectly much further up the body to the pelvis, lower back and, arguably, as far up as the shoulders and neck.

Foot orthoses come in many shapes and forms. The most basic form is a simple ethyl vinyl acetate (EVA) wedge, whereas some are pre-made contoured devices, which may or may not need some form of modification to the patient's prescription. Lockard (1988) highlighted the fact that there are many classification systems used to describe shoe inserts. These range from the description of the properties of the materials used (i.e. soft, semi-rigid or rigid) to the type of procedure used to construct the appliance (i.e. moulded and non-moulded). Anthony (1991) defines an orthosis as 'an orthopaedic device which is designed to promote the structural integrity of the joints of the foot and lower limb, by resisting GRFs that cause abnormal skeletal motion to occur during the stance phase of gait'. Root and colleagues (1977) suggest they 'assist in controlling foot geometry and force direction, stabilising joints and reducing muscle contractions'.

The foot is an extremely complex system of articulating segments. Therefore, the movements of the foot and ankle cannot be completely explained by rotations about a single plane, but by a combination of movements in all three planes. This makes the assessment of the foot and the action of foot orthoses one of the most complex biomechanical systems in the body. This

282

is compounded by the fact that the analysis of the foot, up until fairly recently, has only been considered as a single segment due to restrictions in movement analysis technology.

It is highly likely that in the coming years our knowledge of foot function and the effects of foot orthotic management will be greatly expanded. These next sections consider both the direct action of several types of commonly used foot orthoses on the foot and also their indirect action on the lower limb and pelvis. This section includes summaries from key papers and clinical case studies; however, I would advise any reader to remain up to date with advances in this field: in particular, by looking at the current research literature.

12.1.1 The Assessment of Leg-Length Discrepancy

Leg-length discrepancies of 2 cm or less are very common. Inequalities in leg length greater than 1.5 cm have been linked with low back pain and abnormal gait with higher energy consumption and early fatigue. Undetected leg-length discrepancy can often contribute to chronic pain and biomechanical adaptations of the lower limb and pelvis.

If the presence of a leg-length discrepancy needs to be assessed, then this may be conducted by examining for both real and apparent leg-length discrepancy. Real leg-length discrepancies may be assessed with the patient lying supine from a point on the upper pelvis (anterior superior iliac spine) to the medial malleolus, or from the greater trochanter to the lateral malleolus. Leg-length discrepancies may also be apparent. This is where the legs will actually measure the same length, but will function as though they are different because of pelvic obliquity or curvature in the back. Apparent leg-length discrepancy may be measured with the patient lying supine, and a measurement from the navel or sternum to the medial malleoli.

A number of visual clues can be present with individuals with a leg-length discrepancy. These can include dropping the hip on the shorter side, head and shoulder tilting or bending the knee excessively on the contralateral side. There may be also an appearance of vaulting or stepping into a hole, or other gait asymmetries such as unequal step lengths.

12.1.2 Orthotic Treatment of Leg-Length Discrepancy

With leg-length discrepancies we are confronted with 'to treat or not to treat?' that is the question. There are a variety of schools of thought; some clinicians will not treat discrepancies less than 1 cm, whereas others would not treat discrepancies less than 2 cm. Some clinicians would not measure the discrepancy and only treat if there are the associated adaptations of the lower limb and pelvis. These can include hip hiking, where the patient raises the hip on the longer side with associated head and shoulder tilting in the coronal plane. However, most clinicians would agree that treatment should not be considered unless there are associated symptoms of lower back pain.

We will now consider a patient in her mid-20s with chronic lower back pain. The patient was referred for gait analysis after having been found to have a leg-length discrepancy, with the left side being 2 cm shorter than the right. An assessment was carried out looking, in particular, at the movement of the lower limb and pelvis, but also the GRFs. The subject was then fitted with a simple 1 cm heel raise to allow for the leg-length discrepancy and then immediately re-tested with the device.

GRFs with and without the Heel Raise

The results with no heel raise show the shorter side had a marked reduction in the loading response forces in comparison to normal. The movement over the body, the stance limb during midstance and the vertical propulsive force were also affected, with the trough occurring much earlier than normal. The longer limb showed a less smooth and reduced loading response on the right side, a shallow trough indicating a poor progression of the body over the stance limb and the lower propulsive force (Fig. 12.1A, B).

With the heel raise, the shorter limb showed significant improvement in the loading response. The movement of the body over the stance limb shown by the trough, and the vertical propulsive peak, are also improved. The longer limb showed improved movement of the body over the stance limb on the left side; in addition, the smoothness and magnitude of the loading response forces on the unaffected side were

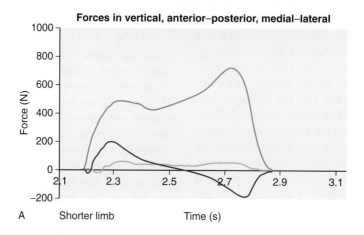

A   Shorter limb   Time (s)

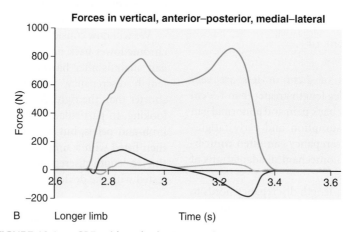

B   Longer limb   Time (s)

FIGURE 12.1 ■ GRFs with no heel raise: (A) shorter limb and (B) longer limb

improved. The magnitudes of the vertical forces show a much better balance between left and right with the heel raise inserted (Fig. 12.2A, B).

Perttunen and colleagues (2004) investigated the gait asymmetries in 25 patients with limb-length discrepancy. They found that the duration of the stance phase was reduced in the short limb, and the vertical GRF during the push-off phase was greater in the long limb. The push-off phase was also initiated earlier in the short leg. These results support this clinical case study where the loading of the longer limb is greater than that of the shorter limb.

But what accounts for this change in GRFs? If we consider the analogy of the patient 'stepping into a hole' then we might expect the forces on the shorter side to, in fact, be greater than the longer, as the foot hits the ground, but this clearly does not happen. It is at this point we need to consider the action of the pelvis. Although the shorter side is descending for longer, the pelvis reaction is to try to adapt to the difference, which has the effect of gently lowering the leg until contact is made. This offers an additional method of deceleration of the shorter limb before the foot reaches the ground, so the foot hits the ground with a lower velocity than normal. If we now consider that this patient was referred for lower back pain, the pelvic involvement becomes very important.

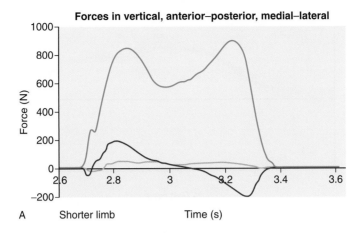

A   Shorter limb   Time (s)

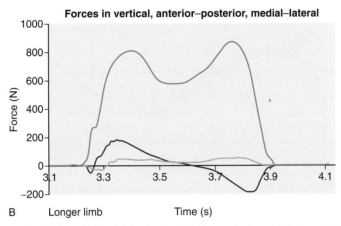

B   Longer limb   Time (s)

FIGURE 12.2 ■ GRFs with heel raise: (A) shorter limb and (B) longer limb

Movement of the Pelvis with and without the Heel Raise

As mentioned previously, some clinicians will examine an individual walking from the coronal plane and examine pelvic, shoulder or head tilting as indicators of functional asymmetry in individuals with suspected or measured leg-length discrepancy. We will now consider the differences in pelvic movement with and without the 1 cm heel raise using movement analysis.

The graphs (Fig. 12.3A, B) show a marked difference in the pelvic movement with and without the heel raise. Without the heel raise, we see an asymmetrical movement pattern, with all the movement being on one side. This relates to the pelvis dropping down on the shorter side, which at no point drops down on the

contralateral side. With the heel raise, the pelvic movement shows a good balance of alternation of pelvic drop during the gait cycle and a more normal pattern.

The improvements in the balance of the forces and the pelvic movement should reduce the strain on the pelvis and lower back, as less compensation will have to be made. In this particular case, the patient's lower back pain did, indeed, reduce when wearing the heel raise. This sounds very good, although whether the use of heel raises for long-term management produces its own compensations and secondary effects is not clear. We also have to be careful with the use of pelvic obliquity as a measure of the efficacy of heel raises. This is highlighted by comments from Wagner (1990) who stated that pelvic tilt (obliquity) is often the

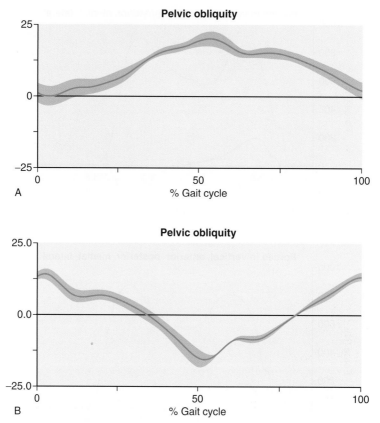

FIGURE 12.3 ■ Pelvic movement in the coronal plane (pelvic obliquity): (A) without heel raise and (B) with heel raise

consequence of a discrepancy in leg length and can be corrected either with orthotic devices or by operative equalization of the leg length. However, pelvic tilt (obliquity) can also occur independently of the leg length in cases of asymmetry of the pelvis, malposition of the hip joint or contracted scoliotic deviation of the spine. In such cases with complex deformities, correction of the pelvis should aim at a balanced body posture rather than necessarily a symmetric level of the iliac crests.

12.1.3 Wedging or Posting of the Rearfoot

The prescription and fitting of medial wedges is commonplace in podiatric practice (Figs 12.4 and 12.5). Wedging aims to control rearfoot motion, which can be responsible for ankle instability and abnormal moments about proximal joints. Foot orthoses have been shown to reduce excessive pronation of the foot.

Nester and colleagues (2003) stated that despite their wide clinical application and success, our understanding of the biomechanical effects of foot orthoses is relatively limited.

Wedging or posting the orthoses can be carried out on both the medial and lateral sides of the foot. There have been many papers considering the biomechanical effect of foot orthoses to control or change rearfoot motion. Nester and colleagues (2003) studied the effect of foot orthoses on the kinematics and kinetics of normal walking gait. This considered not only the effects on the foot and ankle but also the effects at the knee of both medial and lateral wedging. Branthwaite and colleagues (2004) studied the effect of simple insoles on three-dimensional foot motion during normal walking. They found a significantly reduced maximum eversion angle between wearing orthoses and no orthoses.

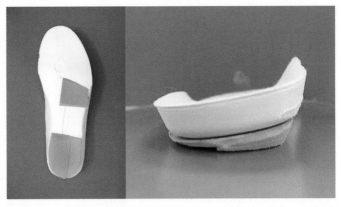

FIGURE 12.4 ■ Medial wedge on a preformed full-length foot orthosis

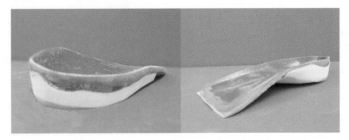

FIGURE 12.5 ■ Bespoke cast orthosis with built-in medial wedge

The underlying theme of all these papers is that the studies were carried out on normal individuals who did not require foot orthoses; yet these may be considered as real effects. But are individuals going to respond in the same way if they have a lack of control in pronation or supination, or have a different foot type, or are either currently suffering or have previously suffered from overuse running injuries? And what amounts of pronation or supination instability are controllable by such devices? What are the limiting factors to their use clinically? We must also consider the modelling of the foot, which is predominantly modelled as a single segment, whereas podiatrists will consider the action of an orthosis on at least three segments of the foot. It is at this point that we still do not have enough literature to refer to when considering the true clinical action of wedging on the foot.

12.1.4 Control of the Line of Action of GRFs

The technique of wedging inside the shoe is sometimes referred to as posting or wedging by a podiatrist. The principal effect of the wedge is to change the orientation of the calcaneus, and, therefore, the subtalar joint when the plantar surface of the heel of the shoe is flat on the ground. A wedge would, therefore, be used if there is some structural deformity that results in the calcaneus not being vertical when the subtalar joint is in neutral. It is assumed clinically that posting the foot induces either a supinatory or pronatory moment on the subtalar joint, and, therefore, will limit the amount of foot pronation.

The action of the orthosis to alter the supinatory or pronatory moments will depend on the point of application and direction of the GRF during the various stages of stance phase. This changes the moments in the coronal plane about the subtalar joint, but may also alter the moments about the knee and hip joint. The diagrams (Fig. 12.6A–C) show a somewhat exaggerated theoretical effect of medial and lateral posting of the rearfoot on the moments about the subtalar joint.

12.1.5 The Effect of Wedging or Posting the Rearfoot During Normal Walking

Rearfoot Motion

The graph in Fig. 12.7 shows how the rearfoot kinematics are changed with the introduction of a medial and lateral rearfoot wedge. These graphs indicate the medial wedging reduces pronation during contact phase, whilst the lateral wedging increases pronation during contact phase.

Medial and Lateral Forces

One way to objectively assess the function of the wedge is to study the GRFs in the frontal plane, with particular attention to the medial–lateral force. This can cause difficulties, as few podiatrists have access to force platforms at this time. Those that do carry out objective

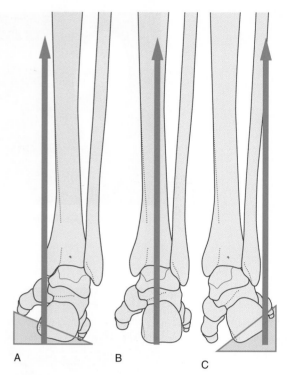

FIGURE 12.6 ■ (A) Medial posting, (B) no posting and (C) lateral posting

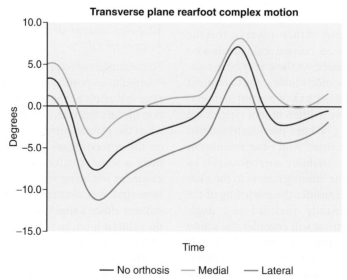

FIGURE 12.7 ■ Rearfoot motion. *(Adapted from Nester et al., 2003.)*

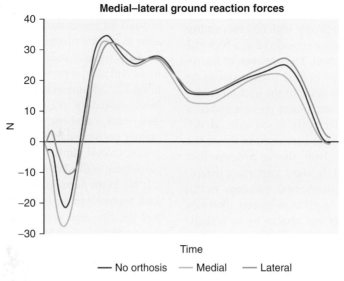

FIGURE 12.8 ■ Medial and lateral forces. *(Adapted from Nester et al., 2003.)*

analysis of force often rely on pressure plate systems that cannot yet measure the forces in any direction other than vertically and are, therefore, not suitable to assess the effect of wedging. The following graphs show that the introduction of a medial rearfoot wedge increases the lateral thrust during loading, whereas a lateral wedge decreases the lateral thrust during loading. In late stance, the medial forces are also slightly affected, with the medial wedge reducing the medial force and the lateral wedge increasing the medial force (Fig. 12.8).

Nester and colleagues (2003) found that medially wedged orthoses decreased rearfoot pronation and increased the laterally directed GRF during the contact phase, suggesting a decrease in the ability to absorb shock and, therefore, a greater shock loading or shock attenuation, whereas laterally wedged orthoses increased rearfoot pronation and decreased the laterally directed GRF during the contact phase, suggesting decreased shock attenuation.

12.1.6 Assessment of Foot Pressure

Body Mass

Whilst this may appear counterintuitive, research has shown that body mass is not a strong predictor of plantar pressures (Cavanagh et al., 1991); this means an increase in body mass does not automatically translate to an increase in plantar pressures. However, research has shown that obese individuals show higher plantar pressures during standing and walking than non-obese individuals (Hills et al., 2001; Birtane and Tuna, 2004). Recent research on children has also found significantly higher plantar pressures and a larger contact area in overweight children when compared to non-overweight children (Gatt et al., 2014).

Gender

Males are known to have longer and broader feet than women, irrespective of height (Wunderlich & Cavanagh, 2001) and, whereas research has shown higher forces under the foot in males when walking, the increased contact area means there appears to be no difference in peak pressures between genders (Putti et al., 2010).

Age

In children by the age of approximately 6 years the foot structure has changed to resemble the adult foot. As the longitudinal arch is not fully developed in young children there is a higher relative load in the midfoot compared to adults (Hennig & Rosenbaum, 1991). The midfoot loading reduces with increasing age, with an increase in the loading in the metatarsals.

The At-Risk Foot

The *at-risk* foot relates to people with diabetes and/or rheumatoid arthritis who are considered at a high risk of developing foot ulceration. The presence of neuropathy and bony deformities in these patient groups are factors that make the soft tissue of the foot more susceptible to damage. Interventions to prevent and treat ulceration in these patients are based around offloading (reducing the plantar pressures) in the at-risk foot during ambulation. Peak plantar pressure data during gait is considered the most important parameter to assess in these patients; peak pressures in this patient group are often reported to exceed 1000 kPa. Unfortunately, there does not appear to be a single threshold pressure value for ulceration, but it is believed that the higher the peak pressure the higher the risk of ulceration (Armstrong et al., 1998). In these patients, it may also be relevant to assess pressure time integral data which provides information on the duration of the pressures during gait.

Case Study – The Effectiveness of a Rocker Sole Diabetic Shoe in Offloading the Forefoot (Healy et al., 2013)

A rocker sole shoe consists of a curved sole at the forefoot to create a rocking effect as the person walks.

The rocker sole shoe has a rigid sole to limit movement of the foot joints during stance, and the rocker sole aims to quickly progress the foot rollover through the forefoot, thereby reducing movement of the plantar tissue and plantar pressures. A 51-year-old male, measuring 171 cm and weighing 116.3 kg, with type 2 diabetes walked in their own footwear (semi-brogue shoe) and a diabetic shoe with a rocker sole while plantar pressure data was recorded using the pedar-x system (Novel GmbH). Data was collected while the participant walked at a self-selected speed (5.22 km/h ± 10%). In the participant's own footwear, their highest peak pressures in the forefoot were found in the hallux and the first and second metatarsal areas, approximately 480, 440 and 360 kPa, respectively (Fig. 12.9A). When the participant walked in the shoe with a rocker sole, these peak pressures were reduced by approximately 40%. The peak pressures for the hallux and the first and second metatarsal areas were 370, 250 and 210 kPa, respectively (Fig. 12.9B).

Footwear/Orthoses

Plantar pressure measurement can also be useful in the assessment of footwear interventions/adaptations and orthoses. In shoe pressure measurement can be a useful tool for clinicians to help assess if the orthotic

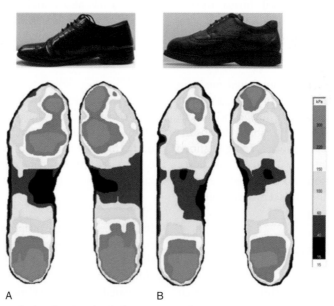

A   B

FIGURE 12.9 ■ Peak stance display showing the peak pressure in the participant's own footwear (A) and the diabetic rocker sole shoe (B) *(Healy et al., 2013.)*

or footwear adaption they have prescribed is achieving the desired outcome: for example, by assessing the peak pressures and the path and velocity of the centre of pressure (CoP). With a vast range of materials now available to clinicians for orthotic prescription, plantar pressure measurement can be used to assist in the selection of the most appropriate material for a patient's orthotic device (Healy et al., 2012).

Fig. 12.10 provides sample data on the effect of different insole materials on peak plantar pressures. In-shoe plantar pressures were measured while a participant walked in flat plimsoles and then the procedure was repeated twice: (1) with the addition of a 3 mm flat insole of medium-density polyurethane (PU) to the plimsole and (2) with the addition of a 3 mm flat EVA insole. Both materials provided a reduction in peak pressure of between 40% and 50% at the hallux and first metatarsal area when compared to having no insole in the shoe, reducing the peak pressure at the hallux from ~290 to ~150 kPa and the first metatarsal area from ~180 to ~90 kPa. Additionally, the PU material reduced the peak pressures at the heel by approximately 20%, reducing from ~300 kPa for the shoe on its own to ~250 kPa with the addition of the PU insole.

This method can also be used to assist in the identification of the most appropriate sports footwear for athletes. Previous research utilized plantar pressure measurement and found significant differences in peak pressures for different running shoe models (Hennig & Milani, 1995).

12.2 MANAGEMENT OF THE ANKLE JOINT USING ORTHOSES

12.2.1 Direct Orthotic Management

Direct orthotic management can be considered to work by three mechanisms: changing the moments directly at a joint, changing shear forces and changing axial forces. In all cases this involves a device being placed around the joint which applies an external system of forces. Most orthoses aim to affect joints and segments directly; however, many orthoses also have additional or secondary effects on a proximal segment or joints, without having any direct contact. In this chapter, we will cover both the direct effects of each of these mechanisms, the effects on the joint being targeted, and also the additional indirect effects on joints not directly targeted by the orthoses.

12.2.2 Modification of Joint Moments with Orthoses

Modification of moments about joints is by far the most common method of direct orthotic

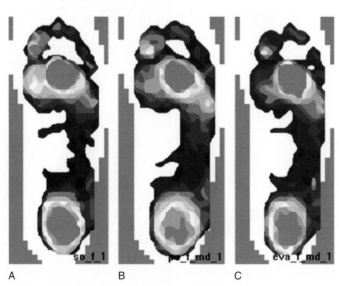

A   B   C

FIGURE 12.10 ■ Peak stance display showing the peak pressure in the plimsole (A), plimsole with the addition of the PU insole (B) and plimsole with the addition of the EVA insole (C) *(Healy et al., 2012.)*

management. These devices are varied and aimed to support and/or control the movement of one or more planes of movement of different joints. In this section, we will cover the theory of the direct orthotic management of the ankle and knee joints with examples of their use on single clinical cases.

12.2.3 Biomechanics of Ankle Foot Orthoses

Plastic ankle foot orthoses (AFOs) have been used to manage weakness and spasticity about the ankle joint for over 40 years. However, the basic design has not changed significantly. The different design options consist of rigid, posterior leaf spring and hinged, although other variations have been tried, including the plastic spiral AFO and, more recently, the introduction of carbon-fibre posterior leaf spring AFO.

Rigid Ankle Foot Orthoses

Rigid AFOs, as the name suggests, are of a completely rigid design which aims to block all movement about the ankle joint and foot in all planes. These are usually made from moulded plastic that extends up the back of the shank and under the foot to the metatarsal–phalangeal joint, although sometimes this is extended over the whole length of the foot (Fig. 12.11).

The rigid design of the orthoses has the effect of supporting a dorsiflexion moment produced by the GRF about the ankle by providing a posteriorly directed force on the anterior tibial strap, which prevents or controls tibial movement over the foot. In this way, the stiffness of the rigid AFO produces a plantarflexion moment that opposes the moment about the ankle from the GRF (Fig. 12.12). Rigid AFOs can also be used to resist knee flexion by setting them into slight plantarflexion; however, if excessive knee flexion needs to be managed, knee orthoses may give more direct control, or if there is ankle involvement, using knee ankle foot orthoses (KAFOs) may be a better option (see Section 12.3).

The clinical guidelines for the use of rigid AFOs are when an individual has: weakness or absence of ankle dorsiflexors and plantarflexors, severe spasticity resulting in equinovarus of the foot during swing and stance phase, weak knee extensors, and proprioceptive sensory loss.

If the plantarflexors are weak, the ankle dorsiflexes too rapidly and results in poor control. This has the additional effect of a rapid movement of the tibia forwards, causing excessive knee flexion and subsequently a crouch gait pattern. If the dorsiflexors are weak then

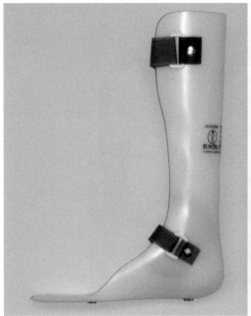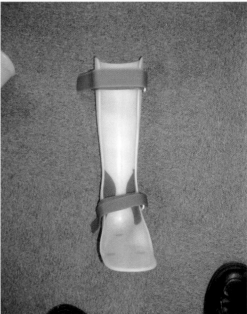

FIGURE 12.11 ■ Rigid ankle foot orthoses

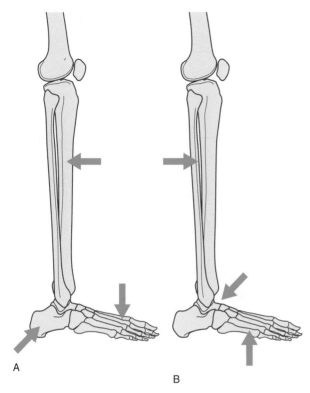

FIGURE 12.12 ■ Force system on a rigid ankle foot orthosis. Stance phase (A), swing phase (B)

this can lead to foot drop during swing phase and foot slap at heel strike. Blocking ankle movement has the effect of supporting plantarflexion/dorsiflexion, pronation/supination and inversion/eversion movement between the metatarsals, cuboid, calcaneus and tibia, so these move as a single segment. In reality, movement of the metatarsal–phalangeal joints still occurs due to reduction in the rigidity of most devices under the forefoot. This blocking effect of the ankle movement can generate problems for the users, forcing them into an early heel lift and the indirect effect of hyperextension of the knee joint. This hyperextension of the knee joint, or at least the movement of the knee away from the flexed position, is one of the key reasons for fitting rigid AFOs, as this aims to reduce the crouch gait position that can be present in cerebral palsy. However, the blocking of the ankle can also make progression on the tibia forward over the stance limb difficult, in essence blocking second rocker of the ankle.

There are several options in the casting and fitting of rigid AFOs, as the orthosis may be set in neutral, plantarflexion or dorsiflexion. If the AFO is set into plantarflexion, then the tibia will be inclined posteriorly during second rocker at the ankle. This will have the effect of reducing the knee flexion moment that may be desirable in crouch gait; however, too much plantarflexion may not be desirable as this will make the progression of the body over the stance limb difficult and can force the knee into hyperextension. The setting into plantarflexion will also have the effect of reducing ground clearance as the foot will also be in a plantarflexed position during swing phase. If the ankle is set into dorsiflexion, then this will increase the moment about the knee during second rocker at the ankle, but would allow for greater ground clearance during swing phase. Therefore, the most important consideration for the amount of plantarflexion or dorsiflexion provided by the orthosis is the degree of knee control the individual patient has.

The Effect of Rigid Ankle Foot Orthoses

The effect of setting a rigid AFO into dorsiflexion allows the tibia to assume a position in front of the ankle joint. This can cause problems for the control of the movement from heel strike to the foot flat position. This is usually achieved by eccentric control into plantarflexion by the ankle dorsiflexors. In particular, this can cause problems when walking down slopes. Restrictions in the plantarflexion during loading and dorsiflexion during the progression of the body over the stance limb can be demonstrated by examining what happens to individuals who are pain and pathology free when wearing a rigid AFO.

Fig. 12.13A and b shows the initial movement into plantarflexion is reduced when wearing a rigid AFO, although some movement is possible. The movement into dorsiflexion is delayed and does not reach the values without the orthosis, although nearly 10° is attained. The nature of where this movement occurs is a point of debate, as most data, including the data presented here, consider the foot as a single segment including the calcaneal, cuboid and metatarsal segments. This does not necessarily tell us the ankle joint movement; however, the fact that movement can occur between the foot and tibia in an individual who is pain and pathology free when wearing a rigid AFO is

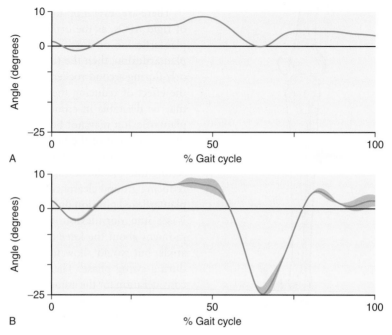

FIGURE 12.13 ■ (A) Normal subject with rigid ankle foot orthosis and (B) normal subject without ankle foot orthosis

interesting and raises the question: exactly how rigid is rigid?

Notwithstanding that some movement can occur, some individuals who wear rigid AFOs do have difficulty in moving their tibia over the foot and subsequently have difficulty in moving the body forward. To aid the progression of the body over the stance limb, or roll over action, a rocker profile is often added to the shoe. This has the effect of allowing the progression of the tibia forwards over the foot without necessarily requiring movement of the ankle joint itself. In fact, this may have the additional benefit of reducing the eccentric work done by the ankle plantarflexors, if this is the intention of the orthosis.

Ankle Foot Orthoses Footwear Combinations

Richardson (1991) reported on the use of rocker-soled shoes in the treatment of subjects with calf claudication. Rocker-soled shoes are believed to reduce the force required at heel off and toe off and, therefore, should reduce the work done by the gastrocnemius and soleus. Richardson found that walking distance was significantly improved, as well as the distance covered before the onset of pain, referred to as 'bothered distance'. The closing comments of this study are

of value. Richardson stated that the clinical value of the use of rocker-soled shoes will not be fully realized until the optimal shoe design and the safety of this shoe have been established. Rocker soles, however, control the progression of the centre of pressure and allow heel lift and, therefore, permit tibial progression in the absence of ankle dorsiflexion. It should at this point be highlighted that little work has been carried out to determine the interaction of rocker soles in conjunction with AFOs, although this is common clinical practice in many countries.

Hullin and colleagues (1992) reported on the kinetics and kinematics during walking with the use of heel raises and rocker soles in subjects with spina bifida. Hullin and colleagues observed that heel raises alter the tibial floor angle, but do not allow tibial progression or heel lift. More recently, Owen (2010) reviewed and summarized evidence and important observations on the significance of shank and thigh kinematics for standing and gait. This focused on the interrelationship between segment kinematics, joint kinematics, and kinetics and their relationship to orthotic design, alignment and tuning. Owen proposed an algorithm for the designing, aligning and tuning of AFO footwear combinations and an algorithm for determining the

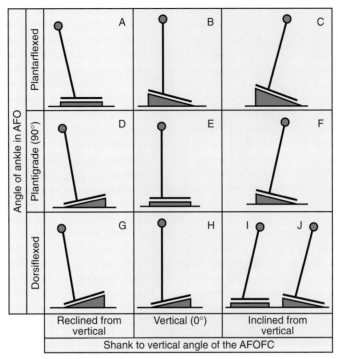

FIGURE 12.14 ■ Nine theoretical configurations of the angle of the ankle in the ankle foot orthosis (AFO) and the shank to vertical angle (SVA) of an ankle foot orthosis footwear combination (AFOFC). Owen (2010) *(Copyright © by International Society for Prosthetics and Orthotics International.)*

sagittal angle of the ankle in an AFO (Fig. 12.14). In providing the correct AFO and footwear combination, it is possible to improve standing balance and gait by improving segment kinematics. This may also prevent or improve bone and joint deformity and offer a normalization of muscle length, strength and stiffness. A recent review focusing on children with cerebral palsy highlighted the paucity of research in this area providing quantitative data on the effects of kinematics, kinetics and energy efficiency of ankle foot orthosis-footwear combination tuning. In addition, there is little data to compare tuned vs. untuned ankle foot orthosis-footwear combinations (Eddison, 2013).

The largest effect is in the restriction of movement into plantarflexion during the power production phase of the ankle motion. This restriction will make it almost impossible for the individual to produce any power about the ankle. This blocking of movement is often used to prevent spastic reactions in the plantarflexors; however, it is now being brought into question whether this is entirely a good thing in most cases, and many clinics are now considering posterior

leaf spring AFOs and hinged AFOs as a viable treatment for cerebral palsy and stroke where spasticity is not so severe.

Posterior Leaf Spring Ankle Foot Orthoses

Posterior leaf spring AFOs, sometimes referred to as flexible plastic shell orthoses, aim to provide dorsiflexion assistance during swing phase, while giving some stability for inversion–eversion of the ankle joint. The clinical guidelines for the use of posterior leaf spring AFOs are for weakness or absence of dorsiflexors; good pronation–supination stability; absence of foot varus or valgus; absence of to moderate spasticity; and good knee stability.

Posterior leaf spring AFOs are usually too flexible to give support in the transverse plane movement of the foot to the tibia, which has been related to pronation and supination (Nester, 2003), although this will in part depend on the 'trim lines': that is, the width and thickness of the material of the posterior leaf spring.

The trim lines of posterior leaf spring AFOs are critical to what these orthoses are capable of.

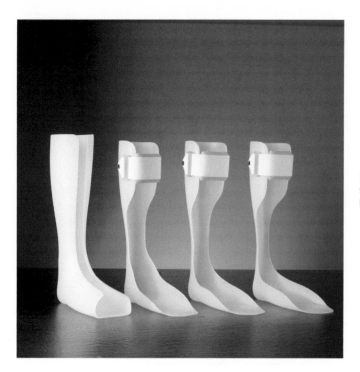

FIGURE 12.15 ■ Otto Bock moulded plastic ankle foot orthoses. *Far left:* rigid; *middle:* alternative trim lines; *far right:* posterior leaf spring

Traditionally, these have been primarily used for dorsiflexion assistance during swing phase to prevent foot drop, in which case not much material will be required to stop the foot plantarflexing as this will have to support the weight of the foot, or, at most, resist spastic muscle activity of the plantarflexors. However, clinically, posterior leaf spring AFOs may also be used to assist the eccentric action of plantarflexors during stance phase. This may be achieved by having wider trim lines. This will provide some resistance to movement into dorsiflexion and, therefore, theoretically improve the control of the movement over the stance limb. This sounds very good; however, the amount of material that should be left will depend on the person's body weight and their available eccentric control of the plantarflexors. This requires the balancing, or tuning, of the orthosis to the individual's needs, which clinically may be difficult to assess; however, the benefits to the individual can be considerable as this allows muscle activity that may reduce muscle atrophy compared with rigid designs. There are several different designs of posterior leaf spring AFOs, which vary from moulded plastic (Fig. 12.15) to carbon fibre (Fig. 12.16).

FIGURE 12.16 ■ Otto Bock WalkOn carbon composite posterior leaf spring ankle foot orthoses for plantarflexion/dorsiflexion assistance

The Effect of Posterior Leaf Spring Ankle Foot Orthoses

The clinical case study of the use of posterior leaf spring AFOs shows a 60-year-old patient who had been suffering from problems with his walking for several years. The patient had got to the point where walking to the local shops was no longer possible.

When the patient walked without the orthosis the ankle motion on the left side showed movement starting from a dorsiflexed position and then moving into further dorsiflexion, indicating poor foot placement and poor control by the posterior muscles of the calf. Poor propulsion was also seen on both left and right, again indicating a deficit in function in the calf group. The knee movement during swing phase showed normal values; however, during stance phase the knee assumed a flexed position on both left and right sides, with the left side never dropping below 40° of flexion. The movement of the thigh on the left and right sides

was also reduced. The left side showed a deficit in flexion during the beginning of stance phase, indicating the limb was not being advanced forwards and, therefore, step length was reduced (Fig. 12.17).

The force patterns seen on both left and right showed similar deficits in the first and second peaks and the trough, indicating poor loading and push off and a poor movement of the body over the stance limb, particularly on the left side (Fig. 12.18A, B).

A posterior leaf spring AFO was fitted to the left side. The aim of the orthosis was to improve the ankle and knee movement, and the progression of the body over the stance limb. This would theoretically offer some resistance to the ankle joint to ensure the ankle was held in neutral position at heel strike, while also

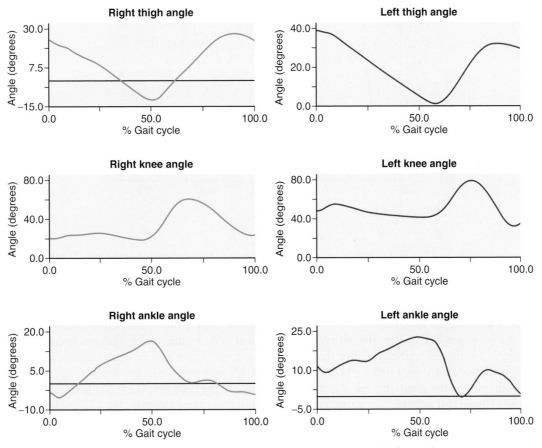

FIGURE 12.17 ■ Movement patterns without a leaf spring ankle foot orthosis

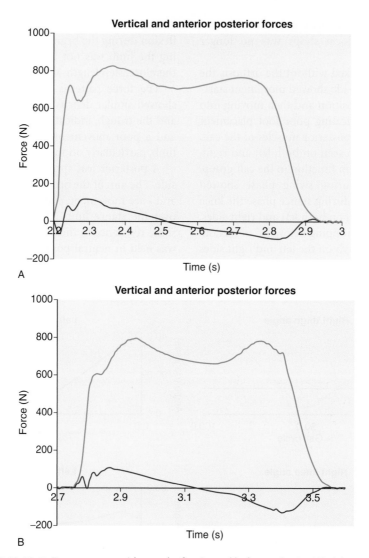

FIGURE 12.18 ■ Force patterns without a leaf spring ankle foot orthosis: (A) right and (B) left

giving some resistance into dorsiflexion to offer more control of the tibia over the ankle and to have an indirect effect of reducing the knee flexion.

With the posterior leaf spring AFO fitted to the ankle, motion for the left and right sides showed a much improved position at heel strike, which enabled the ankle to move into plantarflexion during first rocker. The orthosis also allowed an improved control of the tibial movement over the ankle joint and a reduced knee flexion angle during stance phase. The thigh motion also showed an improved movement

pattern with symmetry between the left and right sides (Fig. 12.19).

The force patterns showed a significant improvement in the loading and propulsion forces, but perhaps most significant was the improvement in the trough, which supports the improvements seen in the movement of the body over the stance limb in the ankle and knee movement patterns (Fig. 12.20A, B).

With a posterior leaf spring AFO fitted, the patient's foot position at heel strike was controlled and he was able to move his body over the stance limb, but with a

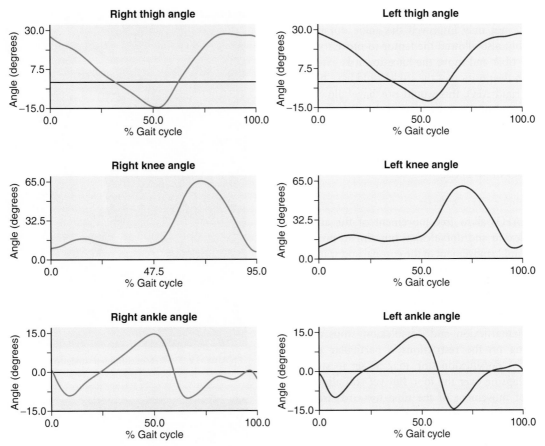

FIGURE 12.19 ■ Movement patterns with a leaf spring ankle foot orthosis

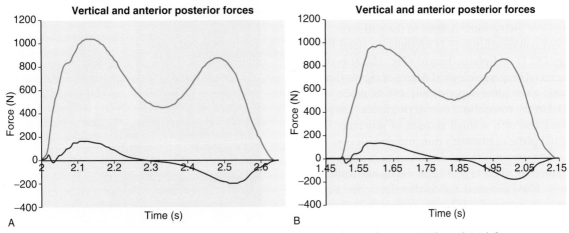

FIGURE 12.20 ■ Force patterns with a leaf spring ankle foot orthosis: (A) right and (B) left

small amount of resistance from the orthosis. This resistance not only improved the ankle dorsiflexion control but also allowed the femur to move forwards over the tibia and move the knee towards extension. Although the position of the ankle could have been set using a rigid AFO, this would not have allowed the same control of the movement of the tibia forwards. The most important aspect in all of this was the patient was again able to walk reasonable distances, his quality of life was much improved and he often joked about the springs in his heels.

Hinged Ankle Foot Orthoses

Hinged AFOs allow free movement of the ankle in plantarflexion and dorsiflexion but aim to provide a block of ankle movement in the coronal and transverse planes, pronation–supination, and inversion–eversion. Although the movement is apparently free in the sagittal plane, the range of motion available is often controlled with plantarflexion and/or dorsiflexion stops. These plantarflexion and dorsiflexion stops are set depending on the restrictions a particular patient requires. A dorsiflexion stop may be set to stop the tibia collapsing over the foot, but yet still allowing a degree of movement of the tibia forwards over the foot. This may be used if the patient has a degree of eccentric control during second rocker, but the prevention of too much movement is required. A plantarflexion stop may be set to prevent foot drop during swing phase or foot slap (uncontrolled movement into the foot flat position) at heel strike. In this way, particular limits can be set to give similar benefits as the casting options in rigid AFOs, but the hinge has the benefit of giving some control to the patient.

Two common hinged AFO designs are metal-hinged (Fig. 12.21) and plastic-hinged (Fig. 12.22) orthoses. The metal hinges give a good degree of rigidity and the plastic gives some support, but not as much as the metal in the coronal and transverse planes. The plastic hinges also offer a small amount of resistance in the sagittal plane in plantarflexion and dorsiflexion.

One reasonably recent development is in the realms of powered AFOs (Fig. 12.23). Ferris and colleagues (2005, 2006) designed and further developed a pneumatically powered AFO with myoelectric (low-pass filtered EMG) control, allowing the artificial muscles to be theoretically controlled from residual muscle

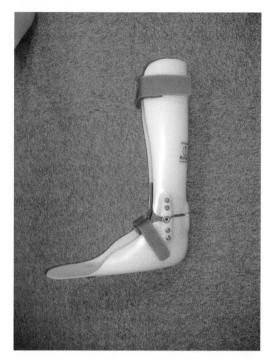

FIGURE 12.21 ■ Metal-hinged ankle foot orthosis

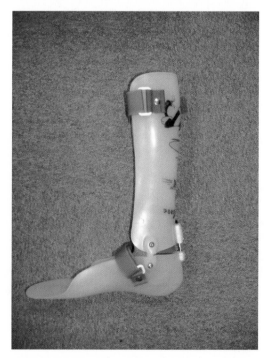

FIGURE 12.22 ■ Plastic-hinged ankle foot orthosis

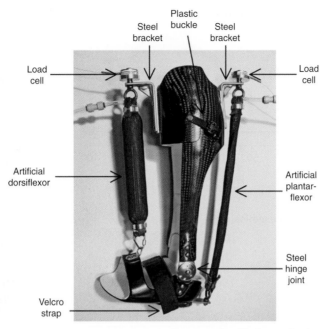

FIGURE 12.23 ■ Pneumatically powered ankle foot orthosis

activity. Robotic exoskeletons and powered orthoses are a very interesting development for the future; however, there is still much to do to improve the portable power supplies, the artificial muscle control systems and the aesthetics to make them a clinically viable option.

The Effect of Hinged Ankle Foot Orthoses

The clinical case study of the use of hinged AFOs shows a young adult with cerebral palsy. This individual was fitted with a metal-hinged AFO to assist his walking. The hinged AFO provided had a dorsiflexor stop at 20°.

When the patient walked without the AFO the ankle began in dorsiflexion and quickly collapsed into further dorsiflexion and internal rotation. This movement of the tibia forward, coupled with internal rotation of the foot, forced the knee into a flexed and apparently valgus position, before eventually moving into extension as the body moved over the stance limb (Fig. 12.24A, B).

With a hinged AFO fitted, the foot started in slight plantarflexion before then moving into dorsiflexion in a more controlled manner. This had the effect of

allowing the knee to be in a more normal position, although the knee moved into a slightly hyperextended position at midstance.

So how can the sagittal plane movement improve when the orthosis allowed movement into plantarflexion and dorsiflexion? The restriction of the dorsiflexion to 20° prevented the collapse into excessive dorsiflexion during second rocker; however, the orthosis also restricted movement in the coronal and transverse planes. This gave added stability and prevented tibial and femoral internal rotation and, therefore, offered a more stable knee position (Fig. 12.25A, B).

Fine Tuning Ankle Foot Orthoses

Singerman and colleagues (1999) investigated the effect of mechanically loading four different designs of AFOs. Perhaps the most notable aspect of this work was the loading and recording of the plantarflexion and dorsiflexion moments in relation to the amount of deflexion. Although this did not consider the effect of the moments on patients, it gives tantalizing evidence to the concept of fine tuning AFOs to individual patient's needs and the effect of the different designs currently available.

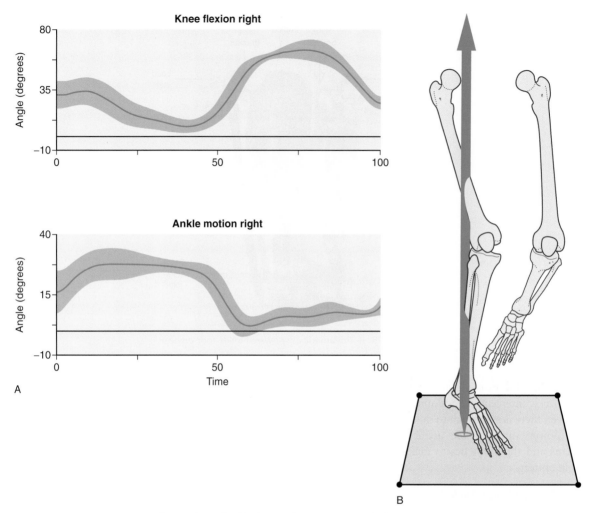

FIGURE 12.24 ■ Movement without a hinged ankle foot orthosis. (A) Sagittal plane movement of the ankle and knee joint. (B) The coronal plane position of the ankle and knee

The solid design showed that a dorsiflexion of 5° produced a resistive moment of approximately 35 Nm into both plantarflexion and dorsiflexion. This is very interesting if we put this in the context of the physiological moment produced in normal adult gait, where at 10° of dorsiflexion, maximum dorsiflexion during second rocker, we would be expecting approximately 80 Nm. Therefore, rigid ankle foot orthoses produce a similar physiological moment. It is also interesting to note the maximum dorsiflexion attained by a normal subject wearing a rigid AFO was nearly 10° with a moment at this point of 85 Nm. This link between the mechanical and biomechanical testing would imply that the eccentric work done by the plantarflexors would be significantly reduced as this is now being controlled almost entirely by the orthosis.

Singerman and colleagues (1999) found that the posterior leaf spring AFO produced a moment of 15 Nm in plantarflexion and dorsiflexion, with a deflection of 10°. This is approximately 20% of the maximum physiological moment during second rocker of ankle movement. This would imply wearing a posterior leaf spring AFO would assist, but not support, the moment about the ankle independently. The flexible plastic

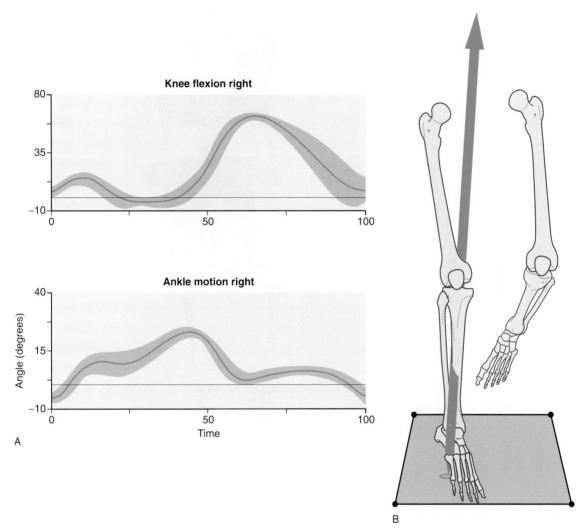

FIGURE 12.25 ■ Movement with a hinged ankle foot orthosis. (A) Sagittal plane movement of the ankle and knee joint. (B) The coronal plane position of the ankle and knee

hinged design also offered a resistance of 10 Nm over a deflection of 12°, which is in the order of 10% of the physiological moment.

There are a number of points of caution to these findings. All the previous comparisons are with normal adult walking and, therefore, the balance between the orthoses and pathological movement and moments will be very different for patients. Therefore, if we consider the use of AFOs in children, the dorsiflexion moments will be substantially different due to their weight, foot length and a variety of other anthropometric factors. So, this brings us back to the importance of clinical assessment and the balancing and tuning of orthoses to achieve the maximum functional benefits, with the minimum of risk of other clinical considerations, such as whether spastic reactions are induced by merely the movement of the joint or by the moments and loads in the muscles, and also the effect of joint contractures on available joint positions. One thing is certain: work into the balance between the

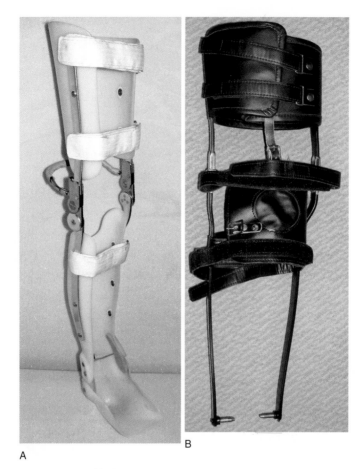

A   B

FIGURE 12.26 ■ Knee ankle foot orthoses: (A) cosmetic KAFO and (B) conventional KAFO

mechanics and biomechanics of AFOs, and the link between this and clinical assessment, requires further interdisciplinary research.

12.3 MANAGEMENT OF THE KNEE, ANKLE AND FOOT USING ORTHOSES

12.3.1 Use of Knee Ankle Foot Orthoses

KAFOs combine the benefits of ankle foot orthoses and knee orthoses. They are generally used when larger moments are required to control the knee, and when there is both substantial lack of control and stability of the ankle and knee joints. The designs of KAFOs can be divided into two categories, conventional and cosmetic. Conventional KAFOs are constructed using leather and metal, with the leather forming the straps

and pads, and the metal forming side steels. Cosmetic, contemporary, or plastic KAFOs are constructed in a similar way to plastic AFOs, with metal hinges joining the sections (Fig. 12.26A, B).

12.3.2 Common Force Systems for Knee Ankle Foot Orthoses

KAFOs can use a three-point force system to offer either resistance to knee flexion moments or knee extension or hyperextension moments (Fig. 12.27A, B, C). However, the nature of the resistive forces will vary with the external moments due to the position of the GRFs with respect to the joints at different stages during the gait cycle.

KAFOs can come in many configurations, with completely rigid joints, to free moving hinges with

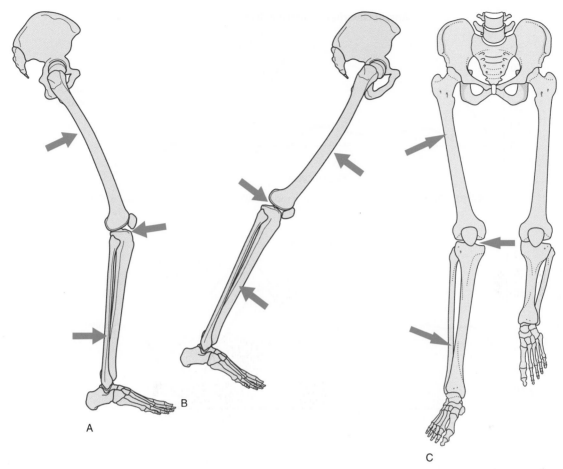

FIGURE 12.27 ■ Three-point force system with a knee ankle foot orthosis: (A) knee flexion, (B) knee extension and (C) knee valgus

hyperextension stops and hinged or leaf spring designed ankle component. The configurations of flexion and extension stops allow different controlling forces at the different stages of the gait cycle by preventing further movement. A similar effect can be achieved by locking the knee, which may be necessary; however, this of course prevents any functional movement.

12.3.3 Clinical Case Study of the Use of Knee Ankle Foot Orthoses

The clinical case study of the use of a KAFO shows a teenager with spina bifida. The orthotic intervention consisted of a pair of stock orthopaedic boots with a compensatory raise to the left, a unilateral cosmetic KAFO, and a rigid AFO section with high lateral walls. The orthotic knee joint provided free motion in the sagittal plane, but was designed to provide lateral stability by the addition of a cosmetic thigh-corset top section.

Without the KAFO fitted, the subject's ankle joint collapsed into a huge amount of dorsiflexion. This would cause the muscles in the posterior compartment of the ankle joint to stretch considerably, which would be very uncomfortable and dangerous as the subject would also be very unstable. With the orthosis fitted, this movement was held, effectively allowing a much more controlled movement over the stance limb. It is

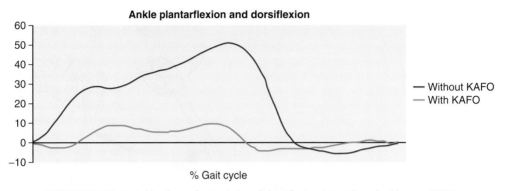

FIGURE 12.28 ■ Ankle plantarflexion (−) and dorsiflexion (+) with and without a KAFO

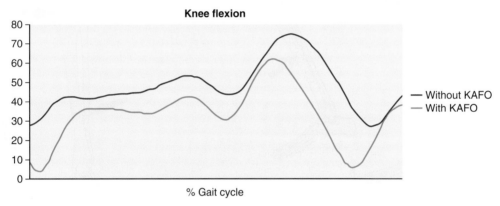

FIGURE 12.29 ■ Knee flexion with and without a KAFO

also interesting to note that the subject was able to move into dorsiflexion, indicating the 'rigid' AFO section is in fact bending (Fig. 12.28).

There is a large difference in the movement patterns at the knee with and without the orthosis (Fig. 12.29). Without the orthosis, the subject's foot strikes the ground with the knee in 28° of knee flexion. With the knee in this position the stride length was significantly reduced, and it also forced the ankle into the excessive dorsiflexion seen previously. After initial contact, the knee should then flex and extend as the load is taken on the front foot and the body moves over the stance limb. However, the subject's knee continued to flex with little recovery, indicating poor quadriceps control and poor movement of the thigh over the tibia. With the orthosis, the knee was initially flexed to 3°, giving a much longer step length; however, the knee was still collapsing into further flexion, but

not to the same degree, giving better movement of the body over the stance limb than without the orthosis. During swing phase the knee flexion without the orthosis was greater than normal, whereas with the orthosis, it was slightly lower than expected. This could be due to the orthosis restricting some motion of the knee during swing phase. The orthosis was clearly offering support and the increase in step length was of considerable benefit to the subject; however, the movement of the body over the knee joint was still an area of concern.

So, could the knee flexion pattern be further improved? If the AFO section were made more rigid, this may reduce the movement of the tibia forwards and bring the femur further forward, thereby reducing the knee flexion during stance phase. Alternatively, the AFO could be set into slight plantarflexion, offering a greater resistance to knee flexion, although this may

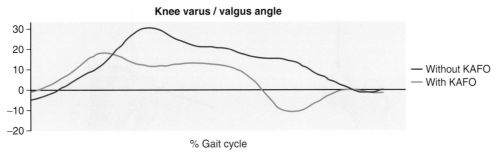

FIGURE 12.30 ■ Varus (+) and valgus (−) angle with and without a KAFO

have a detrimental effect on the foot clearance during the swing phase.

The motion of the knee in the coronal plane was of the most concern. Without the orthosis, the knee collapsed into internal rotation, pushing the knee medially into a valgus position. Without the orthosis, the knee appeared to collapse into 30° of valgus during loading, although this is more likely to be a combination of internal rotation, flexion and abduction of the knee. With the orthosis, this was reduced to 18° valgus (Fig. 12.30).

Although improved, this movement will be extremely detrimental for the patient and, to prevent any further damage/deformity at the knee, this motion needs to be reduced whilst still allowing the knee to move freely in the sagittal plane. This may be achieved by the strengthening of the hinge and side steels, and by applying a force on the medial side of the knee, although the amount of correction needed would require a substantial pressure that may not be tolerable.

12.4 MANAGEMENT OF THE KNEE MOMENTS USING ORTHOSES/BRACES

12.4.1 Biomechanics of Knee Orthoses

The biomechanics of knee orthoses will be considered in relation to their ability to control or change the biomechanics of the knee in the sagittal, coronal and transverse planes. The use of knee orthoses to change the biomechanics of the knee will be examined with examples of varus moments at the knee, shear forces across the knee and the torsional stability of the knee.

12.4.2 Knee Orthoses to Correct Moments

The use of knee orthoses to correct and to support moments about joints is one of the most common uses of direct orthotic management. The theory covered in this section could be applied to many different types of orthotic management including prevention of knee flexion, prevention of hyper-extension or genu recurvatum, and knee valgus and varus instability and deformity, although the exact arrangement of the forces needs to be specific to each condition. These use a variety of types of hinges, joints and locks; however, the example that we will consider is the use of valgus bracing in the management of knee varus in medial compartment osteoarthritis.

The aims of knee valgus braces are to unload the painful compartment through bending moments applied proximally and distally to the knee joint, and reducing the varus deformity (Pollo, 1998). Several studies have been conducted into the use of valgus knee braces for medial compartment osteoarthritis and have reported that patients experience significant pain relief and an improvement in physical function (Hewett et al., 1998; Kirkley et al., 1999; Lindenfeld et al., 1997; Matsumo et al., 1997; Richards et al., 2005) and also a reduction in medial compartment load (Pollo et al., 2002; Jones et al., 2013). But how can a valgus brace reduce the load on the medial compartment of the knee? The answer is that this is a very hard thing to measure directly; however, measures that give an indirect indication of the loading on the medial compartment are a reduction in the knee adduction moments and the varus angle of the knee. We will now consider the biomechanical theory of how knee

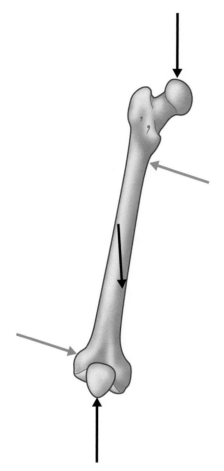

FIGURE 12.31 ■ Proximal segment

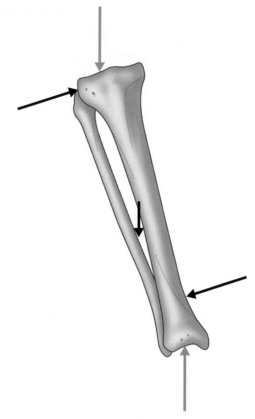

FIGURE 12.32 ■ Distal segment

bracing may be used to change the moments at the knee joint.

12.4.3 Consideration of Individual Segments

When investigating the action of knee bracing to reduce moments we first need to consider the distal (tibial) and proximal (femoral) segments separately. We will first consider the proximal segment. This requires a medial force acting away from the knee joint to deliver a corrective moment, and a lateral force is also required that should act as close to the knee as possible. The lateral force does not create a moment about the knee; however, it is essential as the brace requires an equal and opposite reaction force (Fig. 12.31). In a similar way, the distal segment also requires a medial force away from the knee joint to deliver a

corrective moment and a lateral force as close to the knee as possible (Fig. 12.32).

12.4.4 Consideration of Segments Together

We can now consider the body segments together. When considering the forces on both body segments, we now have two lateral forces acting at the knee. The two lateral forces' optimum position is at the joint centre, therefore, producing no moment, whereas the medial forces theoretically need to be as far away from the joint as possible (Fig. 12.33). The lateral forces can be provided by one point of application, which gives us what is often referred to as a three-point pressure or force system (Fig. 12.34).

12.4.5 Analysis of the Forces Acting on Valgus Bracing

The amount of correction that can be placed on the knee will depend on the bone and cartilage profiles

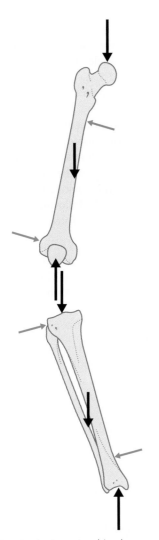

FIGURE 12.33 ■ Combined segments

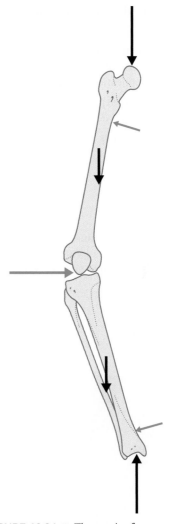

FIGURE 12.34 ■ Three-point force system

within the joint. However, we can consider the supportable angle by considering the mechanics of the orthosis, and the interaction with the body segments. We will first consider the system of forces required to totally support a 5° knee varus angle with no involvement of internal structures and tissues. To do this we again have to consider the proximal and distal segments separately.

Distal Segment

If the total varus angle is 5°, we will consider this to be partly due to the positioning of the proximal and distal

segments. Therefore, we will consider the proximal segment is aligned 2.5° to the vertical:

The length of the shank is 0.45 m

The weight and centre of mass of the shank from the knee are 4.05 kg and 0.25 m, respectively

The orthosis medial is placed 0.25 m away from the knee joint

The angle of inclination is 2.5° to the vertical.

The force on the ankle joint will depend on the GRF and the weight of the foot. If the subject is 90 kg and the mass of the foot is 1.26 kg, then the force from the ankle on the tibia will be 870.54 N:

$$(F_{Ankle} \sin 2.5 \times \text{length of shank}) + (\text{weight of shank} \\ \times \sin 2.5 \times \text{centre of mass}) + (F_{orthosis\ medial} \\ \times \text{point of application of orthosis}) = 0$$

$$(870.54 \times \sin 2.5 \times 0.45) + (4.05 \times 9.81 \times \sin 2.5 \times 0.25) \\ + (F_{orthosis\ medial} \times 0.25) = 0$$

$$17.09 + 0.43 + (F_{orthosis\ medial} \times 0.25) = 0$$

$$F_{orthosis\ medial} \times 0.25 = 16.66$$

$$F_{orthosis\ medial} = \frac{16.66}{0.25}$$

$$F_{orthosis\ medial} = 66.64\ N$$

Therefore, $F_{orthosis\ lateral}$ must also be 66.64 N, which will be acting at the knee joint centre and, therefore, will not provide a moment (Fig. 12.35).

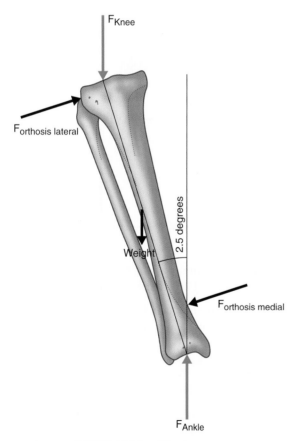

FIGURE 12.35 ■ Distal segment

Proximal Segment

In the same way, we can now consider the distal segment. The total angle at the knee was 5°, with the tibia segment being 2.5° away from the vertical. Therefore, the thigh segment will also be 2.5° away from the vertical.

If we consider the moments acting around the knee joint, then the vertical forces at the hip F_{Hip} and the weight of the thigh segment will provide an adduction moment, whereas force of the orthosis on the thigh segment $F_{orthosis\ medial}$ will provide an abduction moment. Therefore, if we know the following information, then we can find the $F_{orthosis\ medial}$ force:

The length of the thigh is 0.5 m
The weight and centre of mass of the thigh from the knee are 8.64 kg and 0.26 m,
The $F_{orthosis\ medial}$ is placed 0.3 m away from the knee joint
The angle of inclination is 2.5° to the vertical.

The force on the hip joint will depend on the GRF and the weight of the foot, shank and thigh. If the subject is 90 kg and the mass of the foot is 1.26 kg, the mass of the shank is 4.05 kg and the mass of the thigh is 8.64 kg, then the force from the hip onto the head of the femur will be 746.05 N. Both $F_{orthosis\ lateral}$ and F_{Knee} are considered to be acting at the knee joint centre and, therefore, will not provide a moment (Fig. 12.36).

Therefore, if the orthosis is to support the femur, then:

$$(F_{Hip} \sin 2.5 \times \text{length of thigh}) + (mg_{thigh} \times \sin 2.5 \\ \times \text{centre of mass}) - (F_{orthosis\ medial} \\ \times \text{point of application of orthosis}) = 0$$

$$(746.05 \times \sin 2.5 \times 0.5) + (8.64 \times 9.81 \times \sin 2.5 \times 0.26) \\ - (F_{orthosis\ medial} \times 0.3) = 016.27$$

$$+ 0.96 - (F_{orthosis\ medial} \times 0.3) = 0$$

$$17.23 = F_{orthosis\ medial} \times 0.3$$

$$\frac{17.23}{0.3} = F_{orthosis\ medial}$$

$$57.43\ N = F_{orthosis\ medial}$$

So, to support the thigh angulation, a force of 57.4 N will be required from the orthosis. This also means that an equal and opposite force will be required from $F_{orthosis\ lateral}$ to maintain the balance of forces on the

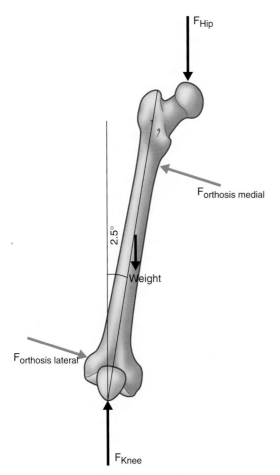

FIGURE 12.36 ▪ Proximal segment

orthosis, although $F_{orthosis\ lateral}$ will not provide a moment about the knee.

However, the largest force will, in fact, be at the knee joint as this must consist of the horizontal components of the lateral forces from **both** the proximal and distal segments as both act on the knee. Therefore, the total lateral knee force may be found by:

$$\text{Lateral knee force} = F_{orthosis\ lateral}\ (\text{femoral}) \times \cos 2.5$$
$$+ F_{orthosis\ lateral}\ (\text{tibial}) \times \cos 2.5$$
$$\text{Lateral knee force} = 57.4 \times \cos 2.5 + 66.64 \times \cos 2.5$$
$$\text{Lateral knee force} = 123.92\ \text{N}$$

The magnitude of this force and the size of the contact area produce a contact pressure that is the limiting factor of the angle which may be supported solely by the orthosis.

12.4.6 Is There a Maximum Supportable Angle Using Valgus Brace?

To find the maximum theoretical angle that may be solely supported by a valgus brace we need first to consider the maximum tolerable contact pressure on the knee.

We now need to consider the maximum size of pad which may be applied to the knee joint in order to find the maximum force that can be applied. A pressure of greater than 32 mmHg or 4.3 kPa has been shown to cause interruption of arteriolar-capillary blood flow (Berjian et al., 1983). However, this is based on long-term pressure application and not intermittent application of pressure that occurs during walking; for instance, pressures on the foot often exceed 250 kPa and stump socket maximum pressure in lower-limb prosthetics have been recorded at 34 kPa for standing and 95 kPa for walking (Lee et al., 1997).

We now need to consider the maximum size of pad which may be applied to the knee joint in order to find the maximum force that can be applied. If we, therefore, consider a pressure of 95 kPa on the proximal and distal pads of the valgus brace, and contact area of 2 cm × 5 cm, then the maximum force that can be applied to the knee may be estimated:

$$\text{Pressure} = \frac{\text{Force}}{\text{Area}}$$
$$\text{Pressure} \times \text{Area} = \text{Force}$$
$$95\,000 \times 0.001\,\text{m}^2 = \text{Force}$$
$$95\,\text{N} = \text{Maximum pad force}$$

If we consider this as the maximum force on both the proximal and distal pads, the maximum supportable joint angle may be found.

Distal Segment

$$(F_{Ankle} \sin \theta \times \text{length of shank}) + (mg_{shank} \times \sin q$$
$$\times \text{centre of mass}) + (F_{orthosis\ medial}$$
$$\times \text{point of application of orthosis}) = 0$$
$$(888.29 \times \sin \theta \times 0.45) + (4 \times 9.81 \times \sin \theta \times 0.25)$$
$$+ (95 \times 0.25) = 0$$
$$399.73 \times \sin \theta + 9.81 \sin \theta + 23.75 = 0$$

$$389.92 \times \sin\theta = 23.75$$
$$\sin\theta = \frac{23.75}{389.92}$$
$$\theta = \sin^{-1} 0.0609$$
$$\theta = 3.49$$

Proximal Segment

$$(F_{Hip}\sin\theta \times \text{length of thigh}) + (mg_{thigh} \times \sin\theta$$
$$\times \text{centre of mass}) - (F_{orthosis\ medial}$$
$$\times \text{point of application of orthosis}) = 0$$
$$(800 \times \sin\theta \times 0.5) + (5 \times 9.81 \times \sin\theta \times 0.26)$$
$$- (95 \times 0.3) = 0$$
$$400\sin\theta + 12.75\sin\theta - 28.5 = 0$$
$$412.75\sin\theta = 28.5$$
$$\sin\theta = \frac{28.5}{412.75}$$
$$\theta = \sin^{-1} 0.0690$$
$$\theta = 3.96$$

However, as before, the largest force will be at the knee joint as this must consist of the horizontal components of the lateral forces from both the proximal and distal segments as both act on the knee. Therefore, the total lateral knee force will be:

$$\text{Lateral knee force} = F_{orthosis\ lateral}\ (\text{femoral}) \times \cos 3.96$$
$$+ F_{orthosis\ lateral}\ (\text{tibial}) \times \cos 3.49$$
$$\text{Lateral knee force} = 95 \times \cos 3.96 + 95 \times \cos 3.49$$
$$\text{Lateral knee force} = 189.6\ \text{N}$$

If the maximum sustainable pressure is 95 kPa then the lateral knee pad size may be found:

$$\text{Pressure} = \frac{\text{Force}}{\text{Area}}$$
$$\text{Area} = \frac{\text{Force}}{\text{Pressure}}$$
$$\text{Area} = \frac{189.6}{95\,000}$$
$$\text{Area} = 0.001996\ \text{m}^2$$

This equates to a circular pad size of radius 0.0252 m or diameter 5.04 cm.

Therefore, the maximum angle that may be solely supported by a valgus brace with a 5.04 cm diameter lateral knee pad is 7.45°. In reality, valgus bracing is useful at much larger knee varus angles as it is very

unlikely that the entire load will be supported by the brace. This supportive element of valgus braces is responsible for the alleviation of pain and the improvement of function often seen with patients suffering from medial compartment osteoarthritis. In the next section, we will consider the biomechanical effects of valgus bracing on individuals with medial compartment osteoarthritis. This angle is specific to this length of brace. Any increase or decrease in the length of the arms of the brace will increase and decrease the maximum angle that can be supported effectively. However, as the length of the brace increases, patient compliance to wear it is likely to decrease.

12.4.7 Valgus Bracing in Medial Compartment Osteoarthritis

It is widely known that knee osteoarthritis is more prevalent in the medial compartment of the knee joint than the lateral compartment, and it has been estimated that during normal gait approximately 60–80% of the load across the knee joint is transmitted to the medial compartment (Prodromos et al., 1985). During walking, individuals have an almost continuous large, external varus moment about their knees throughout stance phase, with the exception of a small valgus moment at initial contact (Johnson et al., 1980; Matsumo et al., 1997). It has been suggested that this varus or adduction moment and the increased loads are a causation factor for the incidence of medial compartment osteoarthritis (Goh et al., 1993). These increasing loads have a degenerative effect on the cartilage in the medial compartment with a narrowing in the joint space between the medial femur and medial tibial plateau. This causes a moment arm increased over that of the unaffected side in a control population (Wang et al., 1990). Increasing disability will arise from the increased moment arm, with pain and functional impairment being the principal complaints of knee osteoarthritis sufferers (Kim et al., 2004), ultimately leading to a reduced quality of life (Fig. 12.37A, B).

Treatment options available to the sufferer are aimed at minimizing these forces at the medial compartment of the knee (Pollo, 1998). Surgical options such as high tibial osteotomy and unicompartmental arthroplasty attempt to unload the medial compartment by removing a portion (wedge) of the tibia, and

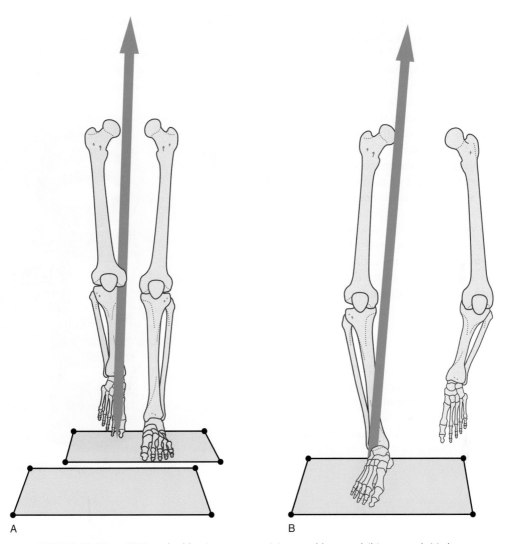

FIGURE 12.37 ▪ GRFs and adduction moment: (a) normal knee and (b) osteoarthritic knee

decrease the loading at the medial compartment by transferring the load to the less affected lateral compartment (Maly et al., 2002; Noyes et al., 1992). However, these types of surgery may not be appropriate for many individuals, such as the younger population, and, therefore, conservative treatment modalities have been introduced in an attempt to reduce this excessive compartmental loading without the need for surgical intervention, and increase the individual's functional independence.

One form of conservative treatment for medial compartment osteoarthritis of the knee is valgus bracing. Valgus braces often claim more than just the ability to support and often claim to offload the painful compartment, correct the varus alignment of the knee and improve quality of life. Various studies have investigated the biomechanical effects and the pain reduction using such devices. This section will consider case study data on the use of valgus bracing in medial compartment osteoarthritis.

Varus Knee Angle

The effect of valgus bracing on knee varus has been a point of debate for some time; however, recent research

(Pollo et al., 2002; Jones, 2013) has shown that bracing can have a direct effect on the knee angulation in the coronal plane. The following data show the immediate effect of an individual walking with and without a valgus brace. The brace fitted in this instance was an OA Adjuster (DJO), which allows the clinician to dial in a 'correction'. In this case, the brace was adjusted until contact was made with the lateral aspect of the knee joint, and then a further 5° was dialled in (Fig. 12.38). This was to first take up the slack in the brace, and then to try to correct by a further 5°. The greatest effect in the varus angle is during loading response from 0 to 20% of the gait cycle (Fig. 12.39). At approximately 10% of the gait cycle, the point of greatest loading, the difference between the braced and unbraced conditions was 3°, indicating that actual correction is in a similar order to the dial-in correction, which in turn will reduce the moment arm of the GRF in the coronal plane.

Knee Adduction Moments

The knee adduction moment ($M = F \times d$) forces the knee outwards (varus) and creates compression on the medial side of the knee joint. Three measures are often taken: peak knee adduction moment during loading (first peak), peak knee adduction moment during push off (second peak) and impulse moment (Fig. 12.40).

Impulse moment adds the extra 'time' dimension, so it is not just how large the moment is, it also matters how long the moment acts for. So, if the person is walking more slowly, the peak moment may be lower but the impulse moment may be higher as the length

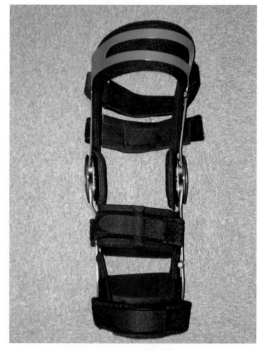

FIGURE 12.38 ■ OA Adjuster, DJO Inc.

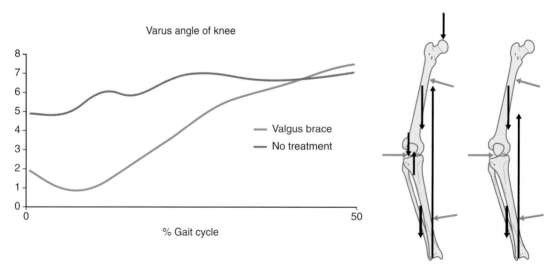

FIGURE 12.39 ■ Varus angle from 0 to 50% of the gait cycle

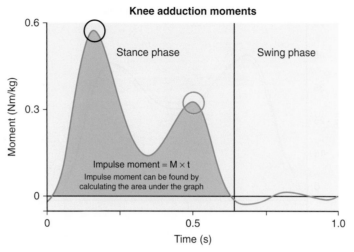

FIGURE 12.40 ■ Knee adduction moments and impulse

of time will be longer, thereby applying the load for longer on the medial side of the knee, which is most often the painful side.

Kim and colleagues (2004) looked at the adduction moment in individuals with and without medial compartment knee osteoarthritis. They found a significant difference in the adduction moment between the osteoarthritis group and an age- and gender-matched normal group, the osteoarthritis group having on average a 50% increase in their adduction moments. Kim also found a correlation between knee adduction moments with the Western Ontario and McMaster Universities Arthritis Index (WOMAC) score. This supports the comments by Goh and colleagues (1993), who suggested that the adduction moment and the increased loads are a causation factor for the incidence of medial compartment osteoarthritis.

The reduction in the moment arm due to the reduction in the varus deformity during loading should in turn lead to a reduction in the adduction moment about the knee joint. In the following section, we see that this is, indeed, the case, with the braced condition reducing the adduction moment by 7% (Jones et al., 2013).

12.4.8 Controlling Moments About the Knee Joint with Shoes and Foot Orthoses

The moments about the knee joint in the coronal plane may also be affected by different footwear and using foot orthoses. This is due to changes in the alignment of the rearfoot in relation to the tibia and changes in the medial–lateral forces acting. The graphs (Fig. 12.41) show an increase in the moments about the knee in the coronal plane with the medial wedge. However, the lateral wedge does not appear to affect the moments about the knee.

The Effect of Lateral Wedging in Medial Compartment Osteoarthritis

We considered the importance of varus or adduction moments in the progression and management of medial compartment osteoarthritis previously when considering the use of a valgus brace to reduce the varus deformity and, therefore, reduce the adduction moments. However, another treatment that has been suggested is the use of lateral wedging of the foot. In the previous section, we saw how, in the application of the GRF, the centre of pressure may be theoretically changed; but can this really be used as an effective way of reducing the adduction moments at the knee? A lateral wedge insole has a thicker lateral border and applies a valgus moment to the heel, attempting to move it into an everted position. It is theorized that by changing the position of the ankle and subtalar joints during weight-bearing (Pollo, 1998) the lateral wedges may apply a valgus moment across the knee, as well as the rearfoot.

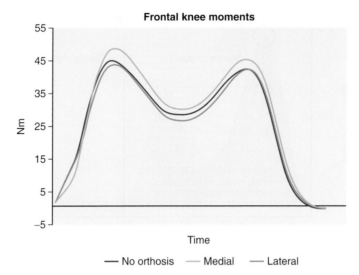

FIGURE 12.41 ■ Moments and the knee joint. *(Adapted from Nester et al., 2003.)*

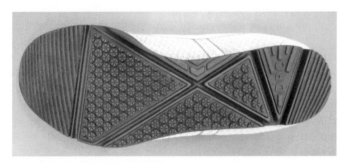

FIGURE 12.42 ■ Mobility shoe for the management of knee osteoarthritis

The knee adduction moment has been shown to be significantly decreased in subjects with medial compartment osteoarthritis using laterally wedged insoles (Jones et al., 2013). Jones and colleagues found that the use of lateral wedging significantly reduced the adduction moments at the knee, which should, in turn, reduce the loading on the medial compartment. It is also clear that such devices give some patients functional improvements during gait. With such indirect management of the knee, many biomechanical factors come into play, including the position of the centre of pressure, the medial GRF, the foot type and also foot contact area.

The Effect of Footwear in Medial Compartment Osteoarthritis

Improvements in the knee adduction moments and patient-reported outcome measures have also been seen when using 'mobility shoes' over standard stable shoes (Shakoor et al., 2013). Subjects with knee osteoarthritis underwent baseline gait analyses under conditions of walking in their own shoes, walking in mobility shoes and walking barefoot. Compared to knee loading at baseline with the participants' own shoes, there was an 18% reduction in the knee adduction moment by 24 weeks with the mobility shoes. In addition, a significant 36% reduction in pain was observed. The mobility shoe achieved this through a modified sole, which was shaped so that the GRF fell more lateral, therefore reducing the knee adduction moments. In addition, the cuts in the sole aimed to guide the movement of the CoP (Fig. 12.42). This investigation provides support for the importance of footwear choice in the management of knee osteoarthritis, and suggests that footwear can be used as a mechanical device to achieve beneficial adaptations and alteration of gait for long-term use.

12.5 MANAGEMENT OF TRANSLATIONAL FORCES AT THE KNEE MOMENTS USING BRACES

12.5.1 Modification of Translational Forces at the Knee with Orthoses

Much has been written on the use of anterior draw testing to determine deficits in the anterior cruciate ligament (ACL). This test looks for translational movement between the tibia and femur.

One treatment that aims to regain translational stability is ACL bracing. ACL braces are sometimes confused with other types of bracing (e.g. valgus bracing in osteoarthritis), which aim to change the moment in the coronal plane; however, the theoretical action of ACL bracing is entirely different. With osteoarthritis bracing we considered that the force system could be reduced to a three-point force system; however, this is not the case with ACL bracing. When walking, we have both vertical and anterior–posterior forces. The anterior–posterior forces will cause shear forces at the ankle and knee joints, which the ACL will, in part, resist and support; however, when the ACL becomes damaged it can no longer support these forces (Fig. 12.43).

To consider the theoretical action of ACL bracing, we first need to consider the GRFs and the action of these forces on the ankle, knee and hip joints. If we consider the forces acting on the tibial and femoral segment separately during late stance phase, at the ankle joint there will be a vertical and anterior force from the foot onto the tibia, which will be resisted by an equal and opposite posterior force at the knee. This posterior force at the knee will subsequently be transmitted to the femoral segment, partly by the action of the ACL. This force will then be balanced by an equal and opposite posterior force at the hip (Fig. 12.44).

Therefore, the anterior force acting on the femur, and the equal and opposite posterior force acting on the tibia, will come from a tensile force in the ACL. So, what happens when the ACL does not function correctly? To consider this, we will imagine a complete absence of the ACL and any other structures that may be able to support these forces in part. Therefore, if these shear forces at the knee are not supported, the anterior force at the ankle will be unopposed and the tibial segment will move in an anterior direction, but

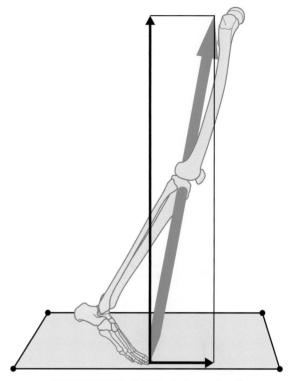

FIGURE 12.43 ■ GRFs during push off

at the same time the posterior force from the hip on the femoral segment will try to move in a posterior direction. This demonstrates the nature of the anterior draw test and how it relates to ACL dysfunction (Fig. 12.45A, B).

Surgical reconstruction of the ACL aims to replace the ligament with either an artificial ligament, sometimes made from Gore-Tex, or a graft harvested from the central third of the patellar tendon or a hamstring tendon. This rebalances the internal forces in the knee and gives translational stability back to the knee. ACL bracing has a similar end functional goal; however, this is achieved by applying an external system of forces, which aims to replace the forces from the ACL and to return translational stability to the knee.

To achieve translational stability to the knee, the brace needs to provide a posteriorly directed force on the anterior proximal aspect of the tibia and an anteriorly directed force acting on the posterior distal aspect of the femur, which will stop its translation posteriorly (Fig. 12.46). Additional forces are also

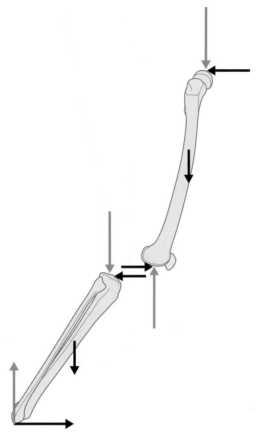

FIGURE 12.44 ■ Normal force system

required to give the brace stability. An anteriorly directed force is also required further distal on the tibia, and a posteriorly directed force acting on the anterior proximal aspect of the femur.

To reduce the turning effect, the forces either side of the knee should be placed as close to the joint centre as possible, therefore minimizing the moment required from the additional forces. Previously we considered braces that can control the moments about a joint where we could use a three-point force system; however, the nature of the direction of the forces required for translational stability is different and a three-point force system would not give support in the correct way. Therefore, a different load system and a different design of brace are required to control shear forces, with four forces now being required to prevent translational movement between body segments. This

is sometimes referred to as a four-point fixation brace or a four-point force system. Movement is possible with such an orthosis if the brace is hinged to allow flexion in the sagittal plane. One example of a hinged four-point ACL brace is the 4Titude Ligament Knee Brace (Fig. 12.47).

12.5.2 The 'Mechanics' of Soft Bracing of the Knee

The focus of many studies on patellofemoral pain has been on the loading limb rather than on the eccentrically controlling limb, although clinically it is most often the eccentrically controlling limb that has most pain. One of the aims of the different treatments is to improve the control of the knee joint; both patellofemoral bracing and taping attempt to do this by realigning the patella. If the patella is being realigned, then this should have greater implications to the coronal and transverse plane mechanics of the knee. Research to date has mainly focused on the sagittal plane, with the mechanics of the torsional and coronal planes attracting very little attention.

Since McConnell's landmark paper in 1986 there has been considerable clinical and research interest in taping techniques for the patellofemoral joint. Patellofemoral taping techniques are now considered as part of standard clinical practice. Although a consensus view that tape is effective at relieving pain is emerging in the literature, there is still an ongoing debate about the mechanism of effect of taping. Recent work has highlighted the importance of the potential proprioceptive effects of taping (Baker et al., 2002; Callaghan et al., 2002).

Compared to taping, there has been much less research on the effect of braces in the management of patellofemoral problems, with only 7% of the recent research literature focusing on this modality (Selfe, 2004). The pain-relieving effects of bracing are reported as occurring through an increased stabilization of the joint, which reduces the force of muscular contraction (Nadler & Nadler, 2001). In particular, patellofemoral braces are designed to 'reduce compression of the patella as well as to prevent excessive lateral shifting' (Nadler & Nadler 2001). Although the results tend to be positive, because of the small number of studies the usefulness of braces remains controversial. The fact that soft braces, by definition, have no rigid

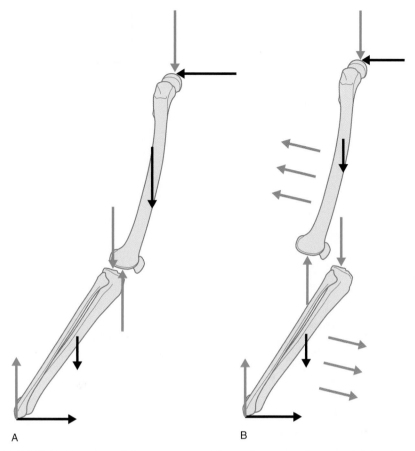

A   B

FIGURE 12.45 ■ (A and B) Force system with anterior cruciate ligament deficient knee

structures has caused much debate over their function. This is highlighted by the fact that in some European countries soft braces are not considered therapeutic and, therefore, cannot be prescribed, whereas their hinged counterparts can be. So, what are the 'mechanics' behind soft patellofemoral bracing and what effect can these have on joint loading and stability?

The System of Forces in Patellofemoral Bracing

The purpose of patellofemoral bracing is to allow full flexion movement but to provide medial–lateral and rotational stability to the knee by realigning the patella. To achieve this the brace is of soft design, which allows the middle section of the brace to move independently of the proximal and distal aspects; therefore, a torsional 'correction' may be produced by tensioning the straps either medially or laterally depending on the patella correction required. On some designs of brace, a semi-circular buttress is fitted to the brace to ensure the force is directed onto the patella (Fig. 12.48).

The Effect of Patellofemoral Bracing on Joint Stability

The vast majority of research into stair climbing has focused on the sagittal plane; however, according to Kowalk and colleagues (1996), although the knee abduction–adduction moment is not in the primary plane of motion, its magnitude should not be ignored when trying to understand the stability and function of the knee during stair-climbing activities. Powers (2003) adds that the knee is designed to absorb rotary forces through its transverse plane motion. Powers also goes on to state that motion of the tibia and femur in

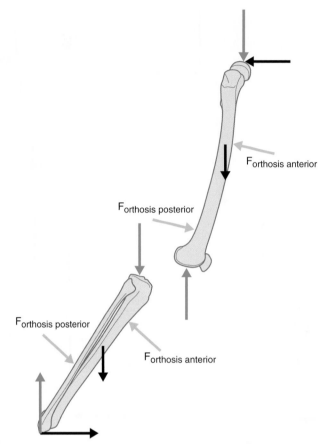

FIGURE 12.46 ■ Four-point force system to achieve translational stability

the frontal and transverse planes can influence patellofemoral mechanics. The comparison of patellar bracing and taping during functional tasks is beginning to attract attention. As the reported purpose of these treatments is to realign the patella, the largest effects should be on the coronal and transverse planes of the knee.

The study of the effect of patellofemoral bracing on joint stability (Selfe et al., 2008, 2011) involved the testing of healthy subjects and patients with patellofemoral pain. They were asked to conduct a slow step-down exercise to assess the control of the knee during a slow eccentric controlled exercise. The step down was conducted under three randomized conditions: (a) no intervention, (b) neutral patella taping and (c) a patellofemoral brace. A step was designed to accommodate one of the plates; the other plate was

embedded in the floor, which arrangement produced a standard step height of 20 cm. The force platforms allowed for the measurement kinetics in the sagittal, coronal and transverse planes. Reflective markers were placed on the foot, shank and thigh using the calibrated anatomical system (CAST) technique (see Chapter 9). The purpose of the step-down exercise was to assess the control of the knee as the body was lowered as slowly as possible from the step. The kinematic and kinetic data about the knee were then quantified from toe off, of the contralateral limb to contact of the supporting, eccentrically controlling limb.

The results from the coronal plane movement and moments show that the patellofemoral brace is having a controlling effect on the mechanics of the knee. This is supported by the results from the coronal and transverse plane knee range of motion experienced at the

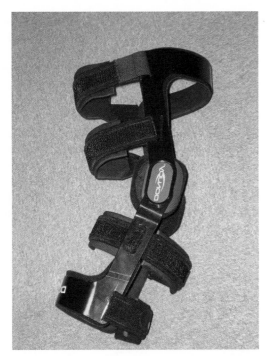

FIGURE 12.47 ■ 4Titude Ligament Knee Brace, DJO Inc.

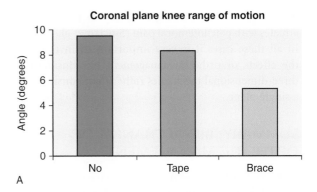

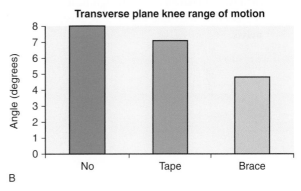

A

B

FIGURE 12.49 ■ Range of movement of the knee during a slow step-down challenge in the (A) coronal and (B) transverse plane *(Adapted from Selfe et al., 2011.)*

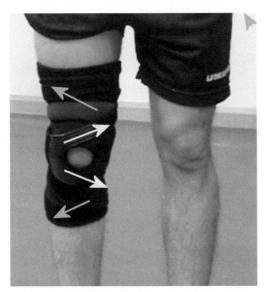

FIGURE 12.48 ■ The system of forces in patellofemoral bracing

knee with the patellofemoral brace during a slow step-down challenge. These results suggest that both taping and bracing are, indeed, having an effect on the coronal and torsional mechanics of the knee, allowing an eccentric descent with considerably more control, with the brace having the largest effect (Fig. 12.49A, B). Although not presented, the sagittal plane did not show any change in the two interventions.

The nature of these results raises the question: how exactly do 'soft' orthoses work? Do they have a direct effect on the mechanics or do they improve proprioception and, therefore, aid joint control? Further studies have shown similar effects in improving the movement patterns and joint control and stability during dynamic sporting tasks (Hanzlíková et al., 2016). In addition, it has been shown that soft proprioceptive knee bracing can also serve to significantly reduce perceived pain and patellofemoral loading during sport-specific tasks, including jog, cut and single leg hop in recreational athletes with patellofemoral pain (Sinclair et al., 2016). In all these cases it is very important to investigate the effects of orthotic management by considering three-dimensional mechanics rather than considering a single plane.

SUMMARY: BIOMECHANICS OF ORTHOTIC MANAGEMENT

- Foot orthoses can have a direct effect on the foot and ankle movement, and the GRFs. Foot orthoses can have a clinically significant effect on the control and function of the knee, hip and pelvis.

- Different configurations of ankle foot orthoses can be used to block movement about the ankle joint, assist with muscle function present, or allow free movement within 'safe' limits. The use of ankle foot orthoses can have clinical effects at the ankle, knee, hip and pelvis.
- Different configurations of knee orthoses or knee brace are required to support moments about a joint, shear and axial forces. Each configuration when correctly prescribed can improve joint function, stability and quality of life.
- The nature of each individual's muscle and joint function dictates which would be the most effective orthotic management.

13

BIOMECHANICS OF THE MANAGEMENT OF LOWER LIMB AMPUTEES

NATALIE VANICEK

AIM

To compare the level and stair walking profiles of lower limb amputees with able-bodied individuals, and also to compare amputees according to their level of amputation (i.e. transtibial vs transfemoral) and prosthetic componentry.

OBJECTIVES

- To discuss symmetry between the affected and intact limbs in lower limb amputees

- To explain the differences in prosthetic componentry available for transtibial and transfemoral amputees

- To describe gait profiles in lower limb amputees during level walking, stair ascent and descent by discussing compensatory kinematic and kinetic strategies.

13.1 CHAPTER INTRODUCTION

A major lower limb amputation is the removal of a limb either through the ankle (Symes), below the knee (transtibial), through the knee (or modified through the knee, i.e. Gritti–Stokes), above the knee (transfemoral) or through the hip (hip disarticulation). Most lower limb amputees learn to ambulate with a prosthesis and can negotiate many daily tasks with relative ease, although they present with distinct temporal-spatial, kinematic and kinetic patterns that will be discussed throughout this chapter.

13.1.1 Amputation Demographics

An amputation can result from trauma, peripheral arterial disease, diabetes, tumour/malignancy or infection. In some rare instances, a skeletal malformation or missing limb may result in a congenital amputation. In the United States, there are approximately 185 000 amputations of the upper or lower extremity performed every year (Owings & Kozak, 1998), whereas in the United Kingdom the annual numbers are significantly smaller, with approximately 5000–6000 new major lower limb amputations (National Amputee Statistical Database [NASDAB], 2009). It is projected that the number of persons living with an amputation will double by 2050 (Ziegler-Graham et al., 2008).

The majority of amputations are a result of disease and therefore the majority of new amputees are adults aged over 50 years. Thus, they must deal with concomitant issues related to their new biomechanical constraints and age-related musculoskeletal alterations. However, thanks to prosthetic technologies, lower limb amputees can relearn to walk with a prosthesis proficiently and develop stereotypical gait patterns relative to their level of amputation.

13.1.2 Symmetry

As the majority of lower limb amputees have a unilateral amputation, they are inherently asymmetric. They have lost the bones, muscles, nervous and connective tissue below or above the knee. These have (partly) been replaced with prosthetic componentry with a range of functional capabilities. Rehabilitation specialists often strive to reduce any between-limb asymmetry in the gait patterns of lower limb amputees. Gross asymmetry may come at a high metabolic cost and may contribute to other debilitating factors. These include increased risk of osteoarthritis on the overloaded joints of the intact side (Lloyd et al., 2010), increased risk of osteoporosis due to extended unloading of the residual limb, especially for trans-femoral amputees (Sherk et al., 2008) and lower back pain (Hammarlund et al., 2011). Conversely, others have argued that an accepted level of asymmetry may be intrinsically natural in lower limb amputees. Winter and Sienko (1988) went so far as to suggest that: 'It is safe to say that any human system with major structural asymmetries in the neuromuscular skeletal system cannot be optimal when the gait is symmetrical. Rather, a new non-symmetrical optimal is probably being sought by the amputee within the constraints of his residual system and the mechanics of his prosthesis.' (Winter and Sienko, 1988, p. 362). This was supported by a previous study that found gait symmetry, measured by stance time and step length, worsened when the inertial properties of the prosthesis were matched to that of the intact limb (Mattes et al., 2000). Moreover, the energetic cost of level walking increased when the inertial properties were matched closely between the two limbs. Therefore, caution should be taken when striving for perfect symmetry between two inherently different limbs.

13.2 TYPES OF PROSTHESES

A prosthetic limb is normally made up of several com-ponents: namely, a socket (interface between the residual limb and the prosthesis), a pylon (providing exoskeleton support), a prosthetic ankle and/or knee, keel for the foot, and a cosmesis for those who prefer the prosthetic leg to look more lifelike. The purpose of a prosthesis is to provide stability during standing and walking activities, and also functionality, especially for individuals fitted with more advanced prosthetic components. A prosthesis should be comfortable and durable to ensure maximum safety. The most techno-logically advanced prosthetic components may not be suitable for everyone, and attention should be given to ensure lower limb amputees are fitted with pros-thetic components to match their activity levels and lifestyle.

13.2.1 Transtibial Amputees

Several types of prostheses are currently available, ranging in performance and cost. At the very basic level, a standard prosthetic foot does not allow for any movement of the prosthetic ankle 'joint' (i.e. non-articulated) and has no energy storing and release capacity as the keel is kept rigid.

Following World War 2, the SACH foot (solid ankle-cushioned heel) was designed with a soft rubber heel that compresses and cushions upon heel strike, attempting to mimic the first heel rocker through plantarflexion, when the (now absent) ankle dorsiflex-ors would normally control the lowering of the foot through eccentric activity. The SACH foot was also designed to aid knee flexion during loading. The SACH foot is still prescribed today, and is generally low maintenance but will limit the wearer's functional per-formance, thus making daily tasks difficult to complete for more active users.

Other prosthetic feet allow movement in one or two planes thanks to a hinge mechanism along a single axis (i.e. passive plantarflexion and dorsiflexion) or multiple axes (i.e. plantarflexion and dorsiflexion and inversion/eversion). They have variable rollover characteristics according to the individual person's alignment. Multi-axial prosthetic feet adjust to the underlying surface, resulting in more fluid movement in challenging environments, such as when ambulat-ing across uneven terrain and negotiating obstacles.

Dynamic response feet, and energy store and release feet, store kinetic energy during loading response and release it as potential energy during push off. In addition to facilitating more natural gait patterns, especially push-off power (Segal et al., 2012), these sophisticated prosthetic designs allow the user to run and jump more comfortably by reducing the loading on the residual limb.

Motor-powered or computerized prosthetic feet use active damping or spring-clutch mechanisms to make the prosthetic ankle-foot adjust to the ground surface (Au & Herr, 2008). Current technology is advancing rapidly in the field, as engineers and bio-mechanists strive to create a prosthesis that mimics the human foot and its mechanical properties without increasing its mass.

13.2.2 Transfemoral Amputees

Broadly speaking, the prosthetic knee for transfemoral amputees can either be categorized as mechanical or microprocessor controlled. Similar to the prosthetic ankle, mechanical knees can move about a single axis or multiple axes. It is generally accepted that the more stable the prosthetic knee joint, the less functional the user. Conversely, the user's safety may be compromised as the prosthetic knee becomes more functional and less stable. Therefore, prosthetic prescription and adjustments are a constant balance between the user's health and physical ability, daily activities and lifestyle goals.

Stance Control Stability

To enhance the stability of the prosthetic knee, the componentry will have some kind of weight-bearing locking mechanism activated during stance. A manual locking system is the most stable as it is locked and unlocked freely by the user. When locked, the knee is prevented from buckling or 'giving way', although this can result is an unnatural-looking gait pattern, which is more tiring and with higher metabolic demands (Dupes, 2005). However, this is often more beneficial for individuals who are otherwise quite unstable or who are negotiating unstable environments. An alternative to the manual locking system is the weight-activated stance control prosthetic knee; this knee is frequently prescribed for less active users. In this case, the prosthetic knee is locked when it is loaded by the user; the knee is then free to bend once the weight-bearing load has been shifted, for example upon initiation of the swing phase.

Motion Control During Swing

To improve the swinging motion, prosthetic knees may use constant or variable friction control or fluid control systems with pneumatic or hydraulic mechanisms. Mechanical friction at the axis of rotation is needed to provide constant friction control at the prosthetic knee joint. Constant friction knees are generally lightweight and durable; the downside is that the level of constant friction is adjusted for a set walking speed and therefore the user has less flexibility of use. A variable friction knee joint is designed in such a way that there is increased friction once the knee starts to bend but it also allows the user to vary their walking speed more freely. However, this type of knee requires more frequent adjustments by prosthetists.

A fluid control system uses fluid dynamics to provide variable resistance for better control during the swing phase. Essentially, they are designed as a piston attached to a pivot within the thigh segment of the prosthesis behind the knee bolt, and a cylinder is connected to a pivot in the leg segment (Muilenburg & Wilson, 1996). A pneumatic system will compress air when the prosthetic knee flexes, storing and then returning energy when the knee is extended again. Instead of air, a hydraulic system uses a viscous fluid (i.e. oil) to accomplish the same task. The benefit of a hydraulic controlled system is that it facilitates a more natural gait pattern, but it is heavier than a pneumatic system and also more expensive.

Mechanical Knee Joints

At the most basic level, a mechanical knee with a single axis acts using a simple hinge mechanism, but this type of knee joint lacks stability during stance as the knee can swing freely. Therefore, the user will adopt compensatory strategies to help their stability during gait. These will be discussed in more detail as this chapter explores knee kinematic and kinetic strategies for lower limb amputees. Mechanical knees will have constant friction control to manage the swinging speed of the prosthetic leg. Some advantages of this type of prosthetic componentry include durability, low cost, low maintenance and light weight.

Multi-axis mechanical knees, also known as polycentric knees, offer more versatility compared to single-axis knees, as the knee can rotate about multiple axes. Multi-axis knees provide stability upon heel strike but are also easier to bend and to facilitate, for example, getting in and out of a chair or a car. Swing control is often provided by hydraulic or pneumatic mechanisms. One distinct advantage to this prosthetic design is that the prosthetic limb shortens in length during the swing phase, which is beneficial for preventing a stumble, trip and possible fall (Dupes, 2005). Some disadvantages to multi-axis prosthetic knees are that they are heavier than their single-axis counterparts and may require more servicing by qualified prosthetists.

Microprocessor-Controlled Knees

Prosthetics technologies are constantly evolving and at a rapid rate. Microprocessor-controlled, or computerized, knee joints have benefitted from significant technological advancements since the 1990s. These are now widely available in many countries, although often at a large financial cost. Microprocessor-controlled prostheses work on the basis that sensors detect the user's movements. These sensors can detect the position of the knee joint during the stance and swing phases, relaying feedback to the computerized knee so that appropriate actions can be executed. These prosthetic components, whether hydraulic, pneumatic or magnetic, are highly sensitive, adjusting the level of stability during stance and controlling the swinging motion easily. Microprocessor-controlled knees allow the user to walk at different speeds and to negotiate more challenging walking tasks such as obstacles, stairs, ramps and uneven surfaces, including grass, rocks and compliant surfaces. One of the most popular microprocessor-controlled knee joints, the C-Leg by Otto Bock, has proven to be especially effective at achieving a good balance between functionality and safety during everyday activities that present a higher risk of trips and stumbles. User satisfaction is also improved (Berry et al., 2009).

The latest prosthetics advancements include bionic prosthetic knee joints, such as the Genium by Otto Bock. Previous research has revealed the bionic knee was more 'in tune' and facilitated a gait pattern that was more similar to an able-bodied individual's thanks to improved physiologic loading of the knee (Bellmann et al., 2012). This characteristic helps the user walk with a reciprocal stair ascent pattern, and walk up and down ramps. It has also been suggested that the gait of transfemoral amputees is enhanced by 'unconsciously controlling their prostheses' (Kahle et al., 2008) and that the bionic technology allows the user to integrate their prosthesis into a positive body image (Bellmann et al., 2012).

13.3 EARLY GAIT RETRAINING

Soon after the amputation of a lower limb, the rehabilitation process typically involves early, partial weight-bearing on the residual limb. This can be achieved with an early walking aid or a preparatory prosthesis. For the majority of transtibial and transfemoral amputees, an early walking aid can be used as early as 7–10 days post-surgery (Redhead et al., 1978). An early walking aid is a generic device consisting of a pneumatic bag inflated around the residual limb and placed within a rigid support frame. Its use helps

reduce oedema in the residual limb, prevents possible post-operative complications and reduces time to fitting of the definitive prosthesis (Redhead et al., 1978; Pollack & Kerstein, 1985). An early walking aid allows patients to practise partial weight-bearing and ambulation between parallel bars before they are cast for their definitive socket and prosthesis. This process improves the cast and fit of the definitive prosthesis. Depending upon the exact rehabilitation protocols provided by the team of healthcare professionals, an alternative to the early walking aid is a preparatory prosthesis. The socket of the preparatory prosthesis is made up of plaster cast or plastic material, which is then attached to a pylon with an artificial foot at the distal end (Muilenburg & Wilson, 1996). In both these cases, a belt and strap around the shoulders are used to secure the prosthesis in place.

A previous study explored whether early mobilization with an articulated (i.e. flexing) early walking aid improved the gait patterns of transtibial amputees during the gait retraining process compared to an early walking aid that kept the biological knee on the affected side locked in extension (Barnett et al., 2009) (Fig. 13.1). These authors hypothesized that walking immediately with a more natural gait pattern would improve longer-term gait. However, the results revealed that neither early walking aid, articulated nor rigid, benefitted patients more than the other, and that patients were being discharged from prosthetic rehabilitation once they achieved a relatively stable gait pattern and their walking speed reached a plateau.

13.4 LEVEL WALKING

13.4.1 Temporal-Spatial Parameters

Broadly speaking, lower limb amputees walk more slowly compared to their age-matched able-bodied counterparts. Walking speed varies according to level of amputation and also with cause of amputation. Amputees with more proximal amputations (i.e. transfemoral amputees) walk more slowly than those with more distal amputations. Older, vascular amputees, lacking the overall fitness levels of younger, traumatic amputees, will also exhibit significantly slower walking speeds.

The average comfortable walking speed of able-bodied adults, aged 20–60 years, is approximately

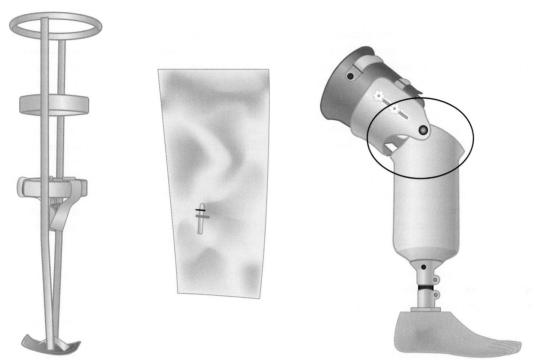

FIGURE 13.1 ■ The rigid early walking aid (Pneumatic Post-Amputation Mobility Aid with inflatable residuum bag, *left*) and the articulated early walking aid (Amputee Mobility Aid, with hinge mechanism knee circled, *right*). Both early walking aids allow transtibial amputees to practise early mobilization and partial weight-bearing prior to being cast and fit for their definitive prosthesis. Transfemoral amputees may use the Pneumatic Post-Amputation Mobility Aid or an alternative generic prosthetic device with a simple prosthetic knee joint such as the Femurett. *(Images taken from Barnett et al., 2009.)*

1.40 m/s or 5 km/h (Bohannon, 1997). Although somewhat variable, most research with lower limb amputees has demonstrated reduced speed. Comfortable walking speeds for established, unilateral transtibial amputees are approximately 1.0–1.2 m/s (De Asha & Buckley, 2015; Vanicek et al., 2009; Wong et al., 2015), whereas transfemoral amputees walk even more slowly, with average speeds closer to 0.85 m/s (Schaarschmidt et al., 2012), although they can achieve faster speeds approaching those of transtibial amputees (Starholm et al., 2016). Recent amputees, still adapting to their new biomechanical constraints, initially walk very slowly, ranging from 0.49 to 0.72 m/s during rehabilitation (Barnett et al., 2009), but they can increase their speed by 0.15–0.20 m/s during the first 6 months following discharge from prosthetic rehabilitation (Barnett et al., 2014a, 2014b).

13.4.2 Stance and Swing Times

From a biomechanical perspective, and as measured through gait analysis, one of the most consistent differences between able-bodied individuals and lower limb amputees is the duration of stance and swing times. Whereas able-bodied individuals consistently exhibit stance to swing ratios of 60:40% of the gait cycle, these values are often significantly different in lower limb amputees, who also demonstrate a marked asymmetry between the intact and affected limbs. Lower limb amputees load their intact limb for longer than their affected side and therefore the stance duration on the intact limb is greater than the affected side. Conversely, the swing duration is longer for the affected limb. Amputees also tend to have longer double support phases (initial and terminal), as maintaining both feet on the ground enhances dynamic stability.

Reasons for decreased stance duration on the affected side include discomfort of weight-bearing on the residual limb and inadequate prosthetic confidence, with sensations of the knee buckling and 'giving way'. This is particularly relevant in transfemoral amputees and will be discussed in more detail shortly. At their preferred walking speed, lower limb amputees may spend 66% of the gait cycle in stance on their intact limb and only 62% on their affected limb (Vanicek et al., 2009; Eshraghi et al., 2014), whereas other research has found intact stance duration as high as 71% (Wentink et al., 2013). Double support increases from 20% of the gait cycle in able-bodied individuals to 30% in amputees (Vanicek et al., 2009). Stance duration is negatively correlated with walking speed, so it is unsurprising that even the intact limb demonstrates a longer stance phase compared to faster walking able-bodied individuals. This also explains why very new amputees have very long stance durations of 75–80% on the intact side, and even 72–73% on the affected side, as they spend the majority of their time almost standing still when they are relearning to walk with their prosthesis between parallel bars (Barnett et al., 2009).

One consequence of a longer swing phase on the affected side is reflected in step length. Lower limb amputees consistently exhibit a longer step length on the affected side compared to the intact side. However, some research has noted that the step length asymmetry may not be detrimental and may, in fact, serve to enhance dynamic stability. Houdijk and colleagues (2014) found that step length asymmetry, notably a shorter step length with the intact limb, caused a larger backwards margin of stability compared to on the affected side. This was thought to be related to the reduced centre of mass velocity (vCoM) at the end of the intact step length, coinciding with reduced push-off power and propulsive force under the affected, prosthetic foot. These authors believed that this gait compensation may help to prevent a backwards fall during gait (Houdijk et al., 2014).

13.4.3 Joint Angular Kinematics

Compared to able-bodied individuals, lower limb amputees show distinct kinematic differences at all three lower limb joints during level walking. Distally, the prosthetic ankle is maintained in a more dorsiflexed

position to facilitate toe clearance during swing. It does not actively plantarflex following toe off and does not dorsiflex as much as a biological ankle joint during mid- to terminal stance (Sanderson & Martin, 1997). However, kinematic compensations also exist on the intact side, which also displays less ankle plantarflexion than the ankle of able-bodied individuals (Sanderson & Martin, 1997). Although the intact limb has not lost the plantarflexor musculature to generate adequate push-off power, reduced plantarflexion (and coinciding smaller power generation at the ankle) has been observed in lower limb amputees. This may be due to muscle weakness or may be a compensatory strategy to improve dynamic stability when the prosthetic limb is entering the vulnerable single support phase on the affected side (Vanicek et al., 2009).

Apart from the prosthetic ankle joint, the knee on the affected side, whether it is a prosthetic knee joint for transfemoral amputees or a biological joint for transtibial amputees, displays obvious differences with the intact side and when compared to able-bodied gait profiles. As evidenced by the reduced stance duration on the affected side, lower limb amputees generally attempt to load their prosthetic limb on the affected side less compared with their intact limb. During stereotypical, able-bodied gait, the knee flexes shortly following initial foot contact during the loading response. This shock-absorbing mechanism places increased muscular demands on the eccentrically contracting knee extensor muscles. In transtibial amputees, these muscles are often weakened on the affected side, whereas in transfemoral amputees these muscles are mostly absent. Therefore, to compensate for the lack of muscle strength, amputees demonstrate a reduced loading response, or in the case of transfemoral amputees, an extended and often hyperextended knee position (Fig. 13.2). Although the amount of loading varies according to the prosthetic user and their prosthetic components, this is unsurprising given the stance control stability mechanisms where the knee joint is locked in extension to prevent the prosthetic knee from buckling. The extended knee position presents another distinct characteristic in that the ground reaction force vector (vGRF) is maintained near the knee joint centre if the limb is maintained close to vertical. Thus, with a short moment arm, the external torque about the knee is reduced and can be very close to zero

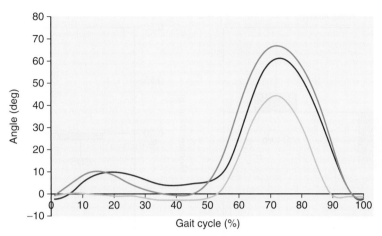

FIGURE 13.2 ■ Prosthetic knee angle (flexion is positive) for a transfemoral amputee (*light green line*) with a mechanical knee joint versus transtibial amputee (*black line*) on the affected side and compared to an able-bodied individual (*dark green line*) walking at their comfortable speed. The transfemoral amputee presents distinct differences compared to the transtibial amputee at the prosthetic knee joint, such as knee extension throughout the entire stance phase and reduced knee flexion in swing

(Winter & Sienko, 1988). In this way, the muscular demands of the knee extensors and flexors can be reduced. This can be advantageous in amputees with weakened knee musculature, but disuse of these muscle groups may also lead to further atrophy. Appropriate strengthening exercises are recommended, especially for transtibial amputees, to help attenuate any further decline in muscle strength.

13.4.4 Comparison of the Knee Joint Kinematics and Kinetics When Transfemoral Amputees Walk with Different Prosthetic Knee Joints

Kaufman and colleagues (2007) compared the knee joint kinematics and kinetics when transfemoral amputees walked with a mechanical versus micro-processor controlled prosthetic knee joint. They reported that amputees walking with a passive mechanical knee joint (Mauch SNS, Össur) presented with a significantly more hyperextended knee and an internal knee flexor moment, compared with a more stereotypical loading response and corresponding internal knee extensor moment when walking with the microprocessor-controlled knee (C-Leg, Otto Bock) (Fig. 13.3). They found significantly improved postural control when amputees wore the microprocessor-controlled C-Leg versus the mechanical knee joint during static and dynamic balance conditions measured with comput-

erized dynamic posturography (Kaufman et al., 2007). In a separate article, the same authors also reported greater spontaneous physical activity levels (outside of the laboratory environment) among amputees walking with the micro-processor controlled C-Leg compared to the mechanical knee joint (Kaufman et al., 2008). The improvement was related to increased prosthetic satisfaction and more physical movement.

Despite the absent (or reduced) physiologic loading response on the affected side, many lower limb amputees (apart sometimes from transfemoral amputees walking with a mechanical knee joint) still display peak knee flexion of approximately 60° during mid-swing (Powers et al., 1998; Sanderson & Martin, 1997). Adequate knee flexion is important as it raises the prosthetic foot off the ground, facilitating adequate ground clearance and avoiding a possible trip or stumble (also assisted by the prosthetic ankle oriented in dorsiflexion). Moreover, it reduces gait compensations often observed proximally, such as increased pelvic hike or hip circumduction during swing on the affected side. The study by Bellmann and colleagues (2012) compared the knee kinematic profiles of transfemoral amputees walking at various speeds with a microprocessor-controlled C-Leg or a more advanced bionic Genium knee. The differences in knee joint angles are presented in Fig. 13.4 and illustrate a better physiologic loading response with the Genium knee

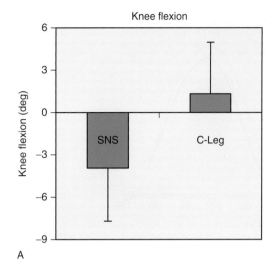

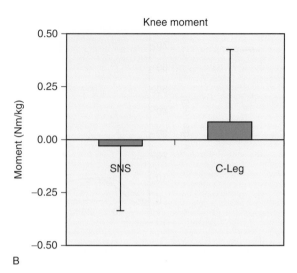

A   B

FIGURE 13.3 ▪ Knee flexion and internal knee moment during the loading response of the gait cycle. (A) The amount of knee hyperextension (negative value in degrees) that was measured for amputees wearing the mechanical SNS (Swing-N-Stance) prosthetic knee versus the amount of flexion (positive value in degrees) for amputees wearing a microprocessor-controlled C-Leg during the loading response of the gait cycle. (B) The virtually absent knee moment for the SNS prosthetic knee versus a more stereotypical internal extensor moment (positive value) during the same gait phase. *(Image taken from Kaufman et al., 2007.)*

compared to the C-Leg. Fig. 13.4 also emphasizes more knee flexion with the Genium knee during swing when walking at slow speeds, which is important for ground clearance.

13.4.5 Compensations at the Hip and Pelvis During Amputee Gait

Compensations for the amputation of the distal portion of the leg occur proximally, particularly at the hip and pelvis. On the affected side, there may be less hip flexion in stance but increased flexion during mid- to terminal swing. This has been considered a foot clearance strategy in the absence of active dorsiflexion at the prosthetic ankle/foot (Sanderson & Martin, 1997). On the intact side, it has been found that hip flexion is greater across the whole gait cycle compared to the affected side (Segal et al., 2006). The hip on both the intact and affected sides generally demonstrates reduced (hyper-)extension. This may be related to the slower walking speed of lower limb amputees, as a faster walking increases hip extension.

At the pelvis, lower limb amputees often demonstrate greater anterior pelvic tilt excursion and range of motion (ROM) compared to able-bodied individuals (Goujon-Pillet et al., 2008) (Fig. 13.5). It is not

uncommon for amputees to use a walking aid for improved dynamic stability, which may cause greater forward trunk lean and anterior pelvic tilt. Amputees with limited knee motion (flexion in particular) still need to ensure ground clearance. Although the prosthetic limb length is often adjusted slightly shorter than the intact limb, adequate knee flexion is important to help clear the ground and avoid a trip or stumble. As a compensation, for example seen among transfemoral amputees, the user may hitch the hip and pelvis upon the affected side, which results in exaggerated pelvic motion in the frontal plane during the swing phase on the ipsilateral limb. No significant differences have been noted between amputee and able-bodied individuals for horizontal plane pelvic kinematics (Goujon-Pillet et al., 2008).

13.4.6 Kinetic Adaptations During Amputee Gait

Ground Reaction Forces (GRFs)

Many of the gait adaptations observed in lower limb amputees are reflected in the kinetic gait profiles, as subtle kinematic differences are amplified in joint moment and power curves. Firstly, both transtibial and transfemoral amputees show reduced anterior

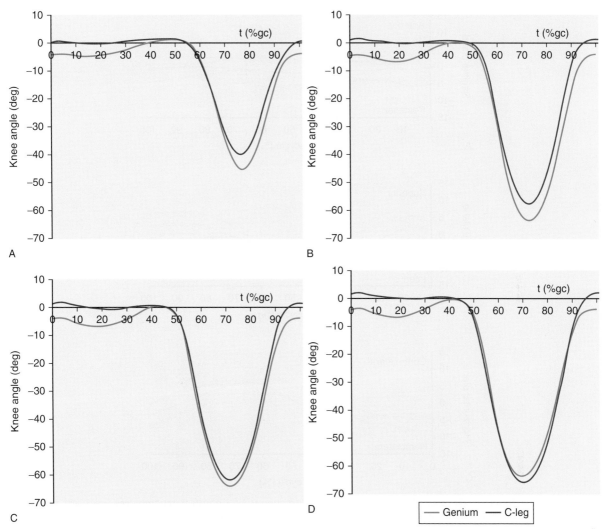

FIGURE 13.4 ■ Knee joint angles for transfemoral amputees walking with the microprocessor-controlled C-Leg versus the bionic Genium. Knee kinematics are compared during level walking at four different speeds: (A) walking with small steps, (B) slow speed, (C) medium speed and (D) fast speed. Negative values (in degrees) indicate knee flexion. *(Image taken from Bellmann et al., 2012.)*

propulsive forces during the second half of stance. This is unsurprising given the absent plantarflexor musculature and lack of energy storing and release prostheses for the majority of lower limb amputees. Vertical GRFs are also altered. Notably, amputees may have a reduced vertical loading force (also known as Fz1) on the affected side and load the prosthetic limb more slowly, resulting in a reduced load rate (Vanicek et al., 2009). They also tend to avoid unloading both limbs much

below body weight, normally seen halfway during stance during single support, and sometimes referred to as the midstance valley and/or Fz2 (Fig. 13.6). This is a vulnerable time in the gait cycle, especially during affected limb stance when the prosthetic foot supports the entire body weight. A greater unloading phase on the affected side involves swinging the intact limb more rapidly, thereby raising the centre of gravity over the stationary prosthetic foot. This may be more

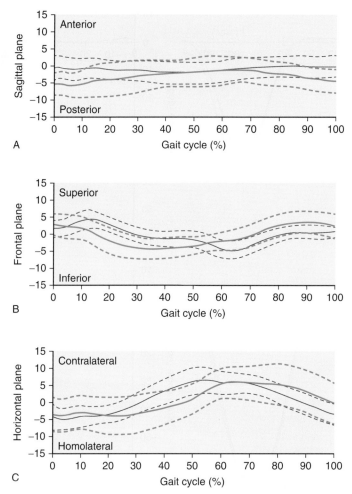

FIGURE 13.5 ■ Mean (± SD) pelvic tilt (A), obliquity (B) and rotation (C) for transfemoral amputees *(green line)* and able-bodied individuals *(grey line)* during level walking. *(Image taken from Goujon-Pillet et al., 2008.)*

difficult to control in lower limb amputees; therefore they may use a strategy to avoid unloading.

Joint Moments and Powers

For lower limb amputees, the biomechanical adaptations observed in joint kinematics and GRFs are amplified in their joint moment and power profiles. Therefore, these set of variables are usually of interest to biomechanists working with engineers, prosthetists and physiotherapists. As joint moments and powers indicate the active muscle groups and type of muscle contraction, respectively, the joint kinetic profiles can help make evidence-based recommendations for targeted and individualized strengthening exercises. All

joint moments discussed herein will be referred to as internal joint moments.

The combination of reduced GRFs in pre-swing (normally push off) and absent prosthetic ankle plantarflexion contribute to a reduced peak plantarflexor moment on the prosthetic side and negligible ankle power generation burst (A2) (Segal et al., 2006; Winter & Sienko, 1988). Absent power generation at the prosthetic knee (Figs 13.2 and 13.7) means the hip must compensate in transfemoral, and often in transtibial, amputees. Consequently, there is increased hip power absorption (H2), then a hip flexor 'pull-off' power generation burst (H3) in pre-swing, adding energy to the swinging limb.

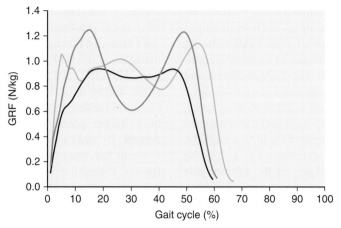

FIGURE 13.6 ■ Vertical GRF (N/kg) for a control participant (*dark green line;* walking speed 1.54 m/s) and for the intact (*light green line*) and prosthetic limb (*black line*) of one transfemoral amputee walking at 1.04 m/s. Note the difference in stance duration of the three limbs

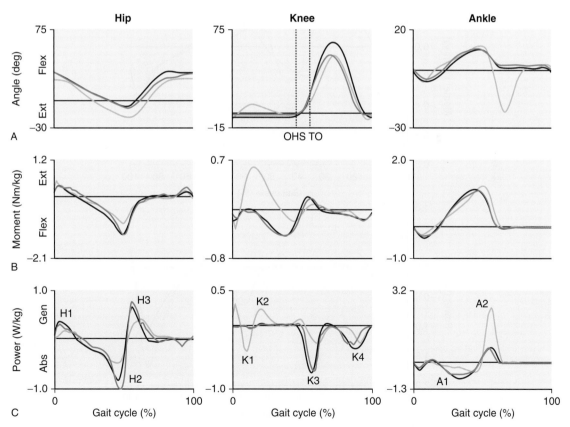

FIGURE 13.7 ■ Hip, knee and ankle angle (A), moment (B) and power (C) profiles of the prosthetic limb for transfemoral amputees wearing a mechanical knee (*black line*), microprocessor-controlled C-Leg (*dark green line*) and compared to an able-bodied individual (*light green line*). *(Image taken from Segal et al., 2006.)*

At the knee, several compensatory mechanisms have been observed, depending on the level of amputation, presence or absence of knee flexion during the loading response, and the user's prosthetic componentry. Fig. 13.8 illustrates the knee moment and power profiles for one transtibial and one transfemoral amputee walking at their self-selected walking speed, similar to walking speeds reported in the literature for their respective levels of amputation. In this example, the transtibial amputee demonstrates a loading response at the biological knee on the affected side (Fig. 13.2), causing a relatively stereotypical knee moment profile for the transtibial amputee that is not too dissimilar to the able-bodied individual (Fig. 13.8).

However, the transfemoral amputee, walking with a mechanical prosthesis, maintains the prosthetic knee joint extended throughout all of stance (Fig. 13.2) and maintains the vGRF in front of the hyperextended knee. Thus, the knee moment remains flexor in orientation during all of the stance (Fig. 13.8). The transtibial amputee demonstrates a slightly delayed K0 power burst (power generation by the knee flexors) as evidenced by the somewhat prolonged knee flexor moment following foot strike, but lacks a distinct K1 (power absorption by the knee extensors) and K2 burst (power generation by the knee extensors). The transfemoral amputee is unable to generate any knee power during stance.

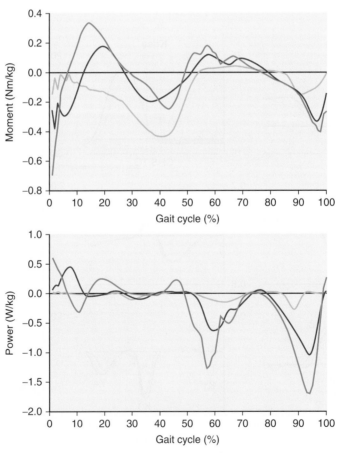

FIGURE 13.8 ■ Prosthetic knee internal moment (extensor is positive) and joint power (generation is positive) for a transfemoral amputee (*light green line*) with a mechanical knee joint versus transtibial amputee (*black line*) on the affected side and compared to an able-bodied individual (*dark green line*) walking at their comfortable speed

13.5 STAIR WALKING

The aim of stair walking is to move the head, arms and trunk (HAT segment) safely in both the vertical and horizontal directions. Kinetic energy is converted into gravitational potential energy during the 'up' phase, whilst potential energy is transferred into kinetic energy during the controlled lowering motion in the 'down' phase. Stair ascent is characterized by concentric muscle activity and stair descent is achieved through eccentric activity.

Stair walking is a challenging locomotor task compared to level walking, as it requires greater joint range of motion and larger joint moments. Both stair ascent and descent rely on muscle power, particularly from the ankle plantarflexors and knee extensors. In transtibial amputees, the ankle plantarflexors are absent and the knee musculature on the affected side, particularly the extensors, is weakened. In transfemoral amputees, both these muscle groups are adversely affected. Therefore, lower limb amputees are likely to present with more compensatory strategies to negotiate stepping up and down.

13.5.1 Phases During Stair Walking

McFadyen and Winter (1988) described several sub-phases for the stance and swing portions of the stride cycle during stair walking with a reciprocal, step over step gait pattern. During stair ascent, the cycle starts with the *weight acceptance phase*, followed by the *pull-up phase* (i.e. contralateral toe off), and then the *forward continuance phase*, when the body has been elevated one step and is continuing to the next step. Swing is initiated with toe off during *foot clearance* and finally *foot placement*. In stair descent, the cycle also starts with *weight acceptance*, followed by the *forward continuance phase*. The body continues its downward path during the *controlled lowering phase*, which is an especially difficult task for lower limb amputees as the prosthetic or affected knee is flexing during single support. In fact, some lower limb amputees adopt very different stair negotiation strategies which will be discussed shortly. Swing begins with the *leg pull through phase* followed by *foot placement* (McFadyen & Winter, 1988).

Temporal-Spatial Parameters

Stair walking places greater functional demands on the prosthesis (Schmalz et al., 2007) and moves the lower limb joints through a greater range of motion than level walking (Lin et al., 2005). The prosthesis may provide structural support but does not allow active plantarflexion. As part of their prosthetic rehabilitation, many amputees are taught to place greater demands on their intact limb by leading with this limb during stair ascent. Conversely, during descent they are taught to lead with their affected, prosthetic limb because the knee musculature on the affected side may not be able to generate sufficient eccentric strength to accomplish the controlled lowering phase safely. In amputees who lack this strength, or who cannot lower the whole body over top of a flexing prosthetic knee, an alternative is to adopt a slower, 'step to' stair descent strategy, described in more detail later.

Lower limb amputees walk more slowly during stair ascent than level walking and also when compared to able-bodied individuals. Powers and colleagues (1997) reported stair ascent speeds of 29.6 m/min (0.49 m/s), whereas Vanicek and colleagues (2010) reported speeds ranging from 0.46 to 0.81 m/s depending upon the lead or trail limb. Transfemoral amputees are likely to walk more slowly than transtibial amputees, although few biomechanical studies actually report stair walking speeds. Lower limb amputees using a 'step to' stair strategy, ascending one vertical step at a time, walk even more slowly at 0.30 m/s (Vanicek et al., 2010). Step lengths are rarely reported as these will be determined and limited in part by the staircase dimensions. Stance durations are similar to level walking ranging from 60 to 66% of the gait cycle for the intact limb and from 56 to 58% for the affected limb (Vanicek et al., 2010, 2015).

As stair descent is more mechanically demanding compared to stair ascent, it could be hypothesized that amputees would demonstrate slower walking speeds during this task, although studies have reported similar speeds to ascent. Powers and colleagues (1997) found average speeds of 0.49 m/s, similar to Ramstrand and Nilsson's (2009) average 0.48 m/s. Vanicek and colleagues (2015) reported speeds ranging from 0.50 to 0.72 m/s for transtibial amputees using a reciprocal stair descent strategy, and as slow as 0.24–0.34 m/s for those amputees using the 'step to' strategy. Wolf and colleagues (2012) found stair descent speeds of 0.42 and 0.45 m/s for transfemoral amputees using a Power Knee and C-Leg, respectively.

Joint Kinematics

There are several studies that have previously investigated the biomechanics of stair walking patterns in lower limb amputees, during stair ascent and descent over several steps (Powers et al., 1997; Schmalz et al., 2007; Yack et al., 1999; Vanicek et al., 2010, 2015; Alimusaj et al., 2009; Hobara et al., 2011). For many amputees, their stair ascent pattern could be described as stepping onto a step with the intact limb and then pulling the prosthetic foot up onto the higher step, or in some cases, onto the same step as the intact foot (i.e. 'step to' pattern). The lack of plantarflexion at the prosthetic ankle negatively affects the pull-up phase on the contralateral limb. Inadequate motion at the prosthetic ankle also affects the ability to push off during the forward continuance phase (i.e. preparation for prosthetic toe off) and compensations must occur proximally, either by greater contralateral knee extensor activity in transtibial amputees and/or exaggerated motion at the pelvis, especially pelvic obliquity, to lift the affected leg up onto the next step. Amputees with very restricted knee motion are likely to adopt a 'step to' gait pattern rather than reciprocal stair ascent. Previous studies have found that sagittal plane hip angle profiles are similar between amputees and able-bodied individuals (Alimusaj et al., 2009; Hobara et al., 2011) but amputees display greater hip and pelvic motion in the frontal plane, with more hip adduction in swing due to weakness in the hip abductor muscles on the affected side (Bae et al., 2009). Yack and colleagues (1999) reported greater reliance on the hip extensor muscles on the affected side during the weight acceptance and pull-up phases.

Stair descent is normally characterized by lowering the body in a controlled manner with eccentric activity in the ankle plantarflexors and knee extensors. However, lower limb amputees display very different joint kinematic profiles. The prosthetic foot makes initial contact during weight acceptance onto a dorsiflexed ankle, whereas the intact ankle is able to land plantarflexed to assist in functionally lengthening (and then lowering) the leg in preparation for contact with the step below. Positioning of the prosthetic foot on the step below is important as some amputees may be able to place the foot on the nose of the step and then use a roll-over technique with the prosthetic foot during the lowering phase. Prosthetic knee components that can generate flexion resistance, such as the C-Leg microprocessor-controlled knee, help to facilitate this gait technique (Schmalz et al., 2007). The roll-over is a more complex movement and additional support from the handrails is usually needed (Vanicek et al., 2015). Because the prosthetic ankle remains dorsiflexed, increased frontal plane motion around the hip and pelvis is necessary to lower the prosthetic foot. Previous research has suggested that proximal compensations are not solely due to insufficiencies of the prosthetic foot and ankle (Vanicek et al., 2015; Alimusaj et al., 2009), but also caused by muscle weakness around the hip (Mian et al., 2007).

The most vulnerable phase in stair descent is the controlled lowering phase, characterized by eccentric knee flexion during single support. Limited knee range of motion on the affected side makes this a challenging task for amputees, who tend to keep that knee more extended (Powers et al., 1997; Alimusaj et al., 2009). Amputees with functional and strength limitations at the knee will adopt the 'step-to' pattern. This strategy reduces the time spent in single support on the affected limb and maintains the knee almost completely extended, reducing the possibility of it buckling or giving way. The controlled lowering phase is substantially shorter for the intact (trail) limb and virtually absent for the affected (lead) limb (Vanicek et al., 2015).

GRFs and Joint Kinetics

Previous research found that vertical and horizontal GRFs are reduced under the prosthetic foot, with more pronounced differences in transfemoral compared to transtibial amputees (Schmalz et al., 2007). However, during stair descent, Schmalz and colleagues (2007) found that the intact limb of both transtibial and transfemoral amputees displayed significantly larger peak vertical forces during weight acceptance compared to able-bodied GRF profiles. This difference, only observed for stair descent and not stair ascent, was attributed to the amputees 'falling' onto the intact limb, and characterizes increased loading of the intact side.

During ascent, as evidenced by the small ankle range of motion and reduced forces under the prosthetic foot, the ankle moment on the affected side is considerably smaller and with concomitant minimal ankle power bursts. The small knee extensor activity during pull-up indicates the lack of power generation by this muscle group, with compensations occurring via increased hip extensor moments and power generation (Fig. 13.9). Previous authors have concluded that lower limb amputees employ a knee-extensor dominant strategy on the intact side (Yack et al., 1999) and a hip-extensor dominant strategy on the affected side (Alimusaj et al., 2009) such that the hip extensors are the main source of power generation during stair

ascent. This hip strategy was still evident, albeit diminished, even when transtibial amputees wore an adaptive microprocessor-controlled prosthetic ankle joint (Proprio-Foot, Össur), adjusted in a dorsiflexing mode versus 'neutral' mode (Alimusaj et al., 2009). These authors hypothesized that a dorsiflexed adjusted prosthetic ankle would enable more physiologic loading at the knee in transtibial amputees.

As previously described, amputees make initial contact with the heel during stair descent (Alimusaj

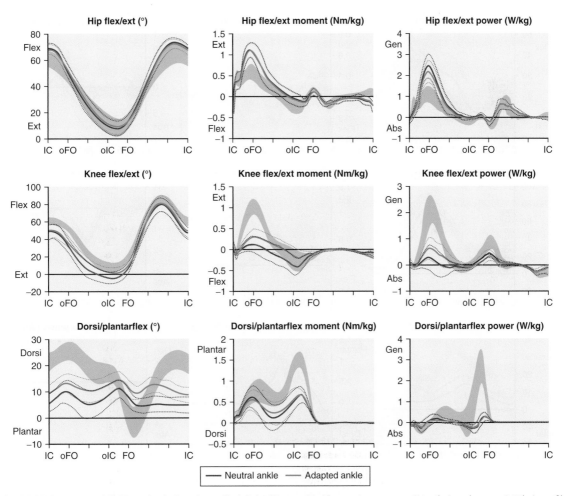

FIGURE 13.9 ■ Mean (±) SD sagittal plane lower limb joint kinematics (degrees), moments (Nm/kg) and power (W/kg) profiles for the affected side of transtibial amputees during stair ascent. The microprocessor-controlled prosthetic Proprio-Foot (Össur) was adjusted in an 'adapted' (dorsiflexed) mode versus neutral mode and compared to able-bodied controls. The stair ascent cycle begins and ends with initial contact (IC). Contralateral gait events are oFO (opposite foot off) and oIC (opposite initial contact). The shaded area represents the mean (± SD) able-bodied control data. *(Image taken from Alimusaj et al., 2009.)*

et al., 2009; Schmalz et al., 2007; Vanicek et al., 2015), causing an initial dorsiflexor moment and very different ankle moment profile compared to the stereotypical 'double hump' profile of able-bodied individuals (Fig. 13.10). Power absorption by the prosthetic ankle following weight acceptance is thereby totally absent, with only a small power burst generated during controlled lowering. Maintaining the knee more extended in stance means the knee moments are very small. Amputees attempt to position the centre of gravity directly above the limb, thereby reducing the moment arm of the GRF vector and minimizing the requirements by the knee extensors (Jones et al., 2006). The most functionally demanding phase during stair descent, the controlled lowering phase, normally characterized by a large knee extensor eccentric power burst during single support (also labelled the K3 power burst (McFadyen & Winter, 1988)), is compensated for by a hip strategy with a large hip extensor moment and hip power absorption (Fig. 13.10).

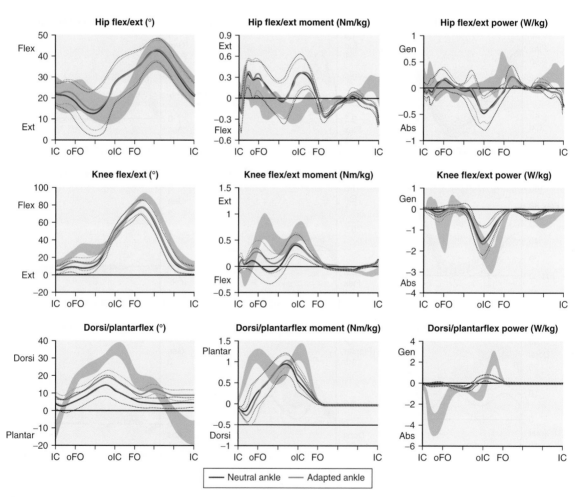

FIGURE 13.10 ■ Mean (±) SD sagittal plane lower limb joint kinematics (degrees), moments (Nm/kg) and power (W/kg) profiles for the affected side of transtibial amputees during stair descent. The microprocessor-controlled prosthetic Proprio-Foot (Össur) was adjusted in an 'adapted' (dorsiflexed) mode versus neutral mode and compared to able-bodied controls. The stair ascent cycle begins and ends with initial contact (IC). Contralateral gait events are oFO (opposite foot off) and oIC (opposite initial contact). The shaded area represents the mean (± SD) able-bodied control data. *(Image taken from Alimusaj et al., 2009.)*

SUMMARY: BIOMECHANICS OF THE MANAGEMENT OF LOWER LIMB AMPUTEES

- Lower limb amputees make many compensatory gait adaptations. They are able to adapt to their unpredictable environments using different prosthetic components. Generally, the more stable prostheses offer less functionality, whereas those that offer improved functional performance are inherently more flexible.

- Lower limb amputees walk more slowly than able-bodied individuals and with greater metabolic requirements. Walking speed is slower and energetic demands are greater for amputees with a more proximal amputation (i.e. transfemoral amputees).

- The prosthetic ankle has restricted ROM. The knee on the affected side is maintained in an extended position to prevent it from feeling as though it might 'give way' or from actually buckling. This also reduces demands on the knee musculature, which is usually weakened in transtibial amputees, and mostly absent in transfemoral amputees. Compensations occur at the hip and pelvis, usually with exaggerated motion in the sagittal and frontal planes, including more pelvic hike and hip adduction.

- Vertical and horizontal GRFs are reduced on the affected side with decreased load rates. Knee joint moments are close to zero, with minimal power bursts during stance. Consequently, hip moments and powers are increased, contributing towards a hip flexor 'pull-off' strategy during level walking and a hip extensor strategy during stair ascent.

- Some amputees, with more significant physical limitations in strength or who are limited by their prosthetic componentry, use a 'step to' stair ascent and descent strategy for enhanced dynamic stability.

REFERENCES

Aagaard, P., Simonsen, E.B., Trolle, M., et al., 1995. Isokinetic hamstring/quadriceps strength ratio: influence from joint angular velocity, gravity correction and contraction mode. Acta Physiol. Scand. 154 (4), 421–427.

Abdel-Aziz, Y.I., Karara, H.M., 1971. Direct linear transformation from comparator coordinates into object space coordinates in close range photogrammetry. In: ASP symposium on close range photogrammetry. American Society of Photogrammetry, Urbana, IL, pp. 1–18.

Alexander, E.J., Andriacchi, T.P., 2001. Correcting for deformation in skin-based marker systems. J. Biomech. 34, 355–361.

Alfredson, H., Pietila, T., Jonsson, P., et al., 1998. Heavy-load eccentric calf muscle training for the treatment of chronic achilles tendinosis. Am. J. Sports Med. 26 (3), 360–366.

Alimusaj, M., Fradet, L., Braatz, F., et al., 2009. Kinematics and kinetics with an adaptive ankle foot system during stair ambulation of transtibial amputees. Gait Posture 30, 356–363.

Al-Majali, M., Solomonidis, S.E., Spence, W., et al., 1993. Design specification of a walk mat system for the measurement of temporal and distance parameters of gait. Gait Posture 1, 119–120.

Andriacchi, T.P., Alexander, E.J., Toney, M.K., et al., 1998. A point cluster method for in vivo motion analysis: applied to a study of knee kinematics. J. Biomech. Eng. 120 (6), 743–749.

Andriacchi, T.P., Anderson, G., Fermier, R.W., et al., 1980. A study of lower limb mechanics during stair climbing. J. Bone Joint Surg. Am. 62, 5–749.

Andriacchi, T.P., Koo, S., Dyrby, C., et al. Rotational changes at the kenn during walking are associated with cartilage thinning following ACL injury. In: 9th International Conference on Orthopaedics, Biomechanics, Sports Rehabilitation. Assisi 2005(Nov 11–13); pp. 99–100.

Andriacchi, T.P., Ogle, J.A., Galante, J.O., 1977. Walking speed as a basis for normal and abnormal gait measurements. J. Biomech. 10, 261–268.

Anthony, R.J., 1991. The manufacture and use of functional foot orthosis. Karger, Basal, Switzerland.

Antonsson, E.K., Mann, R.W., 1989. Automatic 6 DOF kinematic trajectory acquisition and analysis. J. Dyn. Syst. Meas. Control 111, 34–35.

Antonsson, E.K., Mann, R.W., 1985. The frequency of gait. J. Biomech. 18 (1), 39–47.

Archer, S.E., Winter, D.A., Prince, F., 1994. Initiation of gait: a comparison between young, elderly and Parkinson's disease subjects (abstract). Gait Posture 2.

Areblad, M., Nigg, B.M., Ekstrand, J., et al., 1990. Three-dimensional measurement of rearfoot motion during running. J. Biomech. 23 (9), 933–940.

Arenson, J.S., Ishai, G., Bar, A., 1983. A system for monitoring the position and time of feet contact during walking. J. Med. Eng. Technol. 7 (6), 280–284.

Ariel, G., 1974. Method for biomechanical analysis of human performance. Res. Q. 45 (1), 72–79.

Armand, S., Sangeux, M., Baker, R., 2014. Optimal markers' placement on the thorax for clinical gait analysis. Gait Posture 39 (1), 147–153. at http://www.sciencedirect.com/science/article/pii/S0966636213002993. (Accessed 2014).

Armstrong, D.G., Peters, E.J., Athanasiou, K.A., et al., 1998. Is there a critical level of plantar foot pressure to identify patients at risk for neuropathic foot ulceration? J. Foot Ankle Surg. 37 (4), 303–307.

Arts, M.L.J., Bus, S.A., 2011. Twelve steps per foot are recommended for valid and reliable in-shoe plantar pressure data in neuropathic diabetic patients wearing custom made footwear. Clin. Biomech. (Bristol, Avon) 26 (8), 880–884.

Astrand, P., Ryhming, I., 1954. A nomogram for calculation of aerobic capacity (physical fitness) from pulse rate during submaximal work. J. Appl. Physiol. 7, 218–221.

Au, S.K., Herr, H.M., 2008. Powered ankle-foot prosthesis – The importance of series and parallel motor elasticity. IEEE Robot. Autom. Mag. 15, 52–59.

Bae, T.S., Choi, K., Mun, M., 2009. Level walking and stair climbing gait in above-knee amputees. J. Med. Eng. Technol. 33, 130–135.

Bagg, S.D., Forrest, W.J., 1988. A biomechanical analysis of scapular rotation during arm abduction in the scapular plane. Am. J. Phys. Med. Rehabil. 67, 238–245.

Baker, R., 2013. Measuring Walking: A Handbook of Clinical Gait Analysis. MacKeith Press, London.

Baker, V., Bennell, K., Stillman, B., et al., 2002. Abnormal knee joint position sense in individuals with patellofemoral pain syndrome. J. Orthop. Res. 20 (2), 208–214.

Ball, K.A., Pierrynowski, M.R., 1998. Modelling of the pliant surfaces of the thigh and leg during gait. Proc. SPIE-Int. Bio. Opt. Symp. 3254, 435–446.

Ball, K.A., Pierrynowski, M.R. Modelling of the pliant surfaces of the thigh and leg during gait. In: Proc. SPIE-Int. Soc. Opt. Eng., BIOS 1998.

Ball, P., Johnson, G.R., 1993. Reliability of hindfoot goniometry when using a flexible electrogoniometer. Clin. Biomech. (Bristol, Avon) 8, 13–19.

Barnett, C., Vanicek, N., Polman, R., et al., 2009. Kinematic gait adaptations in unilateral transtibial amputees during rehabilitation. Prosthet. Orthot. Int. 33, 135–147.

Barnett, C.T., Polman, R.C.J., Vanicek, N., 2014a. Longitudinal changes in transtibial amputee gait characteristics when negotiating a change in surface height during continuous gait. Clin. Biomech. (Bristol, Avon) 29, 787–793.

Barnett, C.T., Polman, R.C.J., Vanicek, N., 2014b. Longitudinal kinematic and kinetic adaptations to obstacle crossing in recent lower limb amputees. Prosthet. Orthot. Int. 38, 437–446.

Bartlett, R.M., Challis, J.H., Yeadon, M.R., 1992. Cine/video analysis. In: Bartlett, R.M. (Ed.), Biomechanical Analysis of Performance in Sport. British Association of Sports Sciences, Leeds, pp. 8–23.

Batavia, M., Garcia, R.K., 1996. The concurrent validity of a dynamic movement measured by the Ariel Performance Analysis System, the Qualysis MacReflex motion analysis system, and an electrogoniometer (abstract). Phys. Ther. 76, S75.

Batschelet, E., 1981. Circular Statistics in Biology. Academic Press, New York.

Bell, A., Pederson, D., Brand, R., 1990. A comparison of the accuracy of several hip centre location predication methods. J. Biomech. 23 (6), 617–621.

Bell, A.L., Brand, R.A., Pedersen, D.R., 1989. Prediction of hip joint centre location from external landmarks. Hum. Mov. Sci. 8 (1), 3–16.

Bell, F., Ghasemi, M., Rafferty, D., et al., 1995. An holistic approach to gait analysis: Glasgow Caledonian University's CRC. Gait Posture 3, 185.

Bell, F., Shaw, L., Rafferty, D., et al., 1996. Movement analysis technology in clinical practice. Phys. Ther. Rev. 1, 13–22.

Bellmann, M., Schmalz, T., Ludwigs, E., et al., 2012. Immediate effects of a new microprocessor-controlled prosthetic knee joint: a comparative biomechanical evaluation. Arch. Phys. Med. Rehabil. 93, 541–549.

Berjian, R.A., Douglass, H.O., Jr., Holyoke, E.D., et al., 1983. Skin pressure measurements on various mattress surfaces in cancer patients. Am. J. Phys. Med. 62 (5), 217–226.

Bernstein, J., 2003. An overview of MEMS inertial sensing technology. Sensors Weekly 2003, Accessed online at: http://www.sensorsmag.com/sensors/acceleration-vibration/an-overview-mems-inertial-sensing-technology-970.

Berry, D., Olson, M.D., Larntz, K., 2009. Perceived stability, function, and satisfaction among transfemoral amputees using microprocessor and non-microprocessor controlled prosthetic knees: a multicenter survey. J. Prosthet. Orthot. 21, 32–42.

Beynnon, B.D., Ryder, S.H., Konradsen, L., et al., 1999. The effect of anterior cruciate ligament trauma and bracing on knee proprioception. Am. J. Sports Med. 27 (2), 150–155.

Birtane, M., Tuna, H., 2004. The evaluation of plantar pressure distribution in obese and non-obese adults. Clin. Biomech. (Bristol, Avon) 19 (10), 1055–1059.

Blana, D., Hincapie, J.G., Chadwick, E.K., et al., 2008. A musculoskeletal model of the upper extremity for use in the development of neuroprosthetic systems. J. Biomech. 41 (8), 1714–1721.

Bobbert, A.C., 1960. Energy expenditure in level and grade walking. J. Appl. Physiol. 15, 1015–1021.

Bohannon, R.W., 1990. Hand-held compared with isokinetic dynamometry for measurement of static knee extension torque (parallel reliability of dynamometers). Clin. Phys. Physiol. Meas. 11 (3), 217–222.

Bohannon, R.W., 1997. Comfortable and maximum walking speed of adults aged 20–79 years: Reference values and determinants. Age. Ageing 26, 15–19.

Bolsterlee, B., Veeger, D.H.E.J., Chadwick, E.K., 2013. Clinical applications of musculoskeletal modelling for the shoulder and upper limb. Med. Biol. Eng. Comput. 51, 953–963.

Bonacci, J., Vicenzino, B., Spratford, W., et al., 2014. Take your shoes off to reduce patellofemoral joint stress during running. Br. J. Sports Med. 48, 425.

Brand, R.A., Crowninshield, R.D., 1981. Locomotion studies–caves to computers (abstract). J. Biomech. 14 (7), 497.

Branthwaite, H., Chockalingam, N., Greenhalgh, A., 2013. The effect of shoe toe box shape and volume on forefoot interdigital and plantar pressures in healthy females. J. Foot Ankle Res. 6, 28.

Branthwaite, H.R., Payton, C.J., Chockalingam, N., 2004. The effect of simple insoles on three-dimensional foot motion during normal walking. Clin. Biomech. (Bristol, Avon) 19 (9), 972–977.

Braune, W., Fischer, O., 1889. Uber den Schwerpunkt des menschlichen Korpers, mit Rucksicht auf die Ausrustung des deutschen Infanteristen. Abh. Math. Phys. KI. Saechs. Ges. Wiss. 26, 561–672.

Bresler, B., Frankel, J., 1950. The forces and moments in the leg during level walking. Trans. SME 27–36.

Brown, D.C., 1966. Decentering distortion of lenses. Photometric Eng. 32 (3), 444–462.

Bruckner, J. The gait workbook: a practical guide to clinical gait analysis. SLACK Incorporated, 1998.

Brunt, D., Liu, S.M., Trimble, M., et al., 1999. Principles underlying the organization of movement initiation from quiet stance. Gait Posture 10 (2), 121–128.

Bryant, A., Newton, R., Steele, J., 2003. Is tibial acceleration related to knee functionality of ACL deficient and ACL reconstructed patients? J. Sci. Med. Sport 6 (4 Suppl. 1), 31.

Burgess-Limerick, R., Abernethy, B., Neal, J., 1993. Relative phase quantifies interjoint co-ordination. J. Biomech. 26 (1), 91–94.

Bus, S.A., de Lange, A., 2005. A comparison of the 1-step, 2-step, and 3-step protocols for obtaining barefoot plantar pressure data in the diabetic neuropathic foot. Clin. Biomech. (Bristol, Avon) 20 (9), 892–899.

Callaghan, M.J., Selfe, J., 2012. Patellar taping for patellofemoral pain syndrome in adults. Cochrane Library (4), 1–41.

Callaghan, M., Selfe, J., Bagley, P., et al., 2002. Effect of patellar taping on knee joint proprioception. J. Athl. Train. 37 (1), 19–24.

Camomilla, V., Cereatti, A., Vannozzi, G., et al., 2006. An optimized protocol for hip joint centre determination using the functional method. J. Biomech. 39 (6), 1096–1106.

Cappello, A., Cappozzo, A., La Palombara, P.F., et al., 1997. Multiple anatomical landmark calibration for optimal bone pose estimation. Hum. Mov. Sci. 16, 259–274.

Cappozzo, A., Cappello, A., 1997. Surface-marker cluster design criteria for 3-d bone movement reconstruction. IEEE Trans. Biomed. Eng. 40 (12), 1165–1174.

Cappozzo, A., Catani, F., Croce, U.D., et al., 1995. Position and orientation in space of bones during movement: anatomical frame definition and determination. Clin. Biomech. (Bristol, Avon) 10 (4), 171–178.

Cappozzo, A., Catani, F., Leardini, A., et al., 1996. Position and orientation in space of bones during movement: experimental artefacts. Clin. Biomech. (Bristol, Avon) 11 (2), 90–100.

Cappozzo, A., Della Croce, U., Leardini, A., et al., 2005. Human movement analysis using photogrammetry. Part 1: theoretical background. Gait Posture 21, 186–196.

Cappozzo, A., 1983. The forces and couples in the human trunk during level walking. J. Biomech. 16 (4), 265–277. at: http://www.ncbi.nlm.nih.gov/pubmed/6863342. (Accessed 2016).

Cappozzo, A., 1991. Three dimensional analysis of human walking: experimental methods and associated artefacts. Hum. Mov. Sci. 10, 589–602.

Carson, M.C., Harrington, M.E., Thompson, N., et al. Kinematic analysis of a multi-segment foot model for research and clinical applications: a repeatability analysis. 2001.

Cavagna, G.A., Saibene, F.P., Margaria, R., 1963. External work in walking. J. Appl. Physiol. 18, 1–9.

Cavanagh, P.R., Sims, D.S., Sanders, L.J., 1991. Body mass is a poor predictor of peak pressure in diabetic men. Diabetes Care 14 (8), 750–755.

Cerveri, P., Pedotti, A., Ferrigno, G., 2005. Kinematical models to reduce the effect of skin artifacts on marker-based human motion estimation. J. Biomech. 38 (11), 2228–2236.

Chadwick, E.K., Blana, D., Kirsch, R.F., et al., 2014. Real-time simulation of three-dimensional shoulder girdle and arm dynamics. IEEE Trans. Biomed. Eng. 61 (7), 1947–1956. doi:10.1109/TBME.2014.2309727.

Chandler, R.F., et al. Tech. Report AMRL-TR-74-137. Wright-Patterson Air Force Base, Aerospace Medical Research Laboratories, 1975.

Charlton, I.W., Johnson, G., 2006. A model for the prediction of the forces at the glenohumeral joint. Proc. Inst. Mech. Eng. H 220, 801–812.

Charteris, J., Taves, C., 1978. The process of habituation to treadmill walking: a kinematic analysis. Percept. Mot. Skills 47, 659–666.

Chéze, L., Fregly, B.J., Dimnet, J., 1995. A solidification procedure to facilitate kinematic analyses based on video system data. J. Biomech. 28, 879–884.

Chiari, L., Della Croce, U., Leardini, A., et al., 2005. Human movement analysis using stereophotogrammetry. Part 2: instrumental errors. Gait Posture 21, 197–211.

Chockalingam, N., et al., 2008. Marker placement for movement analysis in scoliotic patients: a critical analysis of existing systems. Stud. Health Technol. Inform. 140, 166–169. at: http://www.ncbi.nlm.nih.gov/pubmed/18810021. (Accessed 2014).

Chockalingam, N., et al., 2002. Study of marker placements in the back for opto-electronic motion analysis. Stud. Health Technol. Inform. 88, 105–109. at: http://www.ncbi.nlm.nih.gov/pubmed/15456012. (Accessed 2014).

Chou, L.S., Draganich, L.F., 1998. Placing the trailing foot closer to an obstacle reduces flexion of the hip, knee, and ankle to increase the risk of tripping. J. Biomech. 31 (8), 685–691.

Chou, R., et al., 2007. Diagnosis and treatment of low back pain: a joint clinical practice guideline from the American College of Physicians and the American Pain Society. Ann. Intern. Med. 147 (7), 478–491. at: http://www.ncbi.nlm.nih.gov/pubmed/17909209. (Accessed 2016).

Chung, M.-J., Wang, M.-J., 2012. Gender and walking speed effects on plantar pressure distribution for adults aged 20–60 years. Ergonomics 55 (2), 194–200.

Clauser, C.E., McConville, J.T., Young, J.W., 1969. Weight, volume, and centre of mass of segments of the human body. AMRL technical report. Wright-Patterson Air Force Base, Ohio.

Cochrane, L., Fergus, K., Arnold, G.P., et al., 2008. A comparative study between two pressure mapping systems: FSA versus Novel Pliance. Clin. Biomech. (Bristol, Avon) 23 (5), 669–670.

Codman, E.A., 1934. Tendinitis of the Short Rotators in the Shoulder: Rupture of the Supraspinatus Tendon and Other Lesions in or About the Subacromial Bursa. Thomas Todd and Co, Boston, MA.

Consant, C.R., Murley, A.H.G., 1987. A clinical method of functional assessment of the shoulder. Clin. Orthop. Relat. Res. 214, 160–164.

Cook, J., Khan, K., 2001. What is the most appropriate treatment for patellar tendinopathy. Br. J. Sports Med. 35 (5), 291–294.

Cook, T.M., Zimmermann, C.L., Lux, K.M., et al., 1992. EMG comparison of lateral step up and stepping machine exercise. J. Orthop. Sports Phys. Ther. 16 (3), 108–113.

Corcoran, P.J., Brengelmann, G.L., 1970. Oxygen uptake in normal and handicapped subjects, in relation to speed of walking beside velocity controlled cart. Arch. Phys. Med. Rehabil. 51, 78–87.

Cotes, J.E., Meade, F., 1960. The energy expenditure and mechanical energy demand in walking. Ergonomics 3, 97–119.

Cowan, S.M., Bennell, K., Hodges, P.W., 2000. The test-retest reliability of the onset of concentric and eccentric vastus medialis obliquus and vastus lateralis electromyographic activity in a stair stepping task. Phys. Ther. Sport 1, 129–136.

Craik, R.L., Oatis, C.A., 1995. Gait analysis theory and action, 1 ed. Mosby, St Louis.

Crosbie, J., Vachalathiti, R., Smith, R., 1997. Age, gender and speed effects on spinal kinematics during walking. Gait Posture 5 (1), 13–20. at: http://www.gaitposture.com/article/S0966636296010685/fulltext. (Accessed 2014).

Crouse, J., Wall, J.C., Marble, A.E., 1987. Measurement of temporal and spatial parameters of gait using a microcomputer based system. J. Biomed. Eng. 9 (1), 64–68.

Crouse, S., Lessard, C., Rhodes, J., et al., 1990. Oxygen consumption and cardiac response of short-leg and long-leg prosthetic ambulation in a patient with bilateral above knee amputation: comparisons with able-bodied men. Arch. Phys. Med. Rehabil. 71, 313–317.

Cubo, E., Leurgans, S., Goetz, C.G., 2004. Short-term and practice effects of metronome pacing in Parkinson's disease patients with gait freezing while in the 'on' state: randomized single blind evaluation. Parkinsonism Relat. Disord. 10 (8), 507–510.

Cutti, A.G., Giovanardi, A., Rocchi, L., et al., 2008. Ambulatory measurement of shoulder and elbow kinematics through inertial and magnetic sensors. Med. Biol. Eng. Comput. 46, 169–178.

Dabnichki, P., Lauder, M., Aritan, S., et al., 1997. Accuracy evaluation of an on-line kinematic system via dynamic tests. J. Med. Eng. Technol. 53–66.

Dahlkvist, N.J., Mayo, P., Seedhom, B.B., 1982. Forces during squatting and rising from a deep squat. Eng. Med. 11 (68), 76.

Davids, J.R., Holland, W.C., Sutherland, D.H., 1993. Significance of the confusion test in cerebral palsy. J. Pediatr. Orthop. 13 (6), 717–721.

Davis, R., Ounpuu, S., Tyburski, D., et al., 1991. A gait data collection and reduction technique. Hum. Movement Sci. 10, 575–587.

De Asha, A.R., Buckley, J.G., 2015. The effects of walking speed on minimum toe clearance and on the temporal relationship between minimum clearance and peak swing-foot velocity in unilateral trans-tibial amputees. Prosthet. Orthot. Int. 39, 120–125.

De Bruin, H., Russell, D.J., Latter, J.E., et al., 1982. Angle-angle diagrams in monitoring and quantification of gait patterns for children with cerebral palsy. Am. J. Phys. Med. 61 (4), 176–192.

de Leva, P., 1996. Adjustments to Zatsiorsky-Seluyanov's segment inertia parameters. J. Biomech. 29 (9), 1223–1230.

della Croce, U., Camomilla, V., Leardini, A., et al., 2003. Femoral anatomical frame: assessment of various definitions. Med. Eng. Phys. 25 (5), 425–431.

della Croce, U., Cappozzo, A., Kerrigan, D.C., 1999. Pelvis and lower limb anatomical landmark calibration precision and its propagation to bone geometry and joint angles. Med. Biol. Eng. Comput. 37 (2), 155–161.

Delp, S.L., Anderson, F.C., Arnold, A.S., et al., 2007. OpenSim: open-source software to create and analyze dynamic simulations of movement. IEEE Trans. Biomed. Eng. 54, 1940–1950.

DeLuca, C.J., Forrest, W.J., 1973. Force Analysis of Individual Muscles Acting Simultaneously on the Shoulder Joint during Isometric Abduction. J. Biomech. 6, 385–393.

Dempster, W.T., Gabel, W.C., Felts, W.J.L., 1959. The anthropometry of manual work space for the seated subject. Am. J. Phys. Anthropol. 17, 289–317.

Dempster, W.T., 1955. Space requirements of the seated operator. WADC Technical Report 55–159. Wright-Patterson Air Force Base, Ohio.

Department of Transport (2004). Inclusive mobility. http://www.ukroads.org/webfiles/inclusivemobility.pdf.

Deutsch, A., Altchek, D., Schwartz, E., et al., 1996. Radiologic measurement of superior displacement of the humeral head in impingement syndrome. J. Shoulder Elbow Surg. 5 (3), 186–193.

Dibble, L., Nicholson, D., Shultz, B., et al., 2004. Sensory cueing effects on maximal speed gait initiation in persons with Parkinson's disease and healthy elders. Gait Posture 19 (3), 215–225.

Dickerson, C.R., Chaffin, D.B., Hughes, R.E., 2007. A mathematical musculoskeletal shoulder model for proactive ergonomic analysis. Comput. Meth. Biomech. Biomed. Eng. 10, 389–400.

Donovan, S., Lim, C., Diaz, N., et al., 2011. Laser-light cues for gait freezing in Parkinson's disease: an open label study. Parkinsonism Relat. Disord. 17, 240–245.

Doucette, S.A., Child, D.D., 1996. The effect of open and closed chain exercise and knee joint position on patellar tracking in lateral patellar compression syndrome. J. Orthop. Sports Phys. Ther. 23 (2), 104–110.

Dowswell, T., Towner, E., Cryer, C., et al., 1999. Accidental falls: fatalities and injuries ad examination of the data sources and review of the literature on preventive strategies. Department of Trade and Industry, London, p. URN 99/805.

Drezner, J., Staudt, L., Fowler, E., 1994. Examination of intersegmental coordination in spastic cerebral palsy patients before and after selective posterior rhizotomy. Gait Posture 2, 61.

Drillis, R., Contini, R., 1966. Body segment parameters. Report no.1163– 03. Office of Vocational Rehabilitation. Department of health, Education and Welfare, New York.

Dumbleton, T., Buis, A.W., McFayden, A., et al., 2009. Dynamic interface pressure distributions of two transtibial prosthetic socket concepts. J. Rehabil. Res. Dev. 46 (3), 401–415.

Dupes, B. Prosthetic Knee Systems [Online]. Amputee Coalition of America. 2005. Available: http://www.amputee-coalition.org/military-instep/knees.html [Accessed 22 August 2016].

Durie, N.D., Farley, R.L., 1980. An apparatus for step length measurement. J. Biomed. Eng. 2 (1), 38–40.

Dvir, Z., Berme, N., 1978. The shoulder complex in elevation of the arm: a mechanism approach. J. Biomech. 11, 219–225.

Earl, J.E., Schmitz, R.J., Arnold, B.L., 2001. Activation of the VMO and VL during dynamic mini-squat exercises with and without isometric hip adduction. J. Electromyogr. Kinesiol. 11, 381–386.

Eddison, N., Chockalingam, N., 2013. The effect of tuning ankle foot orthoses-footwear combination on the gait parameters of children with cerebral palsy. Prosthet. Orthot. Int. 37 (2), 95–107.

Elble, R.J., Moody, C., Leffler, K., et al., 1994. The initiation of normal walking. Mov. Disord. 9, 139–146.

Elftman, H., 1939. Forces and energy changes in the leg during walking. Am. J. Physiol. 125, 339–356.

Elftman, H.O., 1939. The force exerted by the ground in walking. Arbeitsphysiologie 10, 485–491.

Ellenbecker, S., Davies, G.J., 2001. Closed Kinetic Chain Exercise. Human Kinetics., IL, USA.

Ellis, M., Seedhom, B.B., Wright, V., et al., 1980. An evaluation of the ratio between the tension along the quadriceps tendon and the patellar ligament. Eng. Med. 9 (4), 189–194.

Endo, K., Ikata, T., Katoh, S., et al., 2001. Radiographic assessment of scapular rotational tilt in chronic shoulder impingement syndrome. J. Orthop. Sci. 6 (1), 3–10.

Escamilla, R.F., Fleisig, G.S., Zheng, N., et al., 1998. Biomechanics of the knee during closed kinetic chain and open kinetic chain exercises. Med. Sci. Sports Exerc. 30 (4), 556–569.

Escamilla, R.F., Fleisig, G.S., Zheng, N., et al., 2001. Effects of technique variations on knee biomechanics during the squat and leg press. Med. Sci. Sports Exerc. 33 (9), 1552–1566.

Escamilla, R.F., 2001. Knee biomechanics of the dynamic squat. Med. Sci. Sports Exerc. 33 (1), 127–141.

Eshraghi, A., Abu Osman, N.A., Karimi, M., et al., 2014. Gait biomechanics of individuals with transtibial amputation: effect of suspension system. PLoS ONE 9, 12.

Fayad, F., Roby-Brami, A., Yazbeck, C., et al., 2008. Three-dimensional scapular kinematics and scapulohumeral rhythm in patients with glenohumeral osteoarthritis or frozen shoulder. J. Biomech. 41 (2), 326–332.

Ferrigno, G., Pedotti, A., 1985. ELITE: a digital dedicated hardware system for movement analysis via real-time TV signal processing. IEEE Trans. Biomed. Eng. 32 (11), 943–950.

Ferris, D.P., Czerniecki, J.M., Hannaford, B., 2005. An ankle-foot orthosis powered by artificial pneumatic muscles. J. Appl. Biomech. 21 (2), 189–197.

Ferris, D.P., Gordon, K.E., Sawicki, G.S., et al., 2006. An improved powered ankle-foot orthosis using proportional myoelectric control. Gait Posture 23 (4), 425–428.

Fiolkowski, P., Brunt, D., Bishop, M., et al., 2002. Does postural instability affect the initiation of human gait? Neurosci. Lett. 323 (3), 167–170.

Fong, D.T.-P., Chan, Y.-Y., 2010. The use of wearable inertial motion sensors in human lower limb biomechanics studies: a systematic review. Sensors (Basel) 10, 11556–11565.

Forte, F.C., de Castro, M.P., de Toledo, J.M., et al., 2009. Scapular kinematics and scapulohumeral rhythm during resisted shoulder abduction – Implications for clinical practice. Phys. Ther. Sport 10 (3), 105–111.

Frazzitta, G., Maestri, R., Uccelini, D., et al., 2009. Rehabilitation treatment of gait in patients with Parkinson's disease with freezing: a comparison between two physical therapy protocols using visual and auditory cues with or without treadmill training. Mov. Disord. 24 (8), 1139–1143.

Frigo, C., et al., 2003. The upper body segmental movements during walking by young females. Clin. Biomech. (Bristol, Avon) 18 (5), 419–425. at: http://www.ncbi.nlm.nih.gov/pubmed/12763438. (Accessed 2014).

Fulkerson, J.P., Hungerford, D.S., 1990. Disorders of the Patellofemoral Joint, 2 ed. Williams and Wilkins, Baltimore.

Fuller, J., Liu, L.-J., Murphy, M.C., et al., 1997. A comparison of lower-extremity skeletal kinematics measured using skin- and pin-mounted markers. Hum. Mov. Sci. 16, 219–242.

Gage, J.R., 1994. The clinical use of kinetics for evaluation of pathological gait in cerebral palsy. J. Bone Joint Surg. Am. 76 (4), 622–631.

Gage, J.R., 1994. The role of gait analysis in the treatment of cerebral palsy. J. Pediatr. Orthop. 4 (6), 701–702.

Gardner, G.M., Murray, M.P., 1975. A method of measuring the duration of foot–floor contact during walking. Phys. Ther. 55 (7), 751–756.

Garner, B., Pandy, M., 1999. A kinematic model of the upper limb based on the visible human project (VHP) image dataset. Comput. Methods Biomech. Biomed. Engin. 2 (2), 107–124.

Gatt, A., Spiteri, M., Formosa, C., et al., 2014. Investigation of plantar pressures in overweight and non-overweight children with a neutral foot posture. OA Musculoskeletal Med. 2 (1), 8.

Gerny, K., 1983. A clinical method of quantitative gait analysis. Phys. Ther. 63, 1125–1126.

Giacomozzi, C., Leardini, A., Caravaggi, P., 2014. Correlates between kinematics and baropodometric measurements for an integrated in-vivo assessment of the segmental foot function in gait. J. Biomech. 47 (11), 2654–2659.

Giladi, N., McMahon, D., Przedborski, S., et al., 1992. Motor blocks in Parkinson's disease. Neurology 42, 333–339.

Giladi, N., Nieuwboer, A., 2008. Understanding and treating freezing of gait in Parkinsonism, proposed working definition, and setting the stage. Mov. Disord. 23, 423–425.

Gill, H.S., O'Connor, J.J., 1996. Biarticulating two-dimensional computer model of the human patellofemoral joint. Clin. Biomech. (Bristol, Avon) 11 (2), 81–89.

Goh, J.C., Bose, K., Khoo, B.C., 1993. Gait analysis study on patients with varus osteoarthrosis of the knee. Clin. Orthop. Relat. Res. 294, 223–231.

Goujon-Pillet, H., Sapin, E., Fode, P., et al., 2008. Three-dimensional motions of trunk and pelvis during transfemoral amputee gait. Arch. Phys. Med. Rehabil. 89, 87–94.

Grabiner, M., Owings, T., 2002. EMG differences between concentric and eccentric maximum voluntary contractions are evident prior to movement onset. Exp. Brain Res. 145 (4), 505–511.

Grieve, D.W., 1968. Gait patterns and the speed of walking. Biomed. Eng. 3, 119–122.

Grieve, D., Gear, J., 1966. The relationship between length of stride, step frequency, time of swing and speed of walking for children and adults. Ergonomics 5 (9), 379–399.

Griffin, H.J., Greenlaw, R., Limousin, P., et al., 2011. The effect of real and virtual visual cues on walking in Parkinson's disease. J. Neurol. 258, 991–1000.

Grood, E.S., Suntay, W.J., 1983. A joint coordinate system for the clinical description of three-dimensional motions: application to the knee. J. Biomech. Eng. 105, 136–144.

Growney, E., Cahalan, T., Meglan, D., 1994. Comparison of goniometry and video motion analysis for gait analysis. J. Biomech. 27, 624.

Guido, J., Jr., Voight, M.L., Blackburn, T.A., et al., 1997. The effects of chronic effusion on knee joint proprioception: a case study. J. Orthop. Sports Phys. Ther. 25 (3), 208–212.

Gussoni, M., Margonato, V., Ventura, R., et al., 1990. Energy cost of walking with hip joint imparement. Phys. Ther. 70, 195–301.

Hallen, L.G., Lindahl, O., 1966. The 'screw-home' movement in the knee- joint. Acta Orthop. Scand. 37 (1), 97–106.

Halliday, S.E., Winter, D.A., Frank, J.S., et al., 1998. The initiation of gait in young, elderly, and Parkinson's disease subjects. Gait Posture 8 (1), 8–14.

Hamill, J., Haddad, J.M., McDermott, W.J., 2000. Issues in quantifying variability from a dynamical systems perspective. J. Appl. Biomech. 16, 407–418.

Hammarlund, C.S., Carlstrom, M., Melchior, R., et al., 2011. Prevalence of back pain, its effect on functional ability and health-related quality of life in lower limb amputees secondary to trauma or tumour: a comparison across three levels of amputation. Prosthet. Orthot. Int. 35, 97–105.

Haneline, M.T., et al., 2008. Determining spinal level using the inferior angle of the scapula as a reference landmark: a retrospective analysis of 50 radiographs. J. Can. Chiropr. Assoc. 52 (1), 24–29. at: pmc/articles/PMC2258239/?report=abstract. (Accessed 2016).

Hanzlíková, I., Richards, J., Tomsa, M., et al., 2016. The effect of proprioceptive knee bracing on knee stability during three different sport related movement tasks in healthy subjects and the implications to the management of anterior cruciate ligament (ACL) injuries. Gait Posture 48, 165–170. doi:10.1016/j.gaitpost.2016.05.011. ISSN 0966-6362.

Harrington, M.E., Zavatsky, A.B., Lawson, S.E., et al., 2007. Prediction of the hip joint centre in adults: children, and patients with cerebral palsy based on magnetic resonance imaging. J. Biomech. 40 (3), 595–602.

Haskell, W., Yee, M., Evans, A., et al., 1993. Simultaneous measurement of heart rate and body motion to quantitate physical activity. Med. Sci. Sports Exerc. 25 (1), 109–115.

Hattin, H.C., Pierrynowski, M.R., Ball, K.A., 1989. Effect of load cadence and fatigue on tibiofemoral joint force during a half squat. Med. Sci. Sports Exerc. 21, 613–618.

Hazlewood, M.E., Brown, J.K., Rowe, P.J., et al., 1994. The use of therapeutic electrical stimulation in the treatment of hemiplegic cerebral palsy. Dev. Med. Child Neurol. 36 (8), 661–673.

Hazlewood, M.E., Hillman, S.J., Lawson, A.M., et al., 1997. Marker attachment in gait analysis: on skin or lycra? Gait Posture 6, 265.

Healy, A., Chatzistergos, P., Needham, R., et al., 2013. Comparison of design features in diabetic footwear and their effect on plantar pressure. Footwear Sci. 5 (1), S67–S69.

Healy, A., Dunning, D.N., Chockalingam, N., 2012. Effect of insole material on lower limb kinematics and plantar pressures during treadmill walking. Prosthet. Orthot. Int. 36 (1), 53–62.

Hebert, L.J., Moffet, H., McFadyen, B.J., et al., 2002. Scapular behavior in shoulder impingement syndrome. Arch. Phys. Med. Rehabil. 83 (1), 60–69.

Hennig, E.M., Milani, T.L., 1995. In-shoe pressure distribution for running in various type of footwear. J. Appl. Biomech. 11 (3), 299–310.

Hennig, E.M., Rosenbaum, D., 1991. Pressure distribution patterns under the feet of children in comparison with adults. Foot Ankle 11 (5), 306–311.

Henriksson, M., Hirschfeld, H., 2005. Physically active older adults display alterations in gait initiation. Gait Posture 21 (3), 289–296.

Hershler, C., Milner, M., 1980. Angle–angle diagrams in above-knee amputee and cerebral palsy gait. Am. J. Phys. Med. 59 (4), 165–183.

Hewett, T.E., Noyes, F.R., Barber-Westin, S.D., et al., 1998. Decrease in knee joint pain and increase in function in patients with medial compartment arthrosis: a prospective analysis of valgus bracing. Orthopedics 21, 131–138.

Hewitt, B.A., Refshauge, K.M., Kilbreath, S.L., 2002. Kinesthesia at the knee: the effect of osteoarthritis and bandage application. Arthritis Rheum. 47 (5), 479–483.

Heyrman, L., et al., 2013. Reliability of head and trunk kinematics during gait in children with spastic diplegia. Gait Posture 37 (3), 424–429. at: http://www.ncbi.nlm.nih.gov/pubmed/23062729. (Accessed 2016).

Heyrman, L., et al., 2014. Altered trunk movements during gait in children with spastic diplegia: compensatory or underlying trunk control deficit? Res. Dev. Disabil. 35 (9), 2044–2052. at: http://www.ncbi.nlm.nih.gov/pubmed/24864057. (Accessed 2016).

Hills, A.P., Hennig, E.M., McDonald, M., et al., 2001. Plantar pressure differences between obese and non-obese adults: a biomechanical analysis. Int. J. Obes. 25 (11), 1674–1679.

Hirokawa, S., Matsumura, K., 1987. Gait analysis using a measuring walkway for temporal and distance factors. Med. Biol. Eng. Comput. 25, 577–582.

Hirokawa, S., 1989. Normal gait characteristics under temporal and distance constraints. J. Biomed. Eng. 11, 449–456.

Hobara, H., Kobayashi, Y., Nakamura, T., et al., 2011. Lower extremity joint kinematics of stair ascent in transfemoral amputees. Prosthet. Orthot. Int. 35, 467–472.

Hogfors, C., Peterson, B., Sigholm, G., et al., 1991. Biomechanical model of the human shoulder joint-II. The shoulder rhythm. J. Biomech. 24 (8), 699–709.

Holden, J.P., Orsini, J.A., Lohmann Siegel, K., et al., 1997. Surface movement errors in shank kinematics and knee kinetics during gait. Gait Posture 5, 217–227.

Holden, J.P., Stanhope, S.J., 1998. The effect of variation in knee center location estimates on net knee joint moments. Gait Posture 7 (1), 1–6.

Holzbaur, K.R.S., Murray, W.M., Delp, S.L., 2005. A model of the upper extremity for simulating musculoskeletal surgery and analyzing neuromuscular control. Ann. Biomed. Eng. 33 (6), 829–840.

Hortobagyi, T., Katch, F.I., 1990. Eccentric and concentric torque-velocity relationships during arm flexion and extension. Influence of strength level. Eur. J. Appl. Physiol. Occup. Physiol. 60 (5), 395–401.

Houdijk, H., Hak, L., Beek, P.J., et al., 2014. Step length asymmetry in transtibial amputees: a strategy to regulate gait stability? Gait Posture 39, S84.

Hullin, M.G., Robb, J.E., Loudon, I.R., 1992. Ankle-foot orthosis function in low-level myelomeningocele. J. Pediatr. Orthop. 12, 518–521.

Hurley, G.R., McKenney, R., Robinson, M., et al., 1990. The role of the contralateral limb in below-knee amputee gait. Prosthet. Orthot. Int. 14 (1), 33–42.

Hurmuzlu, Y., Basdogan, C., Carollo, J.J., 1994. Presenting joint kinematics of human locomotion using phase plane portraits and Poincare maps. J. Biomech. 27 (12), 1495–1499.

Hurmuzlu, Y., Basdogan, C., Stoianovici, D., 1996. Kinematics and dynamic stability of the locomotion of post-polio patients. J. Biomed. Eng. 118 (3), 405–411.

Illyés, A., Kiss, R.M., 2006. Kinematic and muscle activity characteristics of multidirectional shoulder joint instability during elevation. Knee Surg. Sports Traumatol. Arthrosc. 14, 673–685.

Imms, F., MacDonald, I., Prestidge, S., 1976. Energy expenditure during walking in patients recovering from fractures of the leg. Scand. J. Rehabil. Med. 8, 1–9.

Inman, V.T., Ralston, H.J., Todd, F., 1981. Human Walking. Williams and Wilkins Company, Baltimore, MD.

Inman, V.T., 1967. Conservation of energy in ambulation. Arch. Phys. Med. Rehabil. 48.

Inman, V.T., 1966. Human locomotion. Can. Med. Assoc. J. 94, 1047–1054.

Inman, V.T., Saunders, J.B., Abbott, L.C., 1996. [1944]. Observations of the function of the shoulder joint. Clin. Orthop. Relat. Res. 330, 3–12.

International Building Code, 2003. ISBN-13: 978-1892395566.

Isaac Newton. Philosophiae Naturalis Principia Mathematica.1687.

Isakov, E., Burger, H., Krajnik, J., et al., 1996. Influence of speed on gait parameters and on symmetry in trans-tibial amputees. Prosthet. Orthot. Int. 20 (3), 153–158.

JAMA, 1992. Evidence Based Practice or Evidenced Based Medicine. The users' guides to evidence-based medicine. J. Am. Med. Assoc.

Jarrett, M.O., Andrews, B.J., Paul, J.P., 1974. Quantitative analysis of locomotion using television. ISPO World Congress, Montreux, Switzerland.

Jerosch, J., Prymka, M., 1996. Knee joint proprioception in patients with posttraumatic recurrent patella dislocation. Knee Surg. Sports Traumatol. Arthrosc. 4, 14–18.

Jevsevar, D.S., Riley, P.O., Hodge, W.A., et al., 1993. Knee kinematics and kinetics during locomotor activities of daily living in subjects with knee arthroplasty and in healthy control subjects. Phys. Ther. 73 (4), 229–242.

Jian, Y., Winter, D.A., Ishac, M.G., et al., 1993. Trajectory of the body COG and COP during initiation and termination of gait. Gait Posture 1, 9–22.

Jiang, Y., Norman, K.E., 2006. Effects of visual and auditory cues on gait initiation in people with Parkinson's disease. Clin. Rehabil. 20 (1), 36–45.

Johnson, F., Leitl, S., Waugh, W., 1980. The distribution of load across the knee. A comparison of static and dynamic measurements. J. Bone Joint Surg. Br. 62 (3), 346–349.

Jones, R.K., Nester, C.J., Kim, W.Y., et al. Direct and indirect orthotic management of medial compartment osteoarthritis of the knee, ESMAC & GCMAS meeting, Amsterdam, 25–30 September, 2006.

Jones, R.K., Nester, C.J., Richards, J.D., et al., 2013. A comparison of the biomechanical effects of valgus knee braces and lateral wedged insoles in patients with knee osteoarthritis. Gait Posture 37 (3), 368–372.

Jones, S.F., Twigg, P.C., Scally, A.J., et al., 2006. The mechanics of landing when stepping down on unilateral lower-limb amputees. Clin. Biomech. (Bristol, Avon) 21, 184–193.

Jonsson, P., Alfredson, H., 2005. Superior results with eccentric compared to concentric quadriceps training in patients with jumper's knee: a prospective randomised study. Br. J. Sports Med. 39, 847–850.

Kadaba, M.P., Ramakrishnan, H.K., Wootten, M.E., et al., 1989. Repeatability of kinematic, kinetic, and electromyographic data in normal adult gait. J. Orthop. Res. 7 (6), 849–860.

Kahle, J.T., Highsmith, M.J., Hubbard, S.L., 2008. Comparison of nonmicroprocessor knee mechanism versus C-Leg on Prosthesis Evaluation Questionnaire, stumbles, falls, walking tests, stair descent and knee preference. J. Rehabil. Res. Dev. 45, 1–13.

Karlsson, D., Peterson, B., 1992. Towards a model for force predictions in the human shoulder. J. Biomech. 25 (2), 189–199.

Kaufman, K.R., Levine, J.A., Brey, R.H., et al., 2007. Gait and balance of transfemoral amputees using passive mechanical and microprocessor-controlled prosthetic knees. Gait Posture 26, 489–493.

Kaufman, K.R., Levine, J.A., Brey, R.H., et al., 2008. Energy expenditure and activity amputees using mechanical and of transfemoral microprocessor-controlled prosthetic knees. Arch. Phys. Med. Rehabil. 89, 1380–1385.

Kavanagh, J.J., Menz, H.B., 2008. Accelerometry: a technique for quantifying movement patterns during walking. Gait Posture 28 (1), 1–15.

Keemink, C.J., Hoek van Dijke, G.A., Snijders, C.J., 1991. Upgrading of efficiency in the tracking of body markers with video techniques. Med. Biol. Eng. Comput. 29 (1), 70–74.

Kellis, E., Baltzopoulos, V., 1996. The effects of normalization on antagonistic activity patterns during eccentric and concentric isokinetic knee extension and flexion. J. Electromyogr. Kinesiol. 6 (4), 235.

Kepple, T.M., Arnold, A.S., Stanhope, S.J., et al., 1994. Assessment of a method to estimate muscle attachments from surface landmarks: a 3D computer graphics approach. J. Biomech. 27 (3), 365–371.

Khan, K., Maffulli, N., Coleman, B., et al., 1998. Patellar tendinopathy: some basic aspects of science and clinical management. Br. J. Sports Med. 32, 346–355.

Kiernan, D., Malone, A., O'Brien, T., et al., 2015. The clinical impact of hip joint centre regression equation error on kinematics and kinetics during paediatric gait. Gait Posture 41 (1), 175–179. doi:10.1016/j.gaitpost.2014.09.026. http://www.sciencedirect.com/science/article/pii/S0966636214007255. ISSN 0966-6362.

Kim, W.Y., Richards, J.D., Jones, R.K., et al., 2004. Single limb stance adduction moment in medial compartment osteoarthritis of the knee. Knee 11, 225–231.

Kingma, I., Toussaint, H.M., Commissaris, D.A.C.M., et al., 1995. Optimising the determination of the body centre of mass. J. Biomech. 28 (9), 1137–1142.

Kirkley, A., Webster-Bogaert, S., Litchfield, R., et al., 1999. The effect of bracing on varus gonarthrosis. J. Bone Joint Surg. Am. 81–A, 539–548.

Klein, P.J., Gabusi, C.A., Brinn, M.B., 1992. Validation of linear and angular displacement estimation by computer-assisted motion analysis. Phys. Ther. 72.

Kowalk, D.L., Duncan, J.A., Vaughan, C.L., 1996. Abduction-adduction moments at the knee during stair ascent and descent. J. Biomech. 29 (3), 383–388.

Ladin, Z., Mansfield, P.K., Murphy, M.C., et al., 1990. Segmental analysis in kinesiological measurements. Image-based motion measurements. SPIE 1356, 110–120.

Lafortune, M.A., 1991. Three-dimensional acceleration of the tibia during walking and running. J. Biomech. 24 (10), 877–886.

Lafortune, M.A., Hennig, V., 1991. Contribution of angular motion and gravity to tibial acceleration. Med. Sci. Sports Exerc. 23 (3), 360–363.

Lafortune, M.A., Hennig, V., Valiant, G.A., 1995. Tibial shock measured with bone and skin mounted transducers. J. Biomech. 28 (8), 989–993.

Lamoth, C.J.C., et al., 2006. Effects of chronic low back pain on trunk coordination and back muscle activity during walking: changes in motor control. Eur. Spine J. 15 (1), 23–40. at: http://www.ncbi.nlm.nih.gov/pubmed/15864670. (Accessed 2016).

Lamoth, C.J.C., et al., 2002. Pelvis-thorax coordination in the transverse plane during walking in persons with nonspecific low back pain. Spine 27 (4), E92–E99. at: http://www.ncbi.nlm.nih.gov/pubmed/11840116. (Accessed 2016).

Larose Chevalier, T., Hodgins, H., Chockalingam, C., 2010. Plantar pressure measurements using an in-shoe system and a pressure platform: a comparison. Gait Posture 31 (3), 397–399.

Laubenthal, K.N., Smidt, G.L., Kettlekamp, D.B., 1972. A quantitative analysis of knee motion during activities of daily living. Phys. Ther. 52 (1), 34–43.

Laudner, K.G., Stanek, J.M., Meister, K., 2006. Assessing posterior shoulder contracture: the reliability and validity of measuring glenohumeral joint horizontal adduction. J. Athl. Train. 41 (4), 375–380.

Leardini, A., Cappozzo, A., Catani, F., et al., 1999. Validation of a functional method for the estimation of hip joint centre location. J. Biomech. 32 (1), 99–103.

Leardini, A., Chiari, L., Della Croce, U., et al., 2005. Human movement analysis using stereophotogrammetry. Part 3. Soft tissue artifact assessment and compensation. Gait Posture 21, 212–225.

Leardini, A., et al., 2011a. Multi-segment trunk kinematics during locomotion and elementary exercises. Clin. Biomech. (Bristol, Avon) 26 (6), 562–571.

Leardini, A., et al., 2011b. Multi-segment trunk kinematics during locomotion and elementary exercises. Clin. Biomech. (Bristol, Avon) 26 (6), 562–571. at: http://www.ncbi.nlm.nih.gov/pubmed/21419535. (Accessed 2014).

Lebold, C., Almeida, Q.J. Evaluating the contributions of dynamic flow to freezing of gait in Parkinson's disease. Sage-Hindawi Access to Research Parkinson's Disease, 2010.

Lee, V.S., Solomonidis, S.E., Spence, W.D., 1997. Stump-socket interface pressure as an aid to socket design in prostheses for transfemoral amputees–a preliminary study. Proc. Inst. Mech. Eng. H 211 (2), 167–180.

Leiper, C.I., Craik, R.L., 1991. Relationship between physical activity and temporal-distance characteristics of walking in elderly women. Phys. Ther. 71 (11), 791–803.

Lephart, S.M., Fu, F.H., 2000. Proprioception and Neuromuscular Control in Joint Stability. Human Kinetics, Champaign, IL, pp. xvii–xxiv.

Lephart, S.M., Henry, T.J., 1995. Functional rehabilitation for the upper and lower extremity. Orthop. Clin. North Am. 26 (3), 579–592.

Lesh, M.D., Mansour, J.M., Simon, S.R., 1979. A gait analysis subsystem for smoothing and differentiation of human motion data. J. Biomech. Eng. 101, 205–212.

Levens, A.S., Inman, V.T., Blosser, J.A., 1948. Transverse rotation of the segments of the lower extremity in locomotion. J. Bone Joint Surg. Am. 30A, 859–872.

Lim, E., Thong-Meng, T., Chee-Seong Seet, R., 2006. Laser assisted device for start hesitation and freezing in Parkinson's disease. Case Rep. Clin. Pract. Rev. 7, 92–95.

Lin, H.C., Lu, T.W., Hsu, H.C., 2005. Comparisons of joint kinetics in the lower extremity between stair ascent and descent. J. Mech. 21, 41–50.

Lin, J.-J., Lim, H.K., Yang, J.-L., 2006. Effect of shoulder tightness on glenohumeral translation, scapular kinematics, and scapulohumeral rhythm in subjects with stiff shoulders. J. Orthop. Res. 24 (5), 1044–1051.

Lindenfeld, T.N., Hewett, T.E., Andriacchi, T.P., 1997. Joint loading with valgus bracing in patients with varus gonarthrosis. Clin. Orthop. 344, 290–297.

Lloyd, C.H., Stanhope, S.J., Davis, I.S., et al., 2010. Strength asymmetry and osteoarthritis risk factors in unilateral trans-tibial, amputee gait. Gait Posture 32, 296–300.

Lockard, M.A., 1988. Foot orthoses. Phys. Ther. 68 (12), 1866–1873.

Lough, J., 1995. Quantifying motor performance in patients with peripheral neuropathy undergoing treatment (abstract). Physiotherapy 81, 745.

Low, J., Reid, A., 1992. Electrotherapy explained. Butterworth Heinemann, Oxford.

Lu, T.W., O'Connor, J.J. Three dimensional computer graphics based modelling and mechanical analysis of the human locomotor system. In: Sixth International Symposium on the 3D Analysis of Human Movement. 1–4 May, 2000.

Lucchetti, L., Cappozzo, A., Cappello, A., et al., 1998. Skin movement artefact assessment and compensation in the estimation of knee-joint kinematics. J. Biomech. 31, 977–984.

Ludewig, P.M., Cook, T.M., 2000. Alterations in shoulder kinematics and associated muscle activity in people with symptoms of shoulder impingement. Phys. Ther. 80 (3), 276–291.

Ludewig, P.M., Hassett, D.R., LaPrade, R.F., et al., 2010. Comparison of scapular local coordinate systems. Clin. Biomech. (Bristol, Avon) 25 (5), 415–421.

Ludewig, P.M., Reynolds, J.R., 2009. The association of scapular kinematics and glenohumeral joint pathologies. J. Orthop. Sports Phys. Ther. 39 (2), 90–104.

Lukaseiwicz, A.C., McClure, P., Michener, L., et al., 1999. Comparison of 3-dimensional scapular position and orientation between subjects with and without shoulder impingement. J. Orthop. Sports Phys. Ther. 29 (10), 574–583.

Lundgren, P., Nester, C., Liu, A., et al., 2008. Invasive in vivo measurement of rear-, mid- and forefoot motion during walking. Gait Posture 28 (1), 93–100.

MacGregor, J., 1979. Rehabilitation ambulatory monitoring. Disability, Strathclyde Bioengineering Seminars. MacMillan, London, pp. 159–172.

MacGregor, J., 1981. The evaluation of patient performance using long- term ambulatory monitoring technique in the domiciliary environment. Physiotherapy 67 (2), 30–33.

MacWilliams, B.A., Cowley, M., Nicholson, D.E., 2003. Foot kinematics and kinetics during adolescent gait. Gait Posture 17 (3), 214–224.

MacWilliams, B.A., et al., 2013. Assessment of three-dimensional lumbar spine vertebral motion during gait with use of indwelling bone pins. J. Bone Joint Surg. Am. 95 (23), e1841–e1848. at: http://www.ncbi.nlm.nih.gov/pubmed/24306707. (Accessed 2014).

Maly, M.R., Culham, E.G., Costigan, P.A., 2002. Static and dynamic biomechanics of foot orthoses in people with medial compartment knee osteoarthritis. Clin. Biomech. (Bristol, Avon) 17 (8), 603–610.

Manal, K., McClay, I., Richards, J., et al., 2002. Knee moment profiles during walking: errors due to soft tissue movement of the shank and the influence of the reference coordinate system. Gait Posture 15, 10–17.

Manal, K., McClay, I., Stanhope, S., et al., 2000. Comparison of surface mounted markers and attachment methods in estimating tibial rotations during walking: an in vivo study. Gait Posture 11, 38–45.

Mann, R.A., Antonsson, E.K., 1983. Gait analysis–precise, rapid, automatic, 3-D position and orientation kinematics and dynamics. Bull. Hosp. Joint Dis. Orthop. Inst. 43 (2), 137–146.

Mann, R.A., Hagey, J.L., White, V., et al., 1979. The initiation of gait. J. Bone Joint Surg. 61–a, 232–239.

Mansour, J.M., Lesh, M.D., Nowak, M.D., et al., 1982. A three dimensional multi-segmental analysis of the energetics of normal and pathological gait. J. Biomech. 15 (1), 51–59.

Marciniak, W., 1973. Design of an electrogoniometer for the examination of the movements of the knee and foot during walking. Chir. Narzadow Ruchu Ortop. Pol. 38 (5), 573–579.

Marey, E.J., 1873. Animal mechanism: a treatise on terrestrial and aerial locomotion. Republished as Vol. XI of the International Scientific Series. Appleton, New York.

Martin, M., Shinberg, M., Kuchibhatla, M., et al., 2002. Gait initiation in community-dwelling adults with Parkinson disease: comparison with older and younger adults without the disease. Phys. Ther. 82 (6), 566–577.

Mason, D.L., et al., 2016. Reproducibility of kinematic measures of the thoracic spine, lumbar spine and pelvis during fast running. Gait Posture 43, 96–100.

Matias, R., Pascoal, A.G., 2006. The unstable shoulder in arm elevation: a three-dimensional and electromyographic study in subjects with glenohumeral instability. Clin. Biomech. (Bristol, Avon) 21, S52–S58.

Matsumo, H., Kadowaki, K., Tsuji, H., 1997. Generation II knee bracing for severe medial compartment osteoarthritis of the knee. Arch. Phys. Med. Rehabil. 78, 745–749.

Mattes, S.J., Martin, P.E., Royer, T.D., 2000. Walking symmetry and energy cost in persons with unilateral transtibial amputations: matching prosthetic and intact limb inertial properties. Arch. Phys. Med. Rehabil. 81, 561–568.

Maurel, W., Thalmann, D., 1999. A case study on human upper limb modelling for dynamic simulation. Comput. Methods Biomech. Biomech. Engin. 1 (2), 1–17.

Maurel, W., Thalmann, D., 2000. Human shoulder modelling including scapulothoracic constraint and joint sinus cones. Comput. Graphics 24, 203–218.

Mell, A.G., LaScalza, S., Guffey, P., et al., 2005. Effect of rotator cuff pathology on shoulder rhythm. J. Shoulder Elbow Surg. 14 (1 Suppl.S), 58S–64S.

McCandless, P.J., Evans, B.J., Janssen, J., et al., 2016. Effect of three cueing devices for people with Parkinson's disease with gait initiation difficulties. Gait Posture 44, 7–11.

McClay, I., Manal, K., 1998. A comparison of 3D lower extremity kinematics during running between excessive pronators and normals. Clin. Biomech. (Bristol, Avon) 13 (3), 195–203.

McClure, P.W., Michener, L.A., Karduna, A.R., 2006. Shoulder function and 3-dimensional scapular kinematics in people with and without shoulder impingement syndrome. Phys. Ther. 86, 1075–1090.

McConnell, J., 2002. The physical therapist's approach to patellofemoral disorders. Clin. Sports Med. 21 (3), 363–387.

McDonough, A.L., Batavia, M., Chen, F.C., et al., 2001. The validity and reliability of the GAITRite system's measurements: a preliminary evaluation. Arch. Phys. Med. Rehabil. 82 (3), 419–425.

McFadyen, B., Winter, D.A., 1988. An integrated biomechanical analysis of normal stair ascent and descent. J. Biomech. 21 (9), 733–744.

McGinty, G., Irrgang, J.J., Pezullo, D., 2000. Biomechanical considerations for rehabilitation of the knee. Clin. Biomech. (Bristol, Avon) 15, 160–166.

McGrath, D., Judkins, T.N., Pipinos, I.I., et al., 2012. Peripheral arterial disease affects the frequency response of ground reaction forces during walking. Clin. Biomech. (Bristol, Avon) 27 (10), 1058–1063. doi:10.1016/j.clinbiomech.2012.08.004. [Epub 2012 Sep 9].

Medical Progress Through Technology, 1993. Med. Prog. Technol. 19 (2), 61–81.

Messier, S.P., Loeser, R.F., Hoover, J.L., et al., 1992. Osteoarthritis of the knee: effects on gait, strength, and flexibility. Arch. Phys. Med. Rehabil. 73 (1), 29–36.

Mian, O.S., Thom, J.M., Narici, M.V., et al., 2007. Kinematics of stair descent in young and older adults and the impact of exercise training. Gait Posture 25, 9–17.

Mickelborough, J., van der Linden, M.L., Tallis, R.C., et al., 2004. Muscle activity during gait initiation in normal elderly people. Gait Posture 19 (1), 50–57.

Miller, C., Verstraete, M., 1996. Determination of the step duration of gait ignition using a mechanical energy analysis. J. Biomech. 29 (9), 1195–1199.

Mizahi, J., Suzak, Z., Heller, L., et al., 1982. Variation of the time distance parameters of the stride as related to clinical gait improvement in hemiplegics. Scand. J. Rehabil. Med. 14, 133–140.

Moore, O., Peretz, C., Giladi, N., 2007. Freezing of gait affects quality of life of peoples with Parkinson's disease beyond its relationships with mobility and gait. Mov. Disord. 22, 2192–2195.

Morag, E., Cavanagh, P.R., 1999. Structural and functional predictors of regional peak pressures under the foot during walking. J. Biomech. 32 (4), 359–370.

Moritani, T., Muramatsu, S., Muro, M., 1987. Activity of motor units during concentric and eccentric contractions. Am. J. Phys. Med. 66 (6), 338–350.

Morrison, J.B., 1970. The mechanics of the knee joint in relation to normal walking. J. Biomech. 3, 51–61.

Muilenburg, A.L., Wilson, A.B., Jr. A manual for above-knee (Trans-Femoral) amputees [Online]. 1996. Available at http://www.oandp.com/resources/patientinfo/manuals/akindex.htm (Accessed 22 August 2016).

Murray, M.P., Drought, A.B., Kory, R.C., 1964. Walking patterns of normal men. J. Bone Joint Surg. Am. 46A, 335–360.

Murray, M.P., Kory, R.C., Clarkson, B.H., et al., 1966. Comparison of free and fast speed walking patterns of normal men. Am. J. Phys. Med. 45, 8–25.

Murray, M.P., Kory, R.C., Sepic, S.B., 1970. Walking patterns of normal women. Arch. Phys. Med. Rehabil. 51 (11), 637–650.

Murray, M.P., Sepic, S.B., Barnard, E.J., 1967. Patterns of sagittal rotation of the upper limbs in walking. Phys. Ther. 47 (4), 272–284.

Murray, M.P., 1967. Gait as a total pattern of movement. Am. J. Phys. Med. 40, 290–333.

Muybridge, E., 1957. Animal locomotion. In: Brown, L.S. (Ed.), Animal in motion. Dover, New York, p. 1887.

Muybridge, E., 1901. The human figure in motion. Chapman and Hall, London.

Nadler, R., Nadler, S., 2001. Assistive devices and lower extremity orthotics in the treatment of osteoarthritis. Physical medicine and rehabilitation. State Art Rev. 15 (1), 57–64.

Naemi, R., Larose Chevalier, T., Healy, A., et al., 2012. The effect of the use of a walkway and the choice of the foot on plantar pressure assessment when using pressure platforms. The Foot. 22 (2), 100–104.

National Amputee Statistical Database for the United Kingdom (NASDAB). 2009.

Needham, R., Naemi, R., Chockalingam, N., 2015. A new coordination pattern classification to assess gait kinematics when utilising a modified vector coding technique. J. Biomech. 48 (12), 3506–3511.

Needham, R., Naemi, R., Healy, A., et al., 2016. Multi-segment kinematic model to assess three-dimensional movement of the spine and back during gait. Prosthet. Orthot. Int. 40 (5), 624–635. at: http://www.ncbi.nlm.nih.gov/pubmed/25991730. (Accessed 2015).

Needham, R., Stebbins, J., Chockalingam, N., 2016. Three-dimensional kinematics of the lumbar spine during gait using marker-based systems: a systematic review. J. Med. Eng. Technol. 40 (4), 172–185. at: http://www.ncbi.nlm.nih.gov/pubmed/27011295. (Accessed 2016).

Needham, R., Naemi, R., Chockalingam, N., 2014. Quantifying lumbar–pelvis coordination during gait using a modified vector coding technique. J. Biomech. 47, 1020–1026.

Needham, R., Naemi, R., Chockalingam, N., 2015. A new coordination pattern classification to assess gait kinematics when utilising a modified vector coding technique. J. Biomech. 48, 3506–3511.

Nelson, R.M., Currier, D.P., 1991. Clinical electrotherapy, 2nd ed. Appleton & Lange, Conneticut.

Nene, A.V., Patrick, J., 1989. Energy cost of paraplegic locomotion with the ORLAU parawalker. Paraplegia 27, 5–18.

Nene, A.V., 1993. Physiological cost index of walking in able-bodied adolescents and adults. Clin. Rehabil. 7 (4), 319–326.

Nester, C.J., van der Linden, M.L., Bowker, P., 2003. Effect of foot orthoses on the kinematics and kinetics of normal walking gait. Gait Posture 17 (2), 180–187.

Ng, G.Y., Zhang, A.Q., Li, C.K., 2008. Biofeedback exercise improved the EMG activity ratio of the medial and lateral vasti muscles in subjects with patellofemoral pain syndrome. J. Electromyogr. Kinesiol. 18 (1), 128–133. [Epub 2006 Oct 27].

Nguyen, T.C., Baker, R., 2004. Two methods of calculating thorax kinematics in children with myelomeningocele. Clin. Biomech. (Bristol, Avon) 19 (10), 1060–1065.

Nicol, A.C., 1987. A flexible electrogoniometer with widespread applications. In: Jonsson, B. (Ed.), Biomechanics XB. Human Kinetics Pub, Illinois, pp. 1029–1033.

Nicol, A.C., 1989. Measurement of joint motion. Clin. Rehabil. 3, 1–9.

Nigg, B.M., Herzog, W., 1994. Biomechanics of the musculo-skeletal system. John Wiley & Sons Ltd.

Nissel, R., Ekholm, J., 1985. Patellar forces during knee extension. Scand. J. Rehabil. Med. 17, 74.

Nisell, R., Ekholm, J., 1986. Joint load during the parallel squat in powerlifting and force analysis of in vivo bilateral quadriceps tendon rupture. Scand. J. Sports Sci. 8, 63–70.

Noyes, F.R., Schipplein, O.D., Andriacchi, T.P., et al., 1992. The anterior cruciate ligament-deficient knee with varus alignment. An analysis of gait adaptations and dynamic joint loadings. Am. J. Sports Med. 20 (6), 707–716.

Ogston, J.B., Ludewig, P.M., 2007. Differences in 3-dimensional shoulder kinematics between persons with multidirectional instability and asymptomatic controls. Am. J. Sports Med. 35, 1361–1370.

Ojima, H., Miyake, S., Kumashiro, M., et al., 1991. Dynamic analysis of wrist circumduction: a new application of the biaxial flexible electrogoniometer. Clin. Biomech. (Bristol, Avon) 6 (4), 221–229.

Olney, S.J., Grondin, R.C., McBride, I.D., 1989. Energy and power considerations in slow walking (abstract). J. Biomech. 22, 1066.

Olney, S.J., Monga, T.N., Costigan, P.A., 1986. Mechanical energy of walking of stroke patients. Arch. Phys. Med. Rehabil. 67, 92–98.

Olney, S.J., Costigan, P.A., Hedden, D.M., 1987. Mechanical energy patterns in gait of cerebral palsied children with hemiplegia. Phys. Ther. 67, 1348–1354.

Olree, K.S., Engsberg, J.R., White, D.K., 1996. Indices of effort and oxygen uptake in children with cerebral palsy. Dev. Med. Child Neurol. 38 (S74), 49–50.

Onishi, H., Yagi, R., Akasaka, K., et al., 2000. Relationship between signals and force in human vastus lateralis muscle using multipolar wire electrodes. J. Electromyogr. Kinesiol. 10, 59–67.

Orthopaedics, Biomechanics and Sports Rehabilitation edn, University of Perugia, 2005.

Osternig, L.R., James, C.R., Bercades, D.T., 1996. Eccentric knee flexor torque following anterior cruciate ligament surgery. Med. Sci. Sports Exerc. 28 (10), 1229–1234.

2000. Outcome of nonoperative and operative management. Am. J. Sports Med. 28 (3), 392–397.

Owen, E., 2010. The importance of being earnest about shank and thigh kinematics especially when using ankle-foot orthoses. Prosthet. Orthot. Int. 34 (3), 254–269.

Owings, M., Kozak, L., 1998. Ambulatory and inpatient procedures in the United States, 1996. Vital Health Stat. 39, 1–119.

Ozaki, J., Nakagawa, Y., Sakurai, G., et al., 1998. Recalcitrant chronic adhesive capsulitis of the shoulder. Role of contracture of the coracohumeral ligament and rotator interval in pathogenesis and treatment. J. Bone Joint Surg. Am. 71 (10), 1511–1515.

Paletta, G.A., Jr., Warner, J.J., Warren, R.F., et al., 1997. Shoulder kinematics with two-plane x-ray evaluation in patients with anterior instability or rotator cuff tearing. J. Shoulder Elbow Surg. 6 (6), 516–527.

Palmitier, R.A., An, K.N., Scott, S.G., et al., 1991. Kinetic chain exercise in knee rehabilitation. Sports Med. 11 (6), 402–413.

Pandy, M.G., 1999. Moment arm of a muscle force. Exerc. Sport Sci. Rev. 27, 79–118.

Panni, A., Tartarone, M., Maffulli, N., 2000. Tendinopathy in athletes. Outcome of nonoperative and operative management. Am. J. Sports Med. 28 (3), 392–397.

Pascoal, A.G., van der Helm, F.F., Pezarat Correia, P., et al., 2000. Effects of different arm external loads on the scapulo-humeral rhythm. Clin. Biomech. (Bristol, Avon) 15, S21–S24.

Passmore, R., Draper, M.H., 1965. Energy metabolism. In: Albance, A. (Ed.), Newer methods of nutritional biochemistry. Academic, New York.

Passmore, R., Durnin, J., 1955. Human energy expenditure. Phys. Rev. 35, 801–840.

Patrick, J., 1991. Gait laboratory investigations to assist decision making. Br. J. Hosp. Med. 45, 35–37.

Paul, J.P., 1967. Forces transmitted by joints in the human body. Proc. Inst. Mech. Eng. 181 (3J), 8–15.

Pearcy, M.J., et al., 1987. Dynamic back movement measured using a three-dimensional television system. J. Biomech. 20 (10), 943–949. at: http://www.ncbi.nlm.nih.gov/pubmed/3693375. (Accessed 2014).

Perry, J., 1992. Gait analysis: Normal and pathological function. SLACK Incorporated, Thorofare, NJ.

Perttunen, J.R., Anttila, E., Sodergard, J., et al., 2004. Gait asymmetry in patients with limb length discrepancy. Scand. J. Med. Sci. Sports 14 (1), 49–56.

Peters, A., et al., 2010. Quantification of soft tissue artifact in lower limb human motion analysis: a systematic review. Gait Posture 31 (1), 1–8. at: http://www.sciencedirect.com/science/article/pii/S0966636209006171. (Accessed 2015).

Petersen, W.A., Brookhart, J.M., Stone, S.A., 1965. A strain-gage platform for force measurements. J. Appl. Physiol. 20, 1095–1097, 8750–7587.

Phadke, V., Braman, J.P., LaPrade, R.F., et al., 2011. Comparison of glenohumeral motion using different rotation sequences. J. Biomech. 44 (4), 700–705.

Pierrynowski, M., Winter, D., Norman, R., 1980. Transfers of mechanical energy within the total body and mechanical efficiency during treadmill walking. Ergonomics 23.

Pierrynowski, M.R., Norman, R.W., Winter, D.A., 1981. Mechanical energy analyses of the human during local carriage on a treadmill. Ergonomics 24 (1), 1–14.

Podsiadlo, D., Richardson, S., 1991. The timed up and go: a test of basic functional mobility for frail elderly persons. J. Am. Geriatr. Soc. 39, 142–148.

Polcyn, A.F., Lipsitz, L.A., Kerrigan, C., et al., 1998. Age-related changes in the initiation of gait: degradation of central mechanisms for momentum generation. Arch. Phys. Med. Rehabil. 79, 1582–1589.

Pollack, C.V., Kerstein, M.D., 1985. Prevention of postoperative complications in the lower-extremity amputee. J. Cardiovasc. Surg. 26, 287–290.

Pollo, F.E., Otis, J.C., Backus, S.I., et al., 2002. Reduction of medial compartment loads with valgus bracing of the osteoarthritis knee. Am. J. Sports Med. 30, 414–421.

Pollo, F.E., 1998. Bracing and heel wedging for unicompartmental osteoarthritis of the knee. Am. J. Knee Surg. 11, 47–50.

Poppen, N.K., Walker, P.S., 1978. Force at the glcnohunieral joint in abduction. Clin. Orthop. 135, 165–170.

Powers, C.M., 2003. The influence of altered lower extremity kinematics on patellofemoral joint dysfunction: a theoretical perspective. J. Orthop. Sports Phys. Ther. 33, 639–646.

Powers, C.M., Boyd, L.A., Torburn, L., et al., 1997. Stair ambulation in persons with transtibial amputation: An analysis of the Seattle LightFoot(TM). J. Rehabil. Res. Dev. 34, 9–18.

Powers, C.M., Rao, S., Perry, J., 1998. Knee kinetics in trans-tibial amputee gait. Gait Posture 8, 1–7.

Prodromos, C.C., Andriacchi, T.P., Galante, J.O., 1985. A relationship between gait and clinical changes following high tibial osteotomy. J. Bone Joint Surg. Am. 67 (8), 1188–1194.

Prymka, M., Schmidt, K., Jerosch, J., 1998. Proprioception in patients suffering from chondropathia patellae. Int. J. Sports Med. 19, S60.

Purdam, C.R., Johnson, P., Alfredson, H., et al., 2004. A pilot study of the eccentric decline squat in the management of painful chronic patellar tendinopathy. Br. J. Sports Med. 38, 395–397.

Putti, A.B., Arnold, G.P., Abboud, R.J., 2010. Foot pressure differences in men and women. Foot Ankle Surg. 16 (1), 21–24.

Quanbury, A., Winter, D., Reimer, G., 1975. Instantaneous power and power flow in body segments during walking. J. Hum. Mov. Stud. 1, 59–67.

Rab, G., Petuskey, K., Bagley, A., 2002. A method for determination of upper extremity kinematics. Gait Posture 15 (2), 113–119. at: http://www.ncbi.nlm.nih.gov/pubmed/11869904. (Accessed 2016).

Rafferty, D., Bell, F., 1995. Gait analysis – a semiautomated approach. Gait Posture 3 (3), 184.

Ralston, H., Lukin, L., 1969. Energy levels of human body segments during level walking. Ergonomics 12 (1), 39–46.

Ralston, H.J., 1958. Energy speed relation and optimal speed during level walking. Int. Z. Angew. Physiol. 17, 277.

Ramstrand, N., Nilsson, K.A., 2009. A comparison of foot placement strategies of transtibial amputees and able-bodied subjects during stair ambulation. Prosthet. Orthot. Int. 33, 348–355.

Redhead, R.G., Davis, B.C., Robinson, K.P., et al., 1978. Post-amputation pneumatic walking aid. Br. J. Surg. 65, 611–612.

Reilly, D.T., Martens, M., 1972. Experimental analysis of the quadriceps muscle force and patellofemoral joint reaction force for various activities. Acta Orthop. Scand. 43, 126–137.

Reinschmidt, C., van den Bogert, T., Nigg, B.M., et al., 1997. Effect of skin movement on the analysis of skeletal knee joint motion during running. J. Biomech. 30 (7), 729–732.

Reinschmidt, C. Three-dimensional tibiocalcaneal and tibiofemoral kinematics during human locomotion – measured with externaland bone markers. Ph.D. thesis. University of Calgary, Calgary, Alberta, 1996.

Rennie, J., Bell, F., Rafferty, D., et al., 1997. Measurement of spatial parameters of gait and velocity in schools and centres for young adults with learning disabilities. Physiotherapy 83 (7), 364.

Richards, J., Jones, R., Kim, W. Biomechanical changes in the conservative treatment of medial compartment osteoarthritis of the knee using valgus bracing. International Cartilage Repair Society, 2006a.

Richards, J.D., Pramanik, A., Sykes, L., et al., 2003. A comparison of knee kinematic characteristics of stroke patients and age-matched healthy volunteers. Clin. Rehabil. 7 (5), 565–571.

Richards, J.D., Sanchez-Ballester, J., Jones, R.K., et al., 2005. A comparison of knee braces during walking for the treatment of osteoarthritis of the medial compartment of the knee. J. Bone Joint Surg. Br. 87 (7), 937–939.

Richards, J., Selfe, J., Kilmurray, S. The biomechanics of step descent under different treatment modalities used in patellofemoral pain. 9th International conference of.

Richards, J., Thewlis, D., Selfe, J., et al. The biomechanics of single limb squats at different decline angles Enkle de Enkle congress 2006b. 2006b.

Richards, J., Thewlis, D., Selfe, J., et al., 2008. A biomechanical investigation of a single-limb squat: implications for lower extremity rehabilitation exercise. J. Athl. Train. 43 (5), 477–482.

Richardson, J.K., 1991. Rocker-soled shoes and walking distance in patients with calf claudication. Arch. Phys. Med. Rehabil. 72 (8), 554–558.

Rigas, C., 1984. Spatial parameters of gait related to the position of the foot on the ground. Prosthet. Orthot. Int. 8 (3), 130–134.

Rine, R.M., Ward, J., Lindeblad, S., 1992. Use of angle–angle diagrams to analyze effects of lower extremity in children with cerebral palsy. Phys. Ther. 72 (Suppl.), S57–S58.

Roetenberg, D., Luinge, H., Slycke, P. Xsens MVN: Full 6DOF Human Motion Tracking Using Miniature Inertial Sensors. XSENS TECHNOLOGIES–VERSION APRIL 3, 2013. http://www.xsens.com/images/stories/PDF/MVN_white_paper.pdf.

Roos, E., Engstrom, M., Lagerquist, A., et al., 2004. Clinical improvement after 6 weeks of eccentric exercise in patients with mid – portion Achilles tendinopathy – a randomized trial with 1-year follow up. Scand. J. Med. Sci. Sports 14, 286–295.

Root, M.L., Orien, W.P., Weed, J.H., 1977. Normal and Abnormal Function of the Foot: Clinical Biomechanics, vol. 2. Clinical Biomechanics Co, Los Angeles.

Rose, J., Gamble, J.G., 1994. Human walking. Williams and Wilkins, Baltimore.

Rosenbaum, D., Hautmann, S., Gold, M., et al., 1994. Effects of walking speed on plantar pressure patterns and hindfoot angular motion. Gait Posture 2 (3), 191–197.

Rosenbaum, D., Westhues, M., Bosch, K., 2013. Effect of gait speed changes on foot loading characteristics in children. Gait Posture 28 (4), 1058–1060.

Rowe, P.J., Nicol, A.C., Kelly, I.G., 1989. Flexible goniometer computer system for the assessment of hip function. Clin. Biomech. (Bristol, Avon) 4, 68–72.

Rowell, L., Taylor, H., Wang, Y., 1964. Limitatiions to prediction of maximal oxygen intake. J. Appl. Physiol. 19 (5), 919–927.

Rozumalski, A., et al., 2008. The in vivo three-dimensional motion of the human lumbar spine during gait. Gait Posture 28 (3), 378–384. at: http://www.ncbi.nlm.nih.gov/pubmed/18585041. (Accessed 2014).

Rundquist, P., 2007. Alterations in scapular kinematics in subjects with idiopathic loss of shoulder range of motion. J. Orthop. Sports Phys. Ther. 37 (1), 19–25.

Rydell, N.W., 1966. Forces acting on the femoral head-prosthesis. A study on strain gauge supplied prostheses in living persons. Acta Orthop. Scand. 37 (Suppl. 88), 1–32.

Salsich, G.B., Brechter, J.H., Farwell, D., et al., 2002. The effects of patellar taping on knee kinetics, kinematics, and vastus lateralis muscle activity during stair ambulation in individuals with patellofemoral pain. J. Orthop. Sports Phys. Ther. 32 (1), 3–10.

Sanderson, D.J., Martin, P.E., 1997. Lower extremity kinematic and kinetic adaptations in unilateral below-knee amputees during walking. Gait Posture 6, 126–136.

Saul, K.R., Vidt, M.E., Gold, G.E., et al., 2015. Upper limb strength and muscle volume in healthy middle-aged adults. J. Appl. Biomech. 31 (6), 484–491.

Saunders, J.B.D.M., Inman, V.T., Eberhart, H.S., 1953. The major determinants in normal and pathological gait. J. Bone Joint Surg. 35A, 543–558.

Schaafsma, J.D., Balash, Y., Gurevich, T., et al., 2003. Characterization of freezing of gait subtypes and the response of each to levodopa in Parkinson's disease. Eur. J. Neurol. 10 (4), 391–398.

Schaarschmidt, M., Lipfert, S.W., Meier-Gratz, C., et al., 2012. Functional gait asymmetry of unilateral transfemoral amputees. Hum. Mov. Sci. 31, 907–917.

Schache, A.G., et al., 2002. Intra-subject repeatability of the three dimensional angular kinematics within the lumbo-pelvic-hip complex during running. Gait Posture 15 (2), 136–145. at: http://www.ncbi.nlm.nih.gov/pubmed/11869907. (Accessed 2014).

Schache, A.G., Wrigley, T.V., Blanch, P.D., et al., 2001. The effect of differing Cardan angle sequences on three dimensional lumbopelvic angular kinematics during running. Med. Eng. Physiother. 23 (7), 493–501.

Schenkman, M., Riley, P.O., Pieper, C., 1996. Sit to stand from progressively lower seat heights – alterations in angular velocity. Clin. Biomech. (Bristol, Avon) 11 (3), 153–158.

Schmalz, T., Blumentritt, S., Marx, B., 2007. Biomechanical analysis of stair ambulation in lower limb amputees. Gait Posture 25, 267–278.

Schwartz, M.H., Rozumalski, A., 2005. A new method for estimating joint parameters from motion data. J. Biomech. 38 (1), 107–116.

Scott, S.H., Winter, D.A., 1990. Internal forces of chronic running injury sites. Med. Sci. Sports Exerc. 22 (2), 357–369.

Seay, J., Selbie, W.S., Hamill, J., 2008. In vivo lumbo-sacral forces and moments during constant speed running at different stride lengths. J. Sports Sci. 26 (14), 1519–1529. at: http://www.ncbi.nlm.nih.gov/pubmed/18937134. (Accessed 2016).

Seay, J.F., Van Emmerik, R.E.A., Hamill, J., 2011. Influence of low back pain status on pelvis-trunk coordination during walking and running. Spine 36 (16), E1070–E1079. at: http://www.ncbi.nlm.nih.gov/pubmed/21304421. (Accessed 2014).

Seedhom, B.B., Takeda, T., Tsubuku, M., et al., 1979. Mechanical factors and patellofemoral osteoarthrosis. Ann. Rheum. Dis. 38, 307–316.

Seel, T., Raisch, J., Schauer, T., 2014. IMU-based joint angle measurement for gait analysis. Sensors (Basel) 14 (4), 6891–6909.

Segal, A.D., Orendurff, M.S., Mute, G.K., et al., 2006. Kinematic and kinetic comparisons of transfemoral amputee gait using C-Leg (R) and Mauch SNS (R) prosthetic knees. J. Rehabil. Res. Dev. 43, 857–869.

Segal, A.D., Zelik, K.E., Klute, G.K., et al., 2012. The effects of a controlled energy storage and return prototype prosthetic foot on transtibial amputee ambulation. Hum. Mov. Sci. 31, 918–931.

Selfe, J., 2000. Peak 5 motion analysis of an eccentric step test performed by 100 normal subjects. Physiotherapy 86 (5), 241–247.

Selfe, J., 2004. The Patellofemoral joint: a review of primary research. Crit. Rev. Phys. Rehabil. Med. 16 (1), 1–30.

Selfe, J., Harper, L., Pedersen, I., et al., 2001a. Four outcome measures for patellofemoral joint problems: part 1 development and validity. Physiotherapy 87 (10), 507–515.

Selfe, J., Harper, L., Pedersen, I., et al., 2001b. Four outcome measures for patellofemoral joint problems: Part 2 reliability and clinical sensitivity. Physiotherapy 87 (10), 516–522.

Selfe, J., Janssen, J., Callaghan, M., et al., 2016. Are there three main subgroups within the patellofemoral pain population? A detailed characterisation study of 127 patients to help develop targeted intervention (TIPPs). Br. J. Sports Med. 50, 873–880.

Selfe, J., Richards, J., Thewlis, D., et al., 2008. The biomechanics of step descent under different treatment modalities used in patellofemoral pain. Gait Posture 27 (2), 258–263.

Selfe, J., Thewlis, D., Hill, S., et al., 2011. A clinical study of the biomechanics of step descent using different treatment modalities for patellofemoral pain. Gait Posture 34 (1), 92–96.

Selles, R.W., et al., 2001. Disorders in trunk rotation during walking in patients with low back pain: a dynamical systems approach. Clin. Biomech. (Bristol, Avon) 6 (3), 175–181. at: http://www.ncbi.nlm.nih.gov/pubmed/11240051. (Accessed 2016).

Seth, A., Matias, R., Veloso, A.P., et al., 2016. A biomechanical model of the scapulothoracic joint to accurately capture scapular kinematics during shoulder movements. PLoS ONE 11 (1), e0141028.

Shaeffer, D.K., 2013. Mems inertial sensors: A tutorial overview. Topics in integrated circuits for communications. IEEE Communications Magazine 100–109.

Shakoor, N., Lidtke, R.H., Wimmer, M.A., et al., 2013. Improvement in knee loading after use of specialized footwear for knee osteoarthritis: results of a six-month pilot investigation. Arthritis Rheum. 65, 1282–1289. doi:10.1002/art.37896.

Sharma, L., Pai, Y.C., Holtkamp, K., et al., 1997. Is knee joint proprioception worse in the arthritic knee versus the unaffected knee in unilateral knee osteoarthritis? Arthritis Rheum. 40 (8), 1518–1525.

Sherk, V., Bemben, M.G., Bemben, D.A., 2008. BMD and bone geometry in transtibial and transfemoral amputees. J. Bone Miner. Res. 23, 1449–1457.

Shinno, N., 1971. Analysis of knee function in ascending and descending stairs. Med. Sport 6, 202–207.

Shumway-Cook, A., Brauer, S., Woollacott, M., 2000. Predicting the probability for falls in community dwelling older adults using the timed up and go test. Phys. Ther. 80 (9), 896–903.

Sidway, B., Heise, G., Schoenfelder-Zohdi, B., 1995. Quantifying the variability of angle–angle plots. J. Hum. Mov. Stud. 29, 181–197.

Simonson, E., Keys, A. Working capacity in patients with orthopaedic handicaps from poliomyelitis, 1947.

Sinclair, J.K., Selfe, J., Taylor, P.J., et al., 2016. Influence of a knee brace intervention on perceived pain and patellofemoral loading in recreational athletes. Clin. Biomech. (Bristol, Avon) 37, 7–12. doi:10.1016/j.clinbiomech.2016.05.002. ISSN 0268-0033.

Singerman, R., Berilla, J., Archdeacon, M., et al., 1999. In vitro forces in the normal and cruciate deficient knee during simulated squatting motion. Trans. ASME 121, 234–242.

Sojka, A.M., Stuberg, W.A., Knutson, L.M., et al., 1995. Kinematic and electromyographic characteristics of children with cerebral palsy who exhibit genu recurvatum. Arch. Phys. Med. Rehabil. 76 (6), 558–565.

Sparrow, W.A., Donovan, E., van Emmerik, R.E.A., et al., 1987. Using relative motion plots to measure changes in intra-limb and inter-limb coordination. J. Mot. Behav. 19, 115–129.

Starholm, I.M., Mirtaheri, P., Kapetanovic, N., et al., 2016. Energy expenditure of transfemoral amputees during floor and treadmill walking with different speeds. Prosthet. Orthot. Int. 40, 336–342.

Stebbins, J., Harrington, M., Thompson, N., 2006. Repeatability of a model for measuring multi-segment foot kinematics in children. Gait Posture 23 (4), 401–410.

Steindler, A., 1955. Kiniesiology of the human body. Charles Thomas, Springfield, Il.

Stergiou, N., Giakas, G., Byrne, J.B., et al., 2002. Frequency domain characteristics of ground reaction forces during walking of young and elderly females. Clin. Biomech. (Bristol, Avon) 17 (8), 615–617.

Struyf, F., Nijs, J., Baeyens, J.P., et al., 2011. Scapular positioning and movement in unimpaired shoulders, shoulder impingement syndrome, and glenohumeral instability. Scand. J. Med. Sci. Sports 21, 352–358.

Stuart, M.J., Meglan, D.A., Lutz, G.E., et al., 1996. Comparison of intersegmental tibiofemoral joint forces and muscle activity during various closed kinetic chain exercise. Am. J. Sports Med. 24, 792–799.

Sutherland, D.H., Cooper, L., 1978. The pathomechanics of progressive crouch gait in spastic diplegia. Orthop. Clin. North Am. 9, 143–154.

Swinnen, E., et al., 2016. Thorax and pelvis kinematics during walking, a comparison between children with and without cerebral palsy: a systematic review. Neurorehabilitation 38 (2), 129–146. at: http://www.medra.org/servlet/aliasResolver?alias=iospress&doi=10.3233/NRE-161303. (Accessed 2016).

Syczewska, M., Oberg, T., Karlsson, D., 1999. Segmental movements of the spine during treadmill walking with normal speed. Clin. Biomech. (Bristol, Avon) 14 (6), 384–388. at: http://www.ncbi.nlm.nih.gov/pubmed/10521619. (Accessed 2014).

Tasi, R.Y. An efficient and accurate camera calibration technique for 3D machine vision. Proceedings of the 1986 IEEE Computer Society Conference on Computer Vision and Pattern Recognition, 1986; pp. 364–374.

Tata, J.A., Quanbury, A.O., Steinke, T.G., et al., 1978. A variable axis electrogoniometer for the measurement of simple plane movement. J. Biomech. 11, 421–425.

Taylor, A.J., Menz, H.B., Keenan, A.-M., 2004. The influence of walking speed on plantar pressure measurements using the two-step gait initiation protocol. Foot 14 (1), 49–55.

Taylor, N.F., Evans, O.M., Goldie, P.A., 1996. Angular movements of the lumbar spine and pelvis can be reliably measured after 4 minutes of treadmill walking. Clin. Biomech. (Bristol, Avon) 11 (8), 484–486. at: http://www.ncbi.nlm.nih.gov/pubmed/11415664. (Accessed 2014).

Taylor, W.R., Heller, M., Bergmann, G., et al., 2004. Tibio femoral loading during human gait and stair climbing. J. Orthop. Res. 22 (3), 625–632.

Thewlis, D., Richards, J., Bower, J., 2008. Discrepancies in knee joint moments using common anatomical frames defined by different palpable landmarks. J. Appl. Biomech. 24 (2), 185–190.

Thewlis, D., Richards, J., Bower, J. Discrepancies in knee joint moments using common anatomical frames defined by different palpable landmarks. Journal of Applied Biomechanics (in press).

Thorstensson, A., et al., 1984. Trunk movements in human locomotion. Acta Physiol. Scand. 121 (1), 9–22. at: http://www.ncbi.nlm.nih.gov/pubmed/6741583. (Accessed 2014).

Thummerer, Y., et al., 2012. Is age or speed the predominant factor in the development of trunk movement in normally developing children? Gait Posture 35 (1), 23–28. at: http://www.sciencedirect.com/science/article/pii/S0966636211002426. (Accessed 2014).

Thurston, A.J., 1982. Repeatability studies of a television/computer system for measuring spinal and pelvic movements. J. Biomed. Eng. 4 (2), 129–132. at: http://www.ncbi.nlm.nih.gov/pubmed/7070066. (Accessed 2014).

Toutoungi, D.E., Lu, T.W., Leardini, A., et al., 2000. Cruciate ligament forces in the human knee during rehabilitation exercises. Clin. Biomech. (Bristol, Avon) 15, 176–187.

Turcot, K., Aissaoui, R., Boivin, K., et al., 2008. Test-retest reliability and minimal clinical change determination for 3-dimensional tibial and femoral accelerations during treadmill walking in knee osteoarthritis patients. Arch. Phys. Med. Rehabil. 89 (4), 732–737.

van der Helm, F., 1994a. A finite element musculoskeletal model of the shoulder mechanism. J. Biomech. 27 (5), 551–569.

Van de Walle, P., et al., 2012. Increased mechanical cost of walking in children with diplegia: the role of the passenger unit cannot be neglected. Res. Dev. Disabil. 33 (6), 1996–2003. at: http://www.ncbi.nlm.nih.gov/pubmed/22750355. (Accessed 2016).

Van Emmerik, R.E.A., et al., 2005. Age-related changes in upper body adaptation to walking speed in human locomotion. Gait Posture 22 (3), 233–239. at: http://www.ncbi.nlm.nih.gov/pubmed/16214663. (Accessed 2016).

van Wegen, E., de Goede, C., Lim, I., et al., 2006. The effect of rhythmic somato-sensory cueing on gait in patients with Parkinson's disease. J. Neurol. Sci. 248 (1-2), 210–214.

Vanicek, N., Strike, S., Mcnaughton, L., et al., 2009. Gait patterns in transtibial amputee fallers vs. non-fallers: Biomechanical differences during level walking. Gait Posture 29, 415–420.

Vanicek, N., Strike, S.C., Mcnaughton, L., et al., 2010. Lower limb kinematic and kinetic differences between transtibial amputee fallers and non-fallers. Prosthet. Orthot. Int. 34, 399–410.

Vanicek, N., Strike, S.C., Polman, R., 2015. Kinematic differences exist between transtibial amputee fallers and non-fallers during downwards step transitioning. Prosthet. Orthot. Int. 39, 322–332.

Vermeulen, H.M., Stokdijk, M., Eilers, P.H., et al., 2002. Measurement of three dimensional shoulder movement patterns with an electromagnetic tracking device in patients with a frozen shoulder. Ann. Rheum. Dis. 61, 115–120.

Viton, J.M., Timsit, M., Mesure, S., et al., 2000. Asymmetry of gait initiation in patients with unilateral knee arthritis. Arch. Phys. Med. Rehabil. 81 (2), 194–200.

Von Eisenhart-Rothe, R., Matsen, F.A., Eckstein, F., et al., 2005. Pathomechanics in atraumatic shoulder instability: scapular positioning correlates with humeral head centering. Clin. Orthop. Relat. Res. 433, 82–89.

Wagner, H., 1990. Pelvic tilt and leg length correction. Orthopade. 19 (5), 273–277.

Wall, J.C., Ashburn, A., 1979. Assessment of gait disability in hemiplegics. Scand. J. Rehabil. Med. 11, 95–103.

Wall, J.C., Charteris, J., Turnbull, G., 1987. Two steps equals one stride equals what? Clin. Biomech. (Bristol, Avon) 2, 119–125.

Wang, J.W., Kuo, K.N., Andriacchi, T.P., et al., 1990. The influence of walking mechanics and time on the results of proximal tibial osteotomy. J. Bone Joint Surg. Am. 72 (6), 905–909.

Warner, J.J., Micheli, L.J., Arslanian, L.E., et al., 1992. Scapulothoracic motion in normal shoulders and shoulders with glenohumeral instability and impingement syndrome: a study using moiré topographic analysis. Clin. Orthop. Relat. Res. 285, 191–199.

Wentink, E.C., Prinsen, E.C., Rietman, J.S., et al., 2013. Comparison of muscle activity patterns of transfemoral amputees and control subjects during walking. J. Neuroengineering Rehabil. 10, 11.

Wiberg, G., 1941. Roentgenographic and anatomic studies on the femoropatellar joint. Acta Orthop. Scand. 12, 319–410.

Wilk, K., Dynamic Muscle strength testing. In: Amundsen, L.T. (Ed.), Muscle strength testing instrumented and non-instrumented systems. Churchill Livingstone, New York, (Chapter 5).

Wilk, K.E., Escamilla, R.F., Fleisig, G.S., et al., 1996. A comparison of tibiofemoral joint forces and electromyographic activity during open and closed kinetic chain exercises. Am. J. Sports Med. 24, 518–527.

Williams, G.N., Krishnan, C., 2007. Articular neurophysiology and sensorimotor control. In: Magee, D.J., Zachewski, J.E., Quillen, W.S. (Eds.), Scientific Foundations and Principles of Practice in Musculoskeletal Rehabilitation. Saunders, Misouri.

Winter, D., 1995. Human balance and posture control during standing and walking. Gait Posture 3 (4), 193–214.

Winter, D.A., 1993. Knowledge base for diagnostic gait assessments. Med. Prog. Technol. 19 (2), 61–81.

Winter, D.A., 1978. Energy assessment in pathological gait. Physiother. Can. 30, 183–191.

Winter, D.A. A. B. C. (Anatomy, Biomechanics, Control) of Balance during Standing and Walking. ISBN : 0-9699420-0-1, 1995.

Winter, D.A. Knowledge base for diagnostic gait assessments.

Winter, D.A., Quanbury, A.O., Reimer, G.D., 1976. Analysis of instantaneous energy of normal gait. J. Biomech. 9 (4), 253–257.

Winter, D.A., Sienko, S.E., 1988. Biomechanics of below-knee amputee gait. J. Biomech. 21, 361–367.

Wolf, E.J., Everding, V.Q., Linberg, A.L., et al., 2012. Assessment of transfemoral amputees using C-Leg and Power Knee for ascending and descending inclines and steps. J. Rehabil. Res. Dev. 49, 831–842.

Wolff, S.L., 1978. Essential considerations in the use of EMG Biofeedback. Phys. Ther. 58, 25.

Woltring, H.J., 1976. Calibration and measurement in 3-dimensional monitoring of human motion by optoelectronic means. II. Experimental results and discussion. Biotelemetry 3 (2), 65–97.

Woltring, H.J., 1980. Planar control in multi-camera calibration for three- dimensional gait studies. J. Biomech. 13 (1), 39–48.

Woltring, H.J., 1982. Estimation and precision of 3D kinematics by analytical photogrametery. Comput. Med. 232–241.

Woltring, H.J., Huiskes, R., de Lange, A., 1985. Finite centroid and helical axis estimation from noisy landmark measurements in the study of human joint kinematics. J. Biomech. 18 (5), 379–389.

Wong, D.W.C., Lam, W.K., Yeung, L.F., et al., 2015. Does longdistance walking improve or deteriorate walking stability of transtibial amputees? Clin. Biomech. (Bristol, Avon) 30, 867–873.

Wood, G.A., Jennings, L.S., 1979. On the use of spline functions for data smoothing. J. Biomech. 12, 477–479.

Woodburn, J., Nelson, K.M., Siegel, K.L., et al., 2004. Multisegment foot motion during gait: proof of concept in rheumatoid arthritis. J. Rheumatol. 31 (10), 1918–1927.

Wretenberg, P., Feng, Y., Arborelius, U.P., 1996. High and low bar squatting techniques during weight training. Med. Sci. Sports Exerc. 22, 218–224.

Wright, F., 1994. Accident prevention and risk taking by elderly people: The need for advice. Age Concern, London.

Wu, G., et al., 2005. ISB recommendation on definitions of joint coordinate systems of various joints for the reporting of human joint motion – Part II: shoulder, elbow, wrist and hand. J. Biomech. 38 (5), 981–992. at: http://www.sciencedirect.com/science/article/pii/S002192900400301X. (Accessed 2014).

Wu, G., Siegler, S., Allard, P., et al., 2002. Standardization and Terminology Committee of the International Society of Biomechanics. ISB recommendation on definitions of joint coordinate system of various joints for the reporting of human joint motion – part I: ankle, hip, and spine. International Society of Biomechanics. J. Biomech. 35 (4), 543–548.

Wu, G., van der Helm, F.C., Veeger, H.E., et al., 2005. International Society of Biomechanics. ISB recommendation on definitions of joint coordinate systems of various joints for the reporting of human joint motion – part II: shoulder, elbow, wrist and hand. J. Biomech. 38 (5), 981–992.

Wunderlich, R.E., Cavanagh, P.R., 2001. Gender differences in adult foot shape: implications for shoe design. Med. Sci. Sports Exerc. 33 (4), 605–611.

Wurdeman, S.R., Huisinga, J.M., Filipi, M., et al., 2011. Multiple sclerosis affects the frequency content in the vertical ground reaction forces during walking. Clin. Biomech. (Bristol, Avon) 26 (2), 207–212. doi:10.1016/j.clinbiomech.2010.09.021. [Epub 2010 Oct 29].

Yack, H.J., Nielsen, D.H., Shurr, D.G., 1999. Kinetic patterns during stair ascent in patients with transtibial amputations using three different prostheses. J. Prosthet. Orthot. 11, 57–62.

Young, M., Cook, J., Purdam, C., et al., 2005. Eccentric decline squat protocol offers superior results at 12 months compared with traditional eccentric protocol for patellar tendinopathy in volleyball players. Br. J. Sports Med. 39, 102–105.

Zatsiorsky, V., Seluyanov, V., 1983. The mass and inertia characteristics of the main segments of the body. In: Matsui, H., Kobayashi, K. (Eds.), Biomechanics VIII–B. Human Kinetics Publishers, Champaign, IL, pp. 1152–1159.

Ziegler-Graham, K., Mackenzie, E.J., Ephraim, P.L., et al., 2008. Estimating the prevalence of limb loss in the United States: 2005 to 2050. Arch. Phys. Med. Rehabil. 89, 422–429.

INDEX

Page numbers followed by '*f*' indicate figures,
'*t*' indicate tables, and '*b*' indicate boxes.